NONLINEAR DIGITAL FILTERS

THE KLUWER INTERNATIONAL SERIES
IN ENGINEERING AND COMPUTER SCIENCE

VLSI, COMPUTER ARCHITECTURE AND
DIGITAL SIGNAL PROCESSING

Consulting Editor
Jonathan Allen

NONLINEAR DIGITAL FILTERS

PRINCIPLES AND APPLICATIONS

by

I. Pitas
Aristotelian University of Thessaloniki, Greece

and

A.N. Venetsanopoulos
University of Toronto, Canada

KLUWER ACADEMIC PUBLISHERS
Boston/Dordrecht/London

Distributors for North America: ~~ANDE~~
Kluwer Academic Publishers
101 Philip Drive
Assinippi Park
Norwell, Massachusetts 02061 USA

Distributors for all other countries:
Kluwer Academic Publishers Group
Distribution Centre
Post Office Box 322
3300 AH Dordrecht, THE NETHERLANDS

Library of Congress Cataloging-in-Publication Data

Pitas, I. (Ioannis)
    Nonliner digital filters : principles and applications / by
I. Pitas and A.N. Venetsanopoulos.
      p. cm. — (The Kluwer international series in engineering and
computer science. VLSI, computer architecture and digital signal
processing)
    Includes bibliographical references.
    ISBN 0-7923-9049-0
    1. Electric filters, Digital. I. Venetsanopoulos, A. N.
(Anastasios N.), 1941–   . II. Title. III. Series.
TK7872.F5P58   1990
621.3815 '324—dc20
                                                89-26769
                                               CIP

Printed in the United States of America

D
621.381532
Pit

TABLE OF CONTENTS

PREFACE

The function of a filter is to transform a signal into another one more suitable for a given purpose. As such, filters find applications in telecommunications, radar, sonar, remote sensing, geophysical signal processing, image processing, and computer vision. Numerous authors have considered deterministic and statistical approaches for the study of passive, active, digital, multidimensional, and adaptive filters. Most of the filters considered were linear although the theory of nonlinear filters is developing rapidly, as it is evident by the numerous research papers and a few specialized monographs now available. Our research interests in this area created opportunity for cooperation and co-authored publications during the past few years in many nonlinear filter families described in this book. As a result of this cooperation and a visit from John Pitas on a research leave at the University of Toronto in September 1988, the idea for this book was first conceived. The difficulty in writing such a monograph was that the area seemed fragmented and no general theory was available to encompass the many different kinds of filters presented in the literature. However, the similarities of some families of nonlinear filters and the need for such a monograph providing a broad overview of the whole area made the project worthwhile. The result is the book now in your hands, typeset at the Department of Electrical Engineering of the University of Toronto during the summer of 1989.

The purpose of this book is to fill an existing gap in the scientific literature. It presents the state of the art in the area of nonlinear filters. It is broad enough to cover various families of nonlinear filters. It is written at a level which can be easily understood by a graduate student in Electrical Engineering or Computer Science. Therefore, it can be used as a textbook, that covers part of a modern graduate digital signal processing course. It can also be used as a textbook for a graduate course on nonlinear digital signal processing. It contains analysis, design algorithms, and digital signal processor architectures for nonlinear filtering. Thus it is useful to a research or development engineer. Furthermore, since it offers an up-to-date survey of the relevant research, it can be used as a reference text by researchers.

The book can be read in many different ways, other than in the order in which it was written. Readers interested in order statistic or median filters may be interested in reading it in the following order: chapters 1, 2, 3, 4, 5, 9, 10, and 11. Others with an interest in morphological filters may choose the order:

chapters 1, 3, 6, and 11. Homomorphic filters can be studied by reading the book as follows: chapters 1, 3, 7. Polynomial (Volterra) type filters can be studied by following chapters 1, 3, 8, 9, and then going to chapter 11.

We are grateful to the following reviewers for their informative inputs and highly valuable suggestions: Prof. M. Omair Ahmad of Concordia University, Prof. Gonzalo Arce of the University of Delaware, Prof. Saleem Kassam of the University of Pennsylvania, Prof. Petros Maragos of Harvard University, and Prof. Giovanni Siccuranza of the University of Trieste, Italy. We would also like to thank our students AL-Shebeily Saleh, Kostas Kotropoulos, Renee Martens, Sunil Sanwalka, and Ziheng Zhou for reading various chapters of the manuscript and contributing to the example and figures of the book, Dr. George Lampropoulos for proofreading the manuscript and Linda Espeut for contributing to the typing of the manuscript.

LIST OF SYMBOLS

Signals and systems

$f(t)$	one-dimensional function
$F(\Omega)$	Fourier transform of an one-dimensional function
x_i, $x(i)$	data, discrete-time sequence
$\mathbf{x}$	data vector
$X(\omega)$	Fourier transform of a sampled sequence
$f(x,y)$	two-dimensional function
(x,y)	image plane coordinates
(ξ,η)	object plane coordinates
(Ω_x,Ω_y)	frequency coordinates
$h(x,y)$	point-spread function
$H(\Omega_x,\Omega_y)$	frequency response of a linear system
x_{ij}, $x(i,j)$	two-dimensional data or sequences
T_1, T_2	sampling periods
(ω_x,ω_y)	frequency coordinates for sampled signals
D_*, $D_\times$	signal space homomorphisms

Probability and statistics

X	random variable
$Pr\{A\}$	probability of an event A
cdf	cumulative distribution function
F,P,G	cumulative distribution functions
pdf	probability density function
f,p,g	probability density functions
Φ,ϕ	cdf and pdf of the unit Gaussian distribution
$N(\mu,\sigma^2)$	Gaussian distribution having mean μ and variance σ^2
$\pi(F,G)$	Prohorov distance of F, G
P_ε	ε neighborhood of a distribution
$E[.]$	expectation operator
μ	mean
σ^2	variance
σ	standard deviation
$R(\tau)$	autocorrelation function
$C(\tau)$	autocovariance function
$r(\tau)$	correlation coefficient

$M(\tau, p)$	third-order moment sequence
$\mathbf{R}$	autocorrelation matrix
$\mathbf{C}$	autocovariance matrix
$S(\omega)$	power spectral density
$B(\omega_1, \omega_2)$	bispectrum
n, z	noise processes
θ	parameter
F_θ	parametric distribution
Θ	parameter range
$\mathbf{X}$	data range
T_n	estimator
$T(G)$	functional of an estimator
$IF(x; T, G)$	influence function
$V(T, F)$	asymptotic variance
$J(F)$	Fisher information
e	absolute asymptotic efficiency
ARE	asymptotic relative efficiency
γ^*	gross error sensitivity
ε^*	breakdown point
MLE	maximum likelihood estimate
$x_{(i)}$	i-th order statistic
$x_{(v+1)}$	median
$med(x_i)$	median filter
A	window of the median filter
$n = 2v+1$	filter length
$med(x, A)$	median filter having window A
$med^\infty(x, A)$	median root having window A
$\bar{x}$	arithmetic mean, moving average filter
y_{L_r}	L_p-mean filter
y_{CH_r}	contraharmonic mean filter
y_H	harmonic mean filter
y_G	geometric mean filter
s	sample standard deviation
MP	midpoint
W	range
ED	extreme deviate
$MAD(x_i)$	median absolute deviation from median
R_i	rank i
μ_i	mean of $x_{(i)}$
σ^2_i	variance of $x_{(i)}$
σ^2_{med}	variance of the median
$\sigma^2_{\bar{x}}$	variance of the arithmetic mean

a	vector of the L-filter parameters
e	unit vector
n	vector of ordered noise variates
$LOMO\,(m)$	locally monotonic sequence
L	length of an one-dimensional data sequence

Morphology

R	real numbers
Z	integer numbers
X, Y, A	set of points representing binary images
B	structuring element
X^c	complement of X
X_b	translate of a set X by a vector b
X^s	symmetric set of X
$m\,(X)$	area of X
$\overline{X}$	adherence of X
$p\,(X)$	power set of X
$\varnothing$	empty set
$\subset, \subseteq$	subsets
$X \cup Y$	set union
$X \cap Y$	set intersection
$X - Y$	set difference
$\psi(X)$	morphological transformation
$X \oplus B^s$	dilation of X by B
$X \ominus B^s$	erosion of X by B
X^B	closing of X by B
X_B	opening of X by B
$X \oplus B\,; Y$	conditional dilation
$B \uparrow X$	B hits X
$X \circledcirc B$	hit-or-miss transformation
$X \bigcirc T$	thinning of X by T
$X \circledcirc T$	thickening of X by T
$U\,(X)$	umbra of X
$T\,(X)$	top surface of X
$SK\,(X)$	skeleton of X
$skf\,(X)$	skeleton function (quench function) of X
$S_n(X)$	skeleton subset of X
$f\,(r)$	pecstrum
g	structuring function
$U\,(g)$	umbra of g

$X_t(f)$ cross-section of function f at level t

$f \oplus g^s$ dilation of function f by g

$f \ominus g^s$ erosion of function f by g

f^g closing of f by g

f_g opening of f by g

CO close-opening filter

OC open-closing filter

$SK(f)$ skeleton of f

$S_n(f)$ skeleton subset of f

Functions

$\Gamma(x)$ Gamma function

$sign(x)$ sign function

$[x]_{-b}^b$ Huber function

$\rho(x), \psi(x)$ functions defining M estimators

$ln(x)$ natural logarithm

$\delta_t, \delta(x-t)$ delta function

Miscellaneous symbols

$\sim$ proportional

$\simeq$ approximately equal

sup supremum

inf infemum

max maximum

min minimum

NONLINEAR DIGITAL FILTERS

CHAPTER 1

INTRODUCTION

1.1 OVERVIEW

In the early development of signal and image processing, linear filters were the primary tools. Their mathematical simplicity and the existence of some desirable properties made them easy to design and implement. Moreover, linear filters offered satisfactory performance in many applications. However, linear filters have poor performance in the presence of noise that is not additive as well as in problems where system nonlinearities or non-Gaussian statistics are encountered. In addition, various criteria, such as the maximum entropy criterion, lead to nonlinear solutions. In image processing applications, linear filters tend to blur the edges, do not remove impulsive noise effectively, and do not perform well in the presence of signal dependent noise. It is also known that, although the exact characteristics of our visual system are not well understood, experimental results indicate that the first processing levels of our visual system possess nonlinear characteristics. For such reasons, nonlinear filtering techniques for signal/image processing were considered as early as 1958 [1].

Nonlinear filtering has had a dynamic development since then. This is indicated by the amount of research presently published and the popularity and widespread use of nonlinear digital filters in a variety of applications, notably in telecommunications, image processing, in geophysical signal processing. Most of the currently available image processing software packages include nonlinear filters (e.g., median filters, morphological filters). Despite this tremendous growth, all relevant literature is scattered in research papers, in some survey articles in edited books, or in specialized monographs [2-10].

The purpose of this monograph is to organize and integrate the material that has been published in scientific publications and books in the area of nonlinear filters. This objective was quite ambitious in view of the enormous number of published papers and the lack of an overall unifying theory in nonlinear filters. The book could have easily become an encyclopedia of filters. To avoid this pitfall only the most significant nonlinear filter families were considered in this book, with an emphasis placed on the description of the similarities and unifying principles behind these families. Table 1.1 shows non-linear filters as part of the objectives and methods of digital image processing. Figure 1.1.1 gives a graphical representation of the various families of nonlinear filters described in this book.

Table 1.1: Objectives and methods of digital image processing (adapted from [7]).

Method Objective	Point & Local Operations	Linear Shift Invariant Filters	Nonlinear Filters	Adaptive Filters
Image Enhance-ment	Intensity mapping (histogram equalization), local operations (selective averaging, unsharp masking), eye modeling, pseudocolor, edge sharpening	Low-pass, High-pass, High-emphasis, Finite impulse response (FIR), Infinite impulse response (IIR), Recursive and non-recursive realizations	Order statistic filters (median) Morpho-logical Nonlinear mean Homo-morphic Polynomial	Heuristic approaches
			Nonlinear adaptive filters	
Image Restoration		Inverse, Pseudo-inverse, Wiener, Power spectrum equalization, Generalized Wiener, Constrained least squares.	Homomorphic, Maximum entropy, Bayesian, Maximum a posteriori probability.	Kalman filters Edge enhancing Kalman filters.

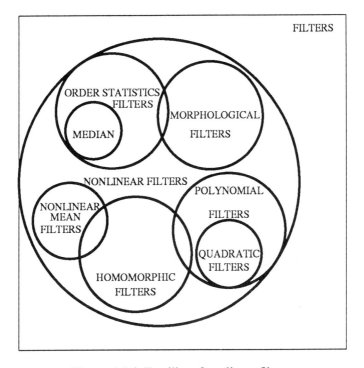

Figure 1.1.1: Families of nonlinear filters.

1.2 LINEAR FILTERS

A filter can be defined in a mathematical sense as an operator $L(\cdot)$, which maps a signal x into a signal y, as follows:

$$y = L(x) \tag{1.2.1}$$

When $L(\cdot)$ is a linear operator, i.e., it satisfies both the superposition and proportionality principles, the filter is linear. When in addition the operator is shift-invariant, the filter is *linear shift-invariant (LSI)*. Due to their mathematical simplicity and their importance, linear filters have been extensively studied in the literature [11-16].

Two-dimensional (2-d) and m-dimensional (m-d) linear filtering is concerned with the extension of 1-d filtering techniques to two and more dimensions. This subject has received considerable attention due to its importance in enhancement and restoration [13,15,16]. One fundamental difficulty in the extension of 1-d techniques to 2-d and m-d is the lack of a Fundamental Theorem of Algebra for polynomials of two or more variables. LSI 2-d digital filters, relate the input, $x(m,n)$, and output, $y(m,n)$, sequences by a linear constant coefficient difference equation:

$$\sum_k \sum_l b(k,l)y(m-k,n-l) = \sum_k \sum_l a(k,l)x(m-k,n-l) \tag{1.2.2}$$

where $a(\cdot)$ and $b(\cdot)$ in (1.2.2) are constant coefficients [12]. In the z transform domain, the corresponding transfer function is

$$H(z_1,z_2) = \frac{A(z_1,z_2)}{B(z_1,z_2)} = \frac{\sum_k \sum_l a(k,l)z_1^{-k}z_2^{-l}}{\sum_k \sum_l b(k,l)z_1^{-k}z_2^{-l}} \tag{1.2.3}$$

The inverse z transform of $H(z_1,z_2)$ is $h(m,n)$, the *impulse response* of the filter. Note that in the case of linear filters due to the existence of linearity it is possible to obtain an expression of the output of the filter, which is the convolution of the input and the impulse response. If $h(m,n)$ has only a finite number of non-zero values, the filter is called a *finite impulse response (FIR)* 2-d filter. Otherwise, it is an *infinite impulse response (IIR)* 2-d filter. If the filter only requires samples of the input image to evaluate the output image, it is called *non-recursive*. If, on the other hand, samples of the input image, together with evaluated output image samples, are required, the filter is called *recursive*.

2-d and 3-d digital filters have been successfully applied to many areas of digital image processing. Some of the main types of filters are:

1) *Low-pass 2-d filters.* These result in the smoothing of the image, reducing high spatial frequency noise components.

2) *High-pass 2-d filters.* These enhance very low contrast features, when they are superimposed on a very dark or very light background.

3) *High-emphasis and band-pass 2-d filters.* These have an effect on the image, which is more subtle than that of the previous two classes. They generally tend to sharpen the edges and enhance small details.

1.3 NONLINEAR FILTERS

Nonlinear filters are also described by (1.2.1). However, the operator $L(\cdot)$ is not linear in this case. The lack of satisfaction of the superposition or proportionality principles or both does not allow us to obtain the output of a nonlinear filter by convolving the input with its impulse response function. For this reason different families of nonlinear filters have been studied, each of which has a different characterization which makes the filters in this class amenable to some form of mathematical analysis.

The simplest possible nonlinear transformation of the form of (1.2.1) corresponds to a memoryless nonlinearity that maps the signal x to y. Such transformations have been used extensively in image processing and are called *gray scale transformations* [15,16]. For example the transformation:

$$y = t(x) \tag{1.3.1}$$

may be used to transform one gray scale x to another y. Another form of intensity mapping is known as *histogram modification* [15,16]. The histogram is a graph of the relative frequency of gray level occurrence in the image. Methods exist allowing us to transform the gray level of an image into another, in order to give the image a specified histogram. A special case of interest is *histogram equalization*. The need for it arises when comparing two images taken under different lighting conditions. The two images must be referred to the same "base", if meaningful comparisons or measurements are to be made. The base that is used as a standard has a uniformly distributed (equalized) histogram. Note that from Information Theory, a uniform histogram signifies maximum information content of the image.

One of the most popular families of nonlinear filters for noise removal are order statistic filters. The median filter which represents one of the simplest such filters was first suggested by Tukey in 1971 [17]. The theoretical basis of order statistics filter is the theory of robust statistics [18-19], which is summarized in chapter 2. There exist several filters which are members of this filter class, e.g., the median filter [3,4,17,20,21], the stack filters [22], the median hybrid filter [23], the α-trimmed mean filter [24], and the L-filters [25]. Order statistics filters have found extensive applications in digital signal and image processing. Their analysis is presented in chapters 4 and 5.

Other approaches attempted to achieve filtering by utilizing geometric rather than analytic features of signals. They originated from the basic set operations for image processing, which were introduced by Matheron [6] and Serra [7-8]. These filters were called morphological filters and found applications in image processing [26-29] and analysis [30-32]. Areas of application include biomedical image processing, shape recognition, nonlinear filtering, edge detection, noise suppression, image enhancement, etc. In recent years, morphological filters have become a hot topic in image processing. Mathematical morphology and its applications are described in chapter 6.

One of the oldest classes of nonlinear filters, which have been used extensively in digital signal and image processing, are the homomorphic filters [11,33,34] and their extensions [34]. This filter class has found applications in image enhancement [33], multiplicative and signal dependent noise removal, as well as in speech processing [34] and in seismic signal processing. It is presented in chapter 7.

In order to generalize nonlinear system theory, the Volterra series have been recognized as a powerful mathematical tool leading to another class of nonlinear filters called polynomial filters. Because of the algebraic complexity in the calculation of high order Volterra kernels, the use of Volterra series remains limited. Recently, some progress was obtained in second order Volterra filter theory (quadratic filters) [2,36-38]. Such filters have been used for nonlinear channel modeling in telecommunications as well as in geophysical signal processing and in image processing. This filter class is described in chapter 8.

Adaptive filtering has also taken advantage of nonlinear filtering techniques. The first applications of adaptive nonlinear filtering were introduced in digital image processing [40-44]. Adaptive Volterra filters were also introduced in echo cancellation and other telecommunication applications recently [45]. Adaptive versions of order statistic filters have been proposed [46,47]. The class of adaptive nonlinear filters is presented in chapter 9.

Finally, a general nonlinear filter structure, that encompasses all these filter classes as special cases is presented in chapter 10. Some new trends in nonlinear filtering, namely neural networks and nonlinear color image processing, are included in this chapter. Chapter 11 presents algorithms and architectures appropriate for the implementation of nonlinear filters.

1.4 METHODOLOGY

To clarify the many issues involved in the design of a nonlinear filter we propose a top down process described by the following five steps.

Step 1: Formulation of the problem

This step is concerned with the clarification of the objective. A satisfactory answer to this question requires an understanding of the signal/image formation process, the signal image degradation, the intended application, and the means and knowledge available. This step may lead to the specification of the signal processing system to be designed.

Step 2: Filter design

This step is concerned with the decision on the method to be used. Once the objective has been decided upon, a course of action is chosen, which may lead either to an analytical solution, or to a computer-aided design solution, or to a heuristic solution.

Step 3: Filter realization

The filter described by step 2 is now realized. This step examines available algorithms and architectures, which may be used for the realization of the system designed, and is concerned with various issues, such as speed, modularity, and the effects of finite precision arithmetic (input quantization, parameter quantization, and roundoff accumulation errors).

Step 4: Filter implementation

The architecture described by step 3 is now implemented in software and/or hardware. This step is concerned with languages and technology. Cost, speed, flexibility, transportability, and accuracy are some of the main issues.

Step 5: Filter verification

The filter obtained is finally verified by utilizing it in the application intended and evaluating its performance. Often it is modified at this step to achieve improved performance.

Although it is convenient to represent the total design process by these five steps, it is important to note that these steps are often interrelated, since a decision made at any one of them affects the others. In addition, not all steps are considered in each signal/image processing problem.

This book reflects the present state of knowledge on the subject of nonlinear filters and emphasizes the material required to respond to the considerations of steps 1 and 2. Special attention is focused on image formation and degradation models, which are described in chapter 3. This special treatment is due to the complexity of the image formation process and to the importance of nonlinear image processing. Chapters 4-9 are mainly concerned with the analysis and design of the various nonlinear filter families (step 2). Finally, chapter 11 deals with realization and implementation problems (steps 3-4).

REFERENCES

[1]   N. Wiener, *Nonlinear problems in random theory*, New York, The Technology Press, MIT and John Wiley and Sons, Inc., 1958.

[2]   M. Schetzen, *The Volterra and Wiener theories of nonlinear filters*, J. Wiley, 1980.

[3]   B.I. Justusson, "Median filtering: statistical properties", in *Two-dimensional digital signal processing II*, T.S. Huang editor, Springer Verlag, 1981.

[4]   G. Tyan, "Median filtering: deterministic properties", in *Two-dimensional digital signal processing II*, T.S. Huang editor, Springer Verlag, 1981.

[5]   G.R. Arce, N.C. Gallagher, T.A. Nodes, "Median filters: theory for one- and two-dimensional filters", in *Advances in computer vision and image processing*, T.S. Huang editor, JAI Press, 1986.

[6]   G. Matheron, *Random sets and integral geometry*, New York, Wiley, 1975.

[7]   J. Serra, *Image analysis and mathematical morphology*, Academic Press, 1982.

[8]   J. Serra (editor), *Image analysis and mathematical morphology: Theoretical advances*, vol. 2, Academic Press, 1989.

[9]   C.R. Giardina, E.R. Dougherty, *Morphological methods in image and signal processing*, Prentice Hall, 1988.

[10] E.R. Dougherty, C.R. Giardina, *Image processing - continuous to discrete*, vol. 1, Prentice-Hall, 1987.

[11] A.V. Oppenheim, R.W. Schafer, *Discrete-time signal processing*, Prentice-Hall, 1989.

[12] D.F. Dudgeon, R.M. Mersereau, *Multidimensional digital signal processing*, Prentice-Hall, 1984.

[13] A.N. Venetsanopoulos, "Digital image processing and analysis", in *Signal processing,* J.L. Lacoume, T.S. Duranni, and R. Stora editors, pp. 573-668, North Holland, 1987.

[14] S.G. Tzafestas (editor), *Progress in multidimensional systems theory*, Marcel Dekker Inc., 1986.

[15] W.K. Pratt, *Digital image processing*, J. Wiley, 1978.

[16] A.K. Jain, *Fundamentals of digital image processing*, Prentice-Hall, 1989.

[17] J.W. Tukey, *Exploratory data analysis*, Addison-Wesley, 1977.

[18] H. .A. David, *Order statistics,* John Wiley, 1980.

[19] P. S. Huber, *Robust statistics,* John Wiley, 1981.

[20] G. Arce, N.C. Gallagher, "State description for the root-signal set of median filters" *IEEE Transactions on Acoustics, Speech and Signal Processing,* vol. ASSP-30, no. 6, pp. 894-902, Dec. 1982.

[21] J.P. Fitch, E.J. Coyle, N.C. Gallagher, "Median filtering by threshold decomposition" *IEEE Transactions on Acoustics, Speech and Signal Processing,* vol. ASSP-32, no. 6, pp. 1183-1188, Dec. 1984.

[22] P.D. Wendt, E.J. Coyle, N.C. Gallagher Jr., "Stack filters", *IEEE Transactions on Acoustics, Speech and Signal Processing,* vol. ASSP-34, no. 4, pp. 898-911, Aug. 1986.

[23] P. Heinonen, Y. Neuvo, "FIR-median hybrid filters", *IEEE Transactions on Acoustics, Speech and Signal Processing*, vol. ASSP-35, no. 6, pp. 832-838, June 1987.

[24] J.B. Bednar, T.L. Watt, "Alpha-trimmed means and their relationship to the median filters", *IEEE Transactions on Acoustics, Speech and Signal Processing*, vol. ASSP-32, no. 1, pp. 145-153, Feb. 1984.

[25] A.C. Bovik, T.S. Huang, D.C. Munson, "A generalization of median filtering using linear combinations of order statistics", *IEEE Transactions on Acoustics, Speech and Signal Processing*, vol. ASSP-31, no. 6, pp. 1342-1349, Dec. 1983.

[26] R.M. Haralick, S.R. Sternberg, X. Zhuang, "Image analysis using mathematical morphology", *IEEE Transactions on Pattern Anal. and Machine Intl.*, vol. PAMI-9, no. 4, pp. 532-550, July 1987.

[27] S.R. Sternberg, "Biological image processing", *Computer*, pp. 22-34, Jan. 1983.

[28] S.R. Sternberg, "Gray scale morphology", *Comput. Vision Graphics and Image Processing*, vol. 35, pp. 333-355, 1986.

[29] P. Maragos, R.W. Schafer, "Morphological filters - Part I: Their set-theoretic analysis and relations to linear shift-invariant filters", *IEEE Transactions on Acoustics, Speech and Signal Processing*, vol. ASSP-35, no. 8, pp. 1153-1168, Aug. 1987.

[30] H. Blum, "A transformation for extracting new descriptors of shape", *Models for the perception of speech and visual Forms*, W. Wathen-Dunn editor, pp. 362-380, MIT Press, 1967.

[31] P.A. Maragos, R.W. Schafer, "Morphological skeleton representation and coding of binary images", *IEEE Transactions on Acoustics, Speech and Signal Processing*, vol. ASSP-34, no. 5, pp. 1228-1244, Oct. 1986.

[32] J.F. Bronskill, A.N. Venetsanopoulos, "Multidimensional shape description and recognition using mathematical morphology", *Journal of Intelligent and Robotic Systems*, vol. 1, pp. 117-143, 1988.

[33] A.V. Oppenheim, R.W. Schafer, T.G. Stockham, "Nonlinear filtering of multiplied and convolved signals", *Proc. of IEEE*, vol.56, pp. 1264-1291, Aug. 1968.

[34] L.R. Rabiner, R.W. Schafer, *Digital processing of speech signals*, Prentice-Hall, 1978.

[35] I. Pitas, A.N. Venetsanopoulos, "Nonlinear mean filters in image processing", *IEEE Transactions on Acoustics, Speech and Signal Processing*, vol. ASSP-34. no. 3, pp. 573-584, June 1986.

[36] T. Koh, E.J. Powers, "Second-order Volterra filter and its application to nonlinear system identification", *IEEE Transactions on Acoustics, Speech and Signal Processing,* vol. ASSP-33, no. 6, pp. 1445-1455, 1985.

[37] H.H. Chiang, C.L. Nikias and A.N. Venetsanopoulos, "Efficient implementations of quadratic digital filters", *IEEE Trans. on Acoustics, Speech and Signal Processing,* vol. ASSP-34, no. 6, pp. 1511-1528, 1986.

[38] B.G. Mertzios, G.L. Sicuranza and A.N. Venetsanopoulos, "Efficient structures for 2-D quadratic filters", *IEEE Trans. on Acoustics, Speech and Signal Processing,* vol. ASSP-37, no. 5, pp. 765-768, May 1989.

[39] I. Pitas and A.N. Venetsanopoulos, "A new filter structure for the implementation of certain classes of image processing operations", *IEEE Trans. on Circuits and Systems,* vol. CAS-35, no. 6, pp. 636-647, June 1988.

[40] J.S. Lee, "Digital image enhancement and noise filtering by local statistics", *IEEE Transactions on Pattern Analysis and Machine Intelligence* , vol. PAMI-2, no. 2, pp. 165-168, March 1980.

[41] N.F. Nahi, A. Habibi, "Decision directed recursive image enhancement", *IEEE Transactions on Circuits and Systems*, vol. CAS-6, pp. 286-293, March 1975.

[42] I. Pitas, A.N. Venetsanopoulos, "Nonlinear order statistics filters for image filtering and edge detection", *Signal Processing* , vol. 10, pp. 395-413, 1986.

[43] X.Z. Sun, A.N. Venetsanopoulos, "Adaptive schemes for noise filtering and edge detection by the use of local statistics", *IEEE Transactions on Circuits and Systems*, vol. CAS-35, no.1, pp. 57-69, Jan. 1988.

[44] R. Bernstein, "Adaptive nonlinear filters for simultaneous removal of different kinds of noise in images" *IEEE Transactions on Circuits and Systems*, vol. CAS-34, no. 11, pp. 1275-1291, Nov. 1987.

[45] G.L. Sicuranza and G. Ramponi, "Adaptive nonlinear filters using distributed arithmetic", *IEEE Trans. on Acoustics, Speech and Signal Processing,* vol. ASSP-34, no. 3, pp. 518-526, June 1986.

[46] F. Palmieri, C.G. Boncelet Jr., "A class of nonlinear adaptive filters", *Proc. IEEE International Conference on Acoustics, Speech and Signal Processing*, pp. 1483-1486, New York, 1988.

[47] I. Pitas, A.N. Venetsanopoulos, "Adaptive *L*-filters", *Proc. European Conference on Circuit Theory and Design*, Brighton, England, 1989.

CHAPTER 2

STATISTICAL PRELIMINARIES

2.1 INTRODUCTION

Many classes of nonlinear filters are based on the field of robust estimation and especially on order statistics. Both of these fields have been developed by statisticians in the last three decades and they have now reached maturity. In this chapter an introduction to robust statistics and order statistics will be given, as a mathematical preliminary to nonlinear filters. The interested reader can find more information in specialized books [1-5].

One of the major approaches to the analysis of statistical data is *parametric data analysis.* This approach makes an assumption about the model underlying the data $\mathbf{x}$. Such a model is usually a probability distribution of the data (e.g., Gaussian, uniform, double exponential distribution, etc.). Frequently, such a distribution is indexed by a parameter vector θ of dimension n, taking values in the Cartesian product $\Theta = \Theta_1 \times \Theta_2 \times \cdots \times \Theta_n$. Such parameters for the Gaussian distribution are the mean μ and the standard deviation σ, i.e., $\theta^T = [\, \mu \; \sigma \,]$. The estimation of the mean μ corresponds to *location estimation*, whereas the estimation of the standard deviation σ corresponds to *scale estimation.* The aim of parametric analysis is to find an estimate of θ. If the model contains only one parameter θ, the estimator is one-dimensional. If the model contains a parameter vector θ and its parameters are estimated simultaneously, the estimator is multidimensional. The use of parametric models is a clever way to perform data reduction, since it reduces the amount of data to the n parameters of the parameter vector θ.

The major problem in parametric estimation (e.g., in maximum likelihood estimation) is the fact that the exact probability distribution of the data $F(\mathbf{x}, \theta)$ must be known. This is seldom the case due to the following reasons [5, p.21]:

1) The existence of the so-called *gross errors* or *outliers,* i.e., data whose distribution is far away from the assumed distribution $F(\mathbf{x}, \theta)$.

2) The presence of rounding and truncation effects. It is very well known that all data processed by computers lie in a limited range due to the finite wordlength of the computer. Therefore the data are truncated or rounded,

if their values exceed the maximum value permissible by the computer.

3)   The probability distribution $F(x,\theta)$ is only an approximation to reality. This distribution usually results from the modeling of the physical phenomenon, which produces the data x. The accuracy of $F(x,\theta)$ depends on the accuracy and the correctness of the underlying physical model.

The most severe deviation from the assumed distribution $F(x,\theta)$ is caused by the presence of outliers. Therefore, this problem will be treated in more detail. Outliers (i.e., the outlying observations) may be caused in three different ways:

1)   **Acquisition or transmission errors.** The data recorded in this case are completely outlying due to errors in the acquisition device (e.g., in the A/D converter), in the recording media (e.g., on a magnetic disk or tape), or in the transmission channel. In image processing applications, such errors produce the so-called *salt-pepper noise* (i.e., positive and negative impulses).

2)   **Long-tailed distributions.** Sometimes moderate outliers are produced by such distributions.

3)   **Model failure.** The assumed model $F(x,\theta)$ is wrong; therefore, some data appear as outliers.

The presence of these outliers renders the classical parametric estimation procedures (e.g., maximum likelihood estimation) unsafe. Mild deviations from normality are catastrophic for classical estimators, e.g., arithmetic mean $\bar{x}$ and sample standard deviation s. Even when the outliers are sparse, their presence causes the efficiency of the classical estimators to deteriorate. Therefore, the two main reasons to develop robust estimation techniques are:

a)   Rejection of outliers, which is necessary to prevent disastrous results.

b)   Development of estimators having high efficiency in the presence of approximate models.

The simplest approach to robustness is to reject outliers by heuristics and then to use classical estimation techniques. This approach, however naive it may be, is still followed by many scientists, for reasons of simplicity. However, it has several disadvantages. First of all, the detection of outliers is not an easy task. Sometimes they are well hidden in the data. In other cases it is not easy to decide if an observation is outlying, because its value is marginal. The second disadvantage is that the efficiency of these techniques is neither optimal, nor can it be estimated, since it employs heuristics. Therefore, the use of the theory of robust estimation is necessary. The following definition can be given to robust estimation [5, p.7]:

Robust estimation is the estimation theory based on approximate parametric models.

Its aims are to find the parameters best fitting to the bulk of the data and to identify outliers. The theory of robust statistics is only three decades old. It was Tukey (1960) who demonstrated the nonrobustness of the mean and investigated various robust estimators. However, the first theory of robust estimation (the so-called *M*-estimators) is due to Huber (1964). Approximately at the same time the so-called *R*-estimators were developed by Hodges and Lehmann (1963). These estimators were originally derived from rank tests. During the same period, the median and the linear combination of order statistics (called *L*-estimators) were proposed and used as robust estimators by Tukey, David, and others. Since then, a whole new field of statistics has been developed for robust estimation and testing. Several books (e.g., [1-5]) have already been published on this subject. Robust estimation techniques, especially the *L*-estimators, have found several applications in time series analysis [19-23] and in digital signal and image processing [6,24-30].

In subsequent sections we shall describe more thoroughly various types of robust estimators. The analysis of their performance will be based on the *influence functions (IF)* [2,5]. These functions have been proven to be very powerful tools for the description of robustness, for the analysis of the performance of robust estimators, and for the design of robust estimators having prescribed efficiency and robustness. Therefore, the next section is devoted to the definition of the influence function.

2.2 INFLUENCE FUNCTION AND MEASURES OF ROBUSTNESS

What is the effect of an outlier on the performance of an estimator? This problem is perhaps the most important for the analysis of an estimator and for the definition of robustness. The influence function (*IF*) is a tool to solve this problem. According to its name, it shows the influence of an additional observation on the estimate. This is a rather heuristic definition of the influence function. A mathematical definition will be given subsequently.

Let us suppose that we have the one-dimensional, independent, and identically distributed (iid) observations $\mathbf{x}^T = [x_1, \ldots, x_n]$, belonging to the sample space $\mathbf{X}$, which is a subset of the real numbers $\mathbf{R}$. Their empirical distribution (histogram) G_n is given by [5, p.82]:

$$G_n = \frac{1}{n} \sum_{i=1}^{n} \delta_{x_i} \qquad (2.2.1)$$

where δ_x is a delta function located at x. When n tends to infinity, the empirical distribution G_n tends to the true distribution G. A parametric model of the observations consists of a family of probability distributions F_θ on the sample space **X**. The probability distribution F_θ depends on the unknown parameter θ, whose parameter space is Θ. In classical parametric data analysis the observations **x** are assumed to be exactly distributed according to F_θ and θ is estimated based on the data **x**. Robust estimation assumes that F_θ is only an approximation to reality. Under this assumption an estimator $T_n = T_n(x_1, \ldots, x_n) = T_n(G_n)$ of θ must be found. In the following the estimators T_n which converge to $T(G)$ in probability will be considered:

$$\lim_{n \to \infty} T_n(x_1,..,x_n) = T(G) \qquad (2.2.2)$$

i.e., the estimators which can be replaced by functionals asymptotically. $T(G)$ is the *asymptotic value* of $\{ T_n; \; n \geq 1 \}$ at G. Furthermore, it is assumed that the functionals $T(G)$ are Fisher consistent:

$$T(F_\theta) = \theta , \qquad \theta \in \Theta \qquad (2.2.3)$$

This means that the estimator $\{ T_n; \; n \geq 1 \}$ asymptotically measures the right quantity at the model distribution F_θ.

The influence function of T at F is given by [2, p.13], [5, p.84]:

$$IF(x;T,F) = \lim_{t \to 0} \frac{T((1-t)F + t\delta_x) - T(F)}{t} \qquad (2.2.4)$$

for those $x \in$ **X**, where the limit exists. The influence function shows the effect of an infinitesimal contamination at point x on the estimator $\{ T_n; \; n \geq 1 \}$ divided by the mass of the contamination. If G_{n-1} is the empirical distribution of $n-1$ observations and $t = 1/n$, the influence function shows approximately the effect of an additional observation at x:

$$IF(x;T,G_{n-1}) \simeq n[T((1-\frac{1}{n})G_{n-1} + \frac{1}{n}\delta_x) - T(G_{n-1})] \qquad (2.2.5)$$

The approximation (2.2.5) is also called *sensitivity curve* [2, p.15], [19]. Because of these facts, the IF approach to robust estimation is also called infinitesimal approach.

The *asymptotic variance* $V(T,F)$ of the estimator T at a distribution F is given by [2, p. 14],[5, p.85]:

$$V(T,F) = \int IF(x;T,F)^2 dF(x) \qquad (2.2.6)$$

In most cases the estimator $\{ T_n; \; n \geq 1 \}$ is asymptotically normal, i.e., for large samples $(n \to \infty)$, the distribution of $\sqrt{n}(T_n - T(F))$ tends weakly to the normal distribution, having zero mean and variance $V(T,F)$. Therefore, the asymptotic variance $V(T,F)$ is a very important measure of the performance of the estimator $\{ T_n; \; n \geq 1 \}$ and (2.2.6) is a practical procedure for its calculation in many cases, as will be seen later on. (2.2.6) can be used to define the *asymptotic relative efficiency ARE* (T,S) of two estimators $T(F)$, $S(F)$:

$$ARE\,(T,S) = \frac{V\,(S,F)}{V\,(T,F)} \tag{2.2.7}$$

If the functional T is Fisher consistent, as is described in (2.2.3), the Cramer-Rao inequality can be stated as follows [5, p.86]:

$$\int IF\,(x;T,F)^2 dF\,(x) \geq \frac{1}{J\,(F)} \tag{2.2.8}$$

where:

$$J\,(F) = \int (\frac{\partial}{\partial \theta}[lnf_\theta(x)])^2 dF = \int (\frac{\partial f_\theta(x)/\partial \theta}{f_\theta(x)})^2 dF \tag{2.2.9}$$

for a specific $\theta \in \Theta$. f_θ is the probability density function (pdf) of F_θ. The equality in (2.2.8) holds if and only if:

$$IF\,(x;T,F) \;\sim\; \frac{\partial}{\partial \theta}[lnf_\theta(x)] \tag{2.2.10}$$

Therefore, the estimator { T_n; $n \geq 1$ } is *asymptotically efficient* if and only if:

$$IF\,(x;T,F) = J\,(F)^{-1} \frac{\partial}{\partial \theta}[lnf_\theta(x)] \tag{2.2.11}$$

Finally, the *absolute asymptotic efficiency* e of an estimator { T_n; $n \geq 1$ } is given by:

$$e = \frac{1}{V\,(T,F)J\,(F)} \tag{2.2.12}$$

The most important robustness measure based on the influence function is the *gross-error sensitivity* γ^* of T at a distribution F:

$$\gamma^* = \sup_x \;|IF\,(x;T,F)| \tag{2.2.13}$$

where *sup* denotes supremum. The gross error sensitivity measures the worst effect of a contamination at any point $x \in X$. If γ^* is finite, T is called *B-robust* at F. B comes from the word bias. An estimator that minimizes γ^* is called *most B-robust* estimator. An estimator that minimizes $V(T,F)$ under the restriction $\gamma^* < c$ is called *optimal B-robust estimator*. Usually the minimization of $V(T,F)$ and γ^* cannot be done simultaneously.

We shall clarify these notions by applying them to the arithmetic mean $\bar{x}$ and to the median $med(x_i)$ as measures of the location, when $X = R$, $\Theta = R$ and the zero mean, unit variance Gaussian distribution is assumed:

$$F_\theta(x) = \Phi(x-\theta) \tag{2.2.14}$$

$$\phi(x) = (2\pi)^{-1/2} exp(-x^2/2)$$

$$\Phi(x) = \int_{-\infty}^{x} \phi(t)dt$$

The arithmetic mean is given by:

$$\bar{x} = T_n = \frac{1}{n} \sum_{i=1}^{n} x_i \qquad (2.2.15)$$

Therefore, the corresponding functional is given by:

$$T(G) = \int u \, dG(u) \qquad (2.2.16)$$

By the combining (2.2.4), (2.2.16) and by using the fact that $\int u d\Phi(u) = 0$, the influence function of $\bar{x}$ is found:

$$IF(x;T,\Phi) = \lim_{t \to 0} \frac{\int ud[(1-t)\Phi+t\delta_x](u) - \int ud\Phi(u)}{t} = x \qquad (2.2.17)$$

Its asymptotic variance $V(T,\Phi)$ is given by:

$$V(T,\Phi) = \int IF(x;T,\Phi)^2 d\Phi(x) = 1 \qquad (2.2.18)$$

Furthermore, it is found that $J(\Phi) = 1$. Therefore the equality holds for the Cramer-Rao bound (2.2.8). Indeed, $IF(x;T,\Phi)$ is proportional to $\frac{\partial}{\partial\theta}[lnf_\theta(x)]$ for $\theta = 0$. The absolute asymptotic efficiency e is found to be $e = 1$, by combining (2.2.12) and (2.2.18). However, $\bar{x}$ is not B-robust, since γ^* is unbounded, as it can be seen from (2.2.17). Its influence function shows that even one distant outlier has catastrophic effects on the arithmetic mean.

The median $med(x_i)$ is simply the middle order statistic, when n is odd. Its functional is given by:

$$T(G) = G^{-1}(\frac{1}{2}) \qquad (2.2.19)$$

(2.2.19) gives an alternative classical definition of the median: it is the point x for which $G(x) = 1/2$. It can be shown [2, p. 57] that its influence function is given by:

$$IF(x;T,F) = \frac{1}{2f(F^{-1}(1/2))} sign(x - F^{-1}(\frac{1}{2})) \qquad (2.2.20)$$

If $F = \Phi$, $\Phi^{-1}(1/2) = 0$ and the influence function becomes:

$$IF(x;T,\Phi) = \frac{sign(x)}{2\phi(0)} \qquad (2.2.21)$$

and it is shown in Figure 2.2.1.

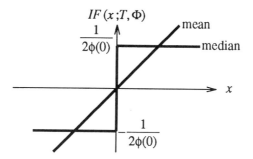

Figure 2.2.1: Influence function of the mean and the median for the normal distribution.

Its absolute asymptotic efficiency is given by:

$$e = \frac{1}{V(T,\Phi)J(\Phi)} = \frac{2}{\pi} \simeq 0.637 \qquad\qquad (2.2.22)$$

therefore the median is worse estimator than $\bar{x}$, for the normal distribution Φ. However, its gross error sensitivity is bounded:

$$\gamma^* = (2\phi(0))^{-1} = (\frac{\pi}{2})^{1/2} \simeq 1.253 \qquad\qquad (2.2.23)$$

and the median is a B-robust estimator.

The influence function gives information about the effect of an infinitesimal contamination at point $x \in X$, i.e., it offers *local* information. It does not give any information about the global reliability of the estimator. Specifically it cannot find the distance of the model distribution beyond which the estimator becomes unreliable. This information is given by the *breakdown point* $\varepsilon^* (0 \leq \varepsilon^* \leq 1)$ [2, p.13],[5, p.97]. If an estimator has breakdown point $\varepsilon^* = 1/2$ (e.g., the median), it is reliable only if less than 50% of the observations outly from the model distribution. The mathematical definition of the breakdown point is based on the notion of the *distance* of two distributions F, G. The *Prohorov distance* of two probability distributions F, G is defined as [5, p.96]:

$$\pi(F,G) = inf \ \{ \ \varepsilon \ ; F(A) \leq G(A^\varepsilon) + \varepsilon \qquad for \ all \ events \ A \ \} \quad (2.2.24)$$

The set A^ε consists of all points whose distance from A is less than ε. If F is the model distribution and G is the data distribution, the term $+\varepsilon$ in (2.2.24) indicates that a fraction ε of the observations may be very differently distributed than the model distribution F. Thus (2.2.24) can take into account the existence of outliers.

The breakdown point ε^* of the estimator $\{T_n; \ n \geq 1\}$ at F for $\Theta = R$ is defined as [5, p.97]:

$$\varepsilon^* = \sup_{n \to \infty} \ \{\varepsilon \leq 1 \ ; \ there \ exists \ r_\varepsilon \ such \ that \ \pi(F,G) < \varepsilon \qquad (2.2.25)$$

$$implies \ G\left(\{|T_n| \leq r_\varepsilon\}\right) \to 1\}$$

The term $\pi(F,G) < \varepsilon$ defines a neighborhood of F. A different neighborhood which can be used in the definition (2.2.25) is $G = (1-\varepsilon)F + \varepsilon H$. This neighborhood denotes that $\varepsilon\%$ of the observations are outliers and belong to the distribution H. It can be proven that the arithmetic mean has breakdown point $\varepsilon^* = 0$, i.e., even one outlier can destroy its results. The median has $\varepsilon^* = 1/2$, i.e., it performs well, even when 50% of the observations are outliers. The α-trimmed mean ($0 \leq \alpha \leq 1/2$) takes away the $[\alpha n]$ smallest and the $[\alpha n]$ largest observations before it calculates the mean. Therefore, it has breakdown point $\varepsilon^* = \alpha$.

2.3 M-ESTIMATORS

The M-estimators were proposed by Huber (1964) as a generalization of the maximum likelihood estimator (MLE). The term M comes from the *generalized Maximum likelihood estimator*. They are defined as the estimators $T_n = T_n(x_1,...,x_n)$, which minimize:

$$\sum_{i=1}^{n} \rho(x_i, T_n) \to min \qquad (2.3.1)$$

where ρ is a function defined on $X \times \Theta$. If ρ possesses the partial derivative $\psi(x,\theta) = \frac{\partial}{\partial \theta}\rho(x,\theta)$, the estimator T_n satisfies the equation:

$$\sum_{i=1}^{n} \psi(x_i, T_n) = 0 \qquad (2.3.2)$$

The maximum likelihood estimator is an M-estimator having $\rho(x,\theta) = -lnf_\theta(x)$ and $\psi(x,\theta) = -\frac{\partial f_\theta(x)/\partial \theta}{f_\theta(x)}$. Definition (2.3.2) is not fully equivalent to (2.3.1). However, it is used to find a solution for the minimization of (2.3.1). The influence function of the M-estimators is given by [2, p. 45],[5, p.101]:

$$IF(x;\psi,F) = -\frac{\psi(x,T(F))}{\int \frac{\partial}{\partial \theta}[\psi(y,\theta)]_{T(F)}dF(y)} \qquad (2.3.3)$$

The asymptotic variance is calculated according to (2.2.6):

$$V(T,F) = \frac{\int \psi^2(x,T(F))dF(x)}{[\int \frac{\partial}{\partial \theta}[\psi(y,\theta)]_{T(F)}dF(y)]^2} \qquad (2.3.4)$$

Both definitions (2.3.1) and (2.3.2) of the *M*-estimators are implicit. Iterative as well as one-step procedures have been implemented for the calculation of the *M*-estimators [5, p. 106]. However, the computational complexity is still high and poses serious difficulties for their calculation in practical cases. This is the main reason for their limited use in digital signal processing and digital image processing applications.

2.4 M-ESTIMATORS OF LOCATION

When the location is estimated and $\Theta = \mathbf{R}$, $\mathbf{X} = \mathbf{R}$, the model has the form:

$$F_{\theta}(x) = F(x-\theta) \tag{2.4.1}$$

Therefore, the ψ-function is chosen to have a similar form:

$$\psi(x, \theta) = \psi(x-\theta) \tag{2.4.2}$$

Furthermore, it must satisfy:

$$\int \psi dF = 0 \tag{2.4.3}$$

so that T is Fisher consistent. The influence function is given by:

$$IF(x;\psi,G) = \frac{\psi(x-T(G))}{\int \psi'(y-T(G))dG(y)} \tag{2.4.4}$$

At the model distribution F, $T(F)=\theta_0=0$ and the influence function becomes:

$$IF(x;\psi,F) = \frac{\psi(x)}{\int \psi'(y)dF(y)} \tag{2.4.5}$$

The asymptotic variance of the *M*-estimator at F is given by:

$$V(\psi,F) = \frac{\int \psi^2(x)dF(x)}{(\int \psi'(y)dF(y))^2} \tag{2.4.6}$$

according to (2.3.4). The Cramer-Rao bound holds for the *M*-estimator with $J(F)$ given by (2.2.9). If the model distribution F is symmetric, the function $\psi(x)$ is chosen to be odd symmetric:

$$\psi(-x) = -\psi(x) \tag{2.4.7}$$

In this case, if ψ is strictly monotone and bounded, the influence function (2.4.5) is bounded and the *M*-estimator is *B*-robust. It has breakdown point $\varepsilon^* = 1/2$.

A special case of the *M*-estimators are the MLE, as has already been stated in the previous section. The MLE estimator possesses the smallest asymptotic variance:

$$V(\psi,F) = \frac{1}{J(F)} = \frac{1}{\int (\frac{\partial f_\theta(x)/\partial\theta}{f_\theta(x)})^2 dF_\theta(x)} \tag{2.4.8}$$

At the normal distribution Φ with $\phi_\theta(x)=(2\pi)^{-1/2}\exp[-(x-\theta)^2/2]$, the MLE minimizes:

$$\sum_{i=1}^{n}[-ln\phi_{T_\bullet}(x_i)] = \frac{1}{2}\sum_{i=1}^{n}(x_i-T_n)^2 + n\,ln(\sqrt{2\pi}) \rightarrow \min \tag{2.4.9}$$

or equivalently satisfies for $\psi(x)=-\dfrac{\partial\phi_\theta(x)/\partial\theta}{\phi_\theta(x)}=-(x-\theta)$:

$$\sum_{i=1}^{n}\psi(x_i,T_n) = -\frac{1}{2}\sum_{i=1}^{n}(x_i-T_n) = 0 \tag{2.4.10}$$

$$T_n = \bar{x} = \frac{1}{n}\sum_{i=1}^{n}x_i \tag{2.4.11}$$

Therefore, the MLE for the normal distribution is the arithmetic mean $\bar{x}$. Its poor robustness properties have already been investigated in section 2.2. Its asymptotic variance is $V(\psi,\Phi)=1$.

At the Laplace distribution $f_\theta(x)=\frac{1}{2}exp(-|x-\theta|)$, the MLE minimizes:

$$\sum_{i=1}^{n}[-lnf_{T_\bullet}(x_i)] = \sum_{i=1}^{n}|x_i-T_n| + n\,ln(2) \rightarrow \min \tag{2.4.12}$$

or equivalently for $\psi(x)=-\dfrac{\partial f_\theta(x)/\partial\theta}{f_\theta(x)}=-sign(x-\theta)$

$$\sum_{i=1}^{n}\psi(x_i,T_n) = -\sum_{i=1}^{n}sign(x_i-T_n) = 0 \tag{2.4.13}$$

(2.4.13) is the implicit definition of the median $T_n=med(x_i)$. The median is a B-robust estimator, as has already been stated. The median minimizes the L_1 error norm, whereas the arithmetic mean is the least mean-square error estimator. At the Laplacian distribution, the asymptotic variance of the median is $V(\psi,F)=J(F)^{-1}=1$.

The first robust M-estimator proposed is the *Huber estimator* [2,7]. Its function $\psi(x)$ for $F=\Phi$ is given by:

$$\psi(x) = [x]_{-b}^{b} = \begin{cases} x & |x|<b \\ b\,sign(x) & |x|\geq b \end{cases} \tag{2.4.14}$$

and it is plotted in Figure 2.4.1.

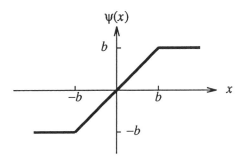

Figure 2.4.1: ψ function of the Huber estimator.

The choice of $\psi(x)$ in (2.4.14) is not arbitrary. It results from the need to find an optimal M-estimator, in the minimax sense, when some observations outly from the symmetric model distribution F. Let us suppose that the outliers follow a symmetric distribution H. The distribution G of the observations lies at the neighborhood P_ε of the distribution F:

$$P_\varepsilon = \{ \ G \ | \ G = (1-\varepsilon)F+\varepsilon H \ \} \tag{2.4.15}$$

where ε is the degree of contamination with outliers. The problem is to find an M-estimator (2.3.2) whose function $\psi(x)$ minimizes the maximal asymptotic variance $V(\psi,G)$ over P_ε [2,5,7]:

$$\min_{\psi} \ \sup_{G \in P_\varepsilon} \ V(\psi,G) \tag{2.4.16}$$

This is obtained by finding a distribution $F_0(x)$ that minimizes the Fisher information matrix $J(F_0)$ over P_ε. The Huber estimator is the MLE for the distribution $F_0(x)$ and its function $\psi(x)$ is given by [2,5,7]:

$$\psi(x) = -\frac{F''_0}{F'_0} \tag{2.4.17}$$

When the model distribution F is Φ, it is found that $F_0(x)$ has the density [2,5,7]:

$$f_0(x) = (1-\varepsilon)(2\pi)^{-1/2} exp\left(-\rho(x)\right) \tag{2.4.18}$$

$$\rho(x) = \int_0^x \psi(t)dt$$

The function $\psi(x)$ is given by (2.4.14) and the constant b satisfies:

$$2\Phi(b)-1+2\frac{\phi(b)}{b} = \frac{1}{1-\varepsilon} \tag{2.4.19}$$

The Huber estimator is B-robust and possesses $\varepsilon^* = 1/2$. When b tends to 0, $\psi(x)$ (2.4.14) tends to $sign(x)$ and the Huber estimator tends to the median . It has been proven [5, p.120] that the Huber estimator is an optimal B-robust estimator for the normal distribution Φ.

2.5 M-ESTIMATORS OF SCALE

The parametric model for scale has the form $F_\theta(x) = F(x/\theta)$, $i.e.$ scale is directly related to the variance of the parametric model. For the case of the normal distribution $\phi(x/\theta)$, the scale parameter θ is essentially the standard deviation σ of the distribution. The parameter space is $\Theta = (0, \infty)$ and $\theta_0 = 1$. The function $\psi(x)$ of the M-estimator has the form:

$$\psi(x, \theta) = \psi(x/\theta) \tag{2.5.1}$$

(2.4.3) is still valid for Fisher consistency. (2.3.3) gives the following influence function for the M-estimators of scale:

$$IF(x; \psi, G) = \frac{\psi(\dfrac{x}{T(G)}) T(G)}{\displaystyle\int \psi'(\dfrac{y}{T(G)}) \dfrac{y}{T(G)} dG(y)} \tag{2.5.2}$$

At the model distribution F, $T(F) = \theta_0 = 1$ and the influence function becomes:

$$IF(x; \psi, F) = \frac{\psi(x)}{\displaystyle\int y \psi'(y) dF(y)} \tag{2.5.3}$$

The Cramer-Rao bound holds. The Fisher information is given by (2.2.9):

$$J(F) = \int (x \frac{f'(x)}{f(x)} + 1)^2 dF(x) \tag{2.5.4}$$

If the distribution F is symmetric, the function $\psi(x)$ is chosen to be even symmetric, $i.e.$ $\psi(x) = \psi(-x)$. If $\psi(x)$ is symmetric, monotone and bounded, the corresponding M-estimator is B-robust and has breakdown point ε^*:

$$\varepsilon^* = -\frac{\psi(0)}{\psi(\infty) - \psi(0)} \leq \frac{1}{2} \tag{2.5.5}$$

If $\psi(x)$ is unbounded the M-estimator of scale is not B-robust.

The maximum likelihood estimate of scale is given by

$$\psi(x) = -x \frac{f'(x)}{f(x)} - 1 \tag{2.5.6}$$

and it has the smallest asymptotic variance $V(\psi, F) = J(F)^{-1}$. At the normal distribution $F = \Phi$, the function $\psi(x)$ is given by:

$$\psi(x) = x^2 - 1 \tag{2.5.7}$$

The M-estimator T_n of standard deviation σ is given by (2.3.2):

$$\sum_{i=1}^{n} [(\frac{x_i}{T_n})^2 - 1] = 0 \tag{2.5.8}$$

$$s = T_n = (\frac{1}{n} \sum_{i=1}^{n} x_i^2)^{1/2} \tag{2.5.9}$$

Therefore, the MLE of scale for the normal distribution Φ is the biased *sample standard deviation* s (2.5.9). The function $\psi(x)$ (2.5.7) is unbounded, thus the sample standard deviation is not B-robust and possesses $\varepsilon^* = 0$.

It has been proven that a robustified version of the MLE of scale is given by [5, p. 122]:

$$\psi(x) = [-x\frac{f'(x)}{f(x)} - 1 - a]_{-b}^{b} \tag{2.5.10}$$

At standard normal distribution $\Phi(x)$, (2.5.10) becomes:

$$\psi(x) = [x^2 - 1 - a]_{-b}^{b} \tag{2.5.11}$$

and its shape is plotted in Figure 2.5.1.

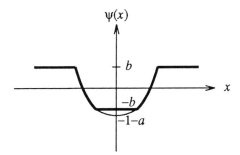

Figure 2.5.1: Shape of the robustified MLE scale estimator for normal distribution.

Estimators (2.5.10-11) are the optimal B-robust M-estimators of scale.

Another robust M-estimator for the normal distribution is described by the equation

$$\psi_{MAD}(x) = sign(|x| - \Phi^{-1}(\frac{3}{4})) \tag{2.5.12}$$

which is the median of the absolute values of the observations multiplied by $1/\Phi^{-1}(3/4) \simeq 1.483$. The multiplication by 1.483 makes the estimator Fisher consistent for $F = \Phi$. If the location is unknown, the estimator (2.5.12) corresponds to the *median of the absolute deviations from the median* (MAD) multiplied by 1.483:

$$T_n = 1.483 \ med\{\,|x_i - med(x_j)|\,\}$$

The influence of (2.5.12) is the following [5, p.107]:

$$IF(x;\psi,\Phi) = \frac{1}{4\Phi^{-1}(3/4)\phi(\Phi^{-1}(3/4))} sign(\,|x|-\Phi^{-1}(3/4)) \qquad (2.5.13)$$

It is easily found from (2.5.13) that:

$$\gamma^* = \frac{1}{4\Phi^{-1}(3/4)\phi(\Phi^{-1}(3/4))} \simeq 1.167 \qquad (2.5.14)$$

Therefore the *M*-estimator (2.5.12) is *B*-robust. In fact, it is proven in [5, p. 142] that the MAD estimator is the most *B*-robust estimator. Its variance is given by $V(\psi,\Phi) = (\gamma^*)^2 \approx 1.361$ and its breakdown point is given by $\varepsilon^* = -\psi(0) / (\psi(\infty) - \psi(0)) = 1/2$.

2.6 ORDER STATISTICS

Order statistics have played an important role in the statistical data analysis and especially in the robust analysis of data contaminated with outlying observations [1-3, 19]. Their robustness properties made them suitable for time series analysis and for digital signal and image applications [6]. All the above-mentioned applications of order statistics will be described in detail in subsequent chapters.

Let the random variables $X_1, X_2, ..., X_n$ be arranged in ascending order of magnitude and then written as

$$X_{(1)} \le X_{(2)} \le \cdots \le X_{(n)} \qquad (2.6.1)$$

$X_{(i)}$ is the so-called *i-th order statistic*. Usually X_i, $i=1,...,n$ are assumed to be independent identically distributed (iid) variables. A very important order statistic is the *median* $X_{(v+1)}$ (for $n=2v+1$) and $(X_{(n/2)} + X_{(n/2+1)})/2$ (for n even). Equally important are the linear combinations of the order statistics, of the form:

$$T_n = \sum_{i=1}^{n} a_i X_{(i)} \qquad (2.6.2)$$

Special cases of T_n are the *range:*

$$W = X_{(n)} - X_{(1)} \tag{2.6.3}$$

and the *extreme deviate* (from the sample mean):

$$ED = X_{(n)} - \bar{x} \tag{2.6.4}$$

In the following, the statistical analysis of the order statistics will be given and their cumulative density functions (cdf), probability density functions (pdf), and expected values and moments will be presented.

If $X_1,...,X_n$ are independent identical distributed (iid) variables, having cdf $F(x)$, the cdf $F_r(x)$ of the r-th order statistic $X_{(r)}$ is given by [1, p.8]:

$$F_r(x) = \sum_{i=r}^{n} \binom{n}{i} F^i(x)[1-F(x)]^{n-i} \tag{2.6.5}$$

Special cases of (2.6.5) are the distribution of the maximum $X_{(n)}$:

$$F_n(x) = F^n(x) \tag{2.6.6}$$

and the distribution of the minimum $X_{(1)}$:

$$F_1(x) = 1-[1-F(x)]^n \tag{2.6.7}$$

The pdf $f_r(x)$ can be derived from (2.6.5) as follows (for continuous random variables):

$$f_r(x) = n\binom{n-1}{r-1} \frac{d}{dx} \int_0^{F(x)} t^{r-1}(1-t)^{n-r} dt = \tag{2.6.8}$$

$$n\binom{n-1}{r-1} F^{r-1}(x)[1-F(x)]^{n-r} f(x)$$

where $f(x) = F'(x)$.

The joint distribution of $X_{(r)}$ and $X_{(s)}$ is denoted by $f_{rs}(x,y),(1 \leq r < s \leq n)$. It is given by the following relation, for $x \leq y$ [1, p.10]:

$$f_{rs}(x,y) = \frac{n!}{(r-1)!(s-r-1)!(n-s)!} \tag{2.6.9}$$
$$\cdot F^{r-1}(x)f(x)[F(y)-F(x)]^{s-r-1}f(y)[1-F(y)]^{n-s}$$

The joint cdf $F_{rs}(x,y)$ can be found by a direct integration of (2.6.9):

$$F_{rs}(x,y) = \sum_{j=s}^{n} \sum_{i=r}^{j} \frac{n!}{i!(j-i)!(n-j)!} \tag{2.6.10}$$
$$\cdot F^i(x)[F(y)-F(x)]^{j-i}[1-F(y)]^{n-j}$$

If $x \geq y, X_{(s)} \leq y$ implies $X_{(r)} \leq x$, thus:

$$F_{rs}(x,y) = F_s(y) \qquad (2.6.11)$$

The conditional pdf of $X_{(s)}$ given $X_{(r)} = x$ for $x \leq y$ is the following:

$$f_{X_{(s)}|X_{(r)}=x}(y) = \frac{(n-r)!}{(s-r-1)!(n-s)!} \qquad (2.6.12)$$

$$\cdot \frac{[F(y)-F(x)]^{s-r-1} f(y)[1-F(y)]^{n-s}}{[1-F(x)]^{n-r}}$$

It can also be shown that for $r < s$ [1, p. 20]:

$$f_{X_{(s)}|X_{(r)}=x_{(r)}, X_{(r-1)}=x_{(r-1)}, \dots, X_{(1)}=x_{(1)}}(y) = f_{X_{(s)}|X_{(r)}=x_{(r)}}(y) \qquad (2.6.13)$$

i.e., the order statistics in a sample from a continuous distribution form a Markov chain.

If $X_1, \dots, X_n$ are independent random variables, and X_i has cdf $F_i(x)$, the cdf of $X_{(r)}$ is given by [1, p.22]:

$$F_r(x, \mathbf{F}) = \sum_{i=r}^{n} \sum_{S_i} \prod_{l=1}^{i} F_{j_i}(x) \prod_{l=i+1}^{n} [1-F_{j_i}(x)] \qquad (2.6.14)$$

where the summation S_i extends over all permutations $(j_1, j_2, \dots, j_n)$ of $1, 2, \dots, n$ for which $j_1 < \cdots < j_i$ and $j_{i+1} < \cdots < j_n$.

If the random independent, identically distributed (iid) variables $X_1, \cdots, X_k$ possess cdf $F_1(x)$ and pdf $f_1(x)$ and the iid variables $X_{k+1}, \dots, X_n$ possess cdf $F_2(x)$ and pdf $f_2(x)$, the pdf $f_r(x)$ of the order statistic $X_{(r)}$ is given by [6, p.176]:

$$f_r(x) = f_1(x) + f_2(x) \qquad (2.6.15)$$

where:

$$f_1(x) = \sum_j k \binom{k-1}{j} \binom{n-k}{r-j-1} f_1(x) F_1(x)^j F_2(x)^{r-j-1} \qquad (2.6.16)$$

$$\cdot [1-F_1(x)]^{k-j-1} [1-F_2(x)]^{n-k-r+j+1}$$

$$f_2(x) = \sum_j (n-k) \binom{k}{j} \binom{n-k-1}{r-j-1} f_2(x) F_1(x)^j F_2(x)^{r-j-1} \qquad (2.6.17)$$

$$\cdot [1-F_1(x)]^{k-j} [1-F_2(x)]^{n-k-r+j}$$

The summations in (2.6.16-17) are to be carried over all natural numbers j for which the coefficients $\binom{p}{q}$ satisfy $p \geq q \geq 0$.

All the previously described formulae (2.6.5-17) for the calculation of the cdf and the pdf of order statistics, although relatively complicated, are numerically tractable. Thus they can be used for the numerical computation of the cdfs and the pdfs. Fast and efficient algorithms for their numerical calculation can also be found in [31].

In the following, some basic formulas on the expected values and moments of order statistics will be given. If X is a continuous random variable, the mean of $X_{(r)}$ is given by:

$$\mu_r = \int_{-\infty}^{\infty} x f_r(x) dx \tag{2.6.18}$$

where $f_r(x)$ is given by (2.6.8). The variance σ_r^2 is given by:

$$\sigma_r^2 = \int_{-\infty}^{\infty} (x - \mu_r)^2 f_r(x) dx \tag{2.6.19}$$

The covariance of $X_{(r)}, X_{(s)}$ is the following:

$$\sigma_{rs} = E[(X_{(r)} - \mu_r)(X_{(s)} - \mu_s)] \tag{2.6.20}$$

and for $r < s$:

$$\sigma_{rs} = \int_{-\infty}^{\infty} \int_{-\infty}^{y} (x - \mu_r)(y - \mu_s) f_{rs}(x,y) dx dy \tag{2.6.21}$$

where $f_{rs}(x,y)$ is given by (2.6.9). Having defined the basics of order statistics, we proceed to the analysis of L-estimators.

2.7 L-ESTIMATORS

L-estimators are based on order statistics. They are explicitly defined and easily calculated. Therefore, they have found extensive applications in digital signal and image processing, as will be shown in subsequent chapters. For example, median filtering is a widely used signal and image filtering technique that is based on L-estimators. In the following, we shall treat again the problem of location and scale estimation by using linear combinations of order statistics, in the general framework of robustness theory and influence functions.

An L-estimator has the following definition:

$$T_n = \sum_{i=1}^{n} a_i x_{(i)} \tag{2.7.1}$$

where $x_{(i)}$ is the i-th order statistic of the observation data. The performance of an L-estimator depends on its weights a_i. A class of location L-estimators is obtained by the following choice [5, p.108]:

$$a_i = \frac{\int_{(i-1)/n}^{i/n} h(\lambda) d\lambda}{\int_0^1 h(\lambda) d\lambda} \tag{2.7.2}$$

where $h(\lambda)$ is a function $[0,1] \to R$ satisfying $\int_0^1 h(\lambda)d\lambda \neq 0$. The corresponding functional is the following:

$$T(G) = \frac{\int xh(G(x))dG(x)}{\int h(F(y))dF(y)} \tag{2.7.3}$$

Its influence function is given by:

$$IF(x;T,F) = \frac{\int_0^x h(F(y))d\lambda(y) - \int_0^t \int h(F(y))d\lambda(y)dF(t)}{\int h(F(y))dF(y)} \tag{2.7.4}$$

The median corresponds to $h(\lambda) = \delta(\lambda - 1/2)$, where $\delta(.)$ is the delta function. Thus $\int_0^1 h(\lambda)d\lambda = 1$. Its coefficients are given by the following relation for $n=2\nu+1$:

$$a_i = \begin{cases} 1 & i=\nu+1 \\ 0 & i\neq\nu+1 \end{cases} \tag{2.7.5}$$

Its functional $T(G)$ is given by (2.7.3)

$$T(G) = \int_0^1 G^{-1}(y)h(y)d\lambda(y) = G^{-1}(1/2) \tag{2.7.6}$$

This has already been stated in (2.2.19). Therefore, its influence function is:

$$IF(x;T,F) = \frac{1}{2f(F^{-1}(1/2))}sign(x-F^{-1}(1/2)) \tag{2.7.7}$$

As it has already been discussed in section 2.2, the median is a B-robust estimator having $\gamma^* = [2f(F^{-1}(1/2))]^{-1}$. It has breakdown point $\varepsilon^* = 1/2$. It is also proven that it is the most B-robust estimator, i.e., it has the minimal gross-error sensitivity γ^* [5, p. 133].

The L-estimators having maximal asymptotic efficiency have influence function proportional to $[lnf(x)]'$ according to (2.2.10). The influence function of the L-estimators satisfies the following equation, according to (2.7.4):

$$\frac{d}{dx}IF(x;T,F) = \frac{h(F(x))}{\int h(F(y))dF(y)} \tag{2.7.8}$$

Therefore, L-estimators having maximal asymptotic efficiency satisfy:

$$h(F(x)) \sim [lnf(x)]'' \tag{2.7.9}$$

At the normal distribution Φ, $[ln\phi(x)]''$ is a constant and after appropriate scaling, (2.7.9) gives $h(\lambda) = 1$, $0 \le \lambda \le 1$. This corresponds to $a_i = 1/n$, $i = 1$, $\cdots$ n. Therefore, the L-estimator having maximal asymptotic efficiency for Φ is the arithmetic mean. Such an L-estimator for the Laplacian distribution is the median. The L-estimator having maximal asymptotic efficiency for the *least favorable distribution* (see 2.4.18-19) is the α-*trimmed mean*:

$$T_n = \frac{1}{n(1-2\alpha)} \sum_{i=\alpha n+1}^{n-\alpha n} x_{(i)}$$

(2.7.10)

Its function $h(\lambda)$ is given by:

$$h(\lambda) = \begin{cases} 1 & \alpha \le \lambda \le 1-\alpha \\ 0 & \lambda < \alpha, \lambda > 1-\alpha \end{cases}$$

(2.7.11)

The α-trimmed mean has the same influence function IF $(x;\psi,F)$ as the Huber estimator (2.4.14) for the distribution F given by (2.4.18) and for $\alpha = F(-b)$. However, these two estimators differ outside this model. The α-trimmed mean has breakdown point $\varepsilon^* = \alpha$, whereas the Huber estimator has $\varepsilon^* = 1/2$, i.e., Huber estimator has better global robustness. It is also proven [5, p.124] that the α-trimmed median is the optimal B-robust estimator for the normal distribution Φ, if $\alpha = \Phi(-b)$ and b satisfies (2.4.19). If b tends to 0, α tends to 1/2 and the α-trimmed mean tends to the median.

Several other L-estimators have been proposed in the literature [1, pp. 159-160, 180]:

a) *Quasi-midrange MR*:

$$MR(i) = \frac{1}{2}(x_{(i)}+x_{(n+1-i)})$$

(2.7.12)

b) *Midpoint MP*:

$$MP = \frac{1}{2}(x_{(1)}+x_{(n)})$$

(2.7.13)

c) *Winsorized mean*:

$$W_n(r) = \begin{cases} \dfrac{(r+1)[x_{(r+1)}+x_{(n-r)}]+ \displaystyle\sum_{i=r+2}^{n-r-1} x_{(i)}}{n} & 0 < r < (n-1)/2 \\ x_{(\frac{n+1}{2})} & r = (n-1)/2, \ n \ odd \end{cases}$$

(2.7.14)

d)   *Linearly weighted means*:

$$
L_n(r) = \begin{cases} \displaystyle\sum_{j=1}^{\frac{n}{2}-r} \frac{(2j-1)[x_{(r+j)}+x_{(n-r+1-j)}]}{2(n/2-r)^2} & n \; even \\[4mm] \displaystyle\sum_{j=1}^{(n-1)/2-r} \frac{(2j-1)[x_{(r+j)}+x_{(n-r+1-j)}]+(n-2r)x_{(\frac{n+1}{2})}}{[(n-1)/2-r\,]^2+[(n+1)/2-r\,]^2} & n \; odd \end{cases}
\tag{2.7.15}
$$

L-estimators (2.7.1) can also be used for scale estimation if the coefficients a_i, $i = 1, ..., n$ are chosen appropriately:

$$
a_i = \frac{\displaystyle\int_{(i-1)/n}^{i/n} h(\lambda)d\lambda}{\displaystyle\int_0^1 h(t)F^{-1}(t)d\lambda(t)}
\tag{2.7.16}
$$

The function $h(\lambda)$: $[0,1] \to R$ is chosen to be odd symmetric for symmetric distributions F. The functional $T(G)$ and the influence functions corresponding to F are given by:

$$
T(G) = \frac{\int x h(G(x))dG(x)}{\int y h(F(y))dF(y)}
\tag{2.7.17}
$$

$$
IF(x;T,F) = \frac{\displaystyle\int_0^x h(F(y))d\lambda(y)-\int\int_0^t h(F(y))d\lambda(y)dF(t)}{\int y h(F(y))dF(y)}
\tag{2.7.18}
$$

A well-known L-estimator of scale is the *t-quantile range* having function $h(\lambda) = \delta(\lambda-1+t) - \delta(\lambda-t)$. The corresponding coefficients a_i are given by:

$$
a_i = \begin{cases} -1 & i= tn \\ 1 & i= (1-t)n \\ 0 & i\neq tn,\ (1-t)n \end{cases}
\tag{2.7.19}
$$

The *interquartile range* ($t = 1/4$) for $F = \Phi$ has the same influence function with the one of MAD(x_i), given by (2.5.13) [5, p.110]. Therefore, it is B-robust and has $\gamma^* \approx 1.167$. Its breakdown point is only $\varepsilon^* = 1/4$ instead of $\varepsilon^* = 1/2$ for MAD(x_i), i.e., MAD has better global robustness than the interquantile range.

The t-quantile range is equivalent to the *quasi-range* for $t \neq 0$:

$$W_{(i)} = x_{(n+1-i)} - x_{(i)}, \qquad 2 \leq i \leq [\frac{n}{2}] \tag{2.7.20}$$

where $[x]$ denotes the integer part of a real number x. A special case of the quasi-range for $i = 1$ is the *range W*:

$$W = x_{(n)} - x_{(1)} \tag{2.7.21}$$

Another scale estimator is the *thickened range*:

$$J_i = W + W_{(2)} + \ldots + W_{(i)} \tag{2.7.22}$$

The range is very sensitive in the presence of outliers, whereas the quasi-range and the thickened range are not. Another unbiased estimator that is not as influenced by outliers is the following [1, p. 191]:

$$\hat{\sigma} = \frac{2\sqrt{\pi}}{n(n-1)} \sum_{i=1}^{n} [i - \frac{1}{2}(n+1)] x_{(i)} \tag{2.7.23}$$

Other interesting variants are also given in [1, p. 192]:

$$G = \sum_{i=1}^{[\frac{n}{2}]} \frac{(n-2i+1)W_{(i)}}{n(n-1)} \tag{2.7.24}$$

$$G = \frac{2}{n(n-1)} \sum_{i=1}^{n-1} i(n-i)(x_{(i+1)} - x_{(i)}) \tag{2.7.25}$$

2.8 R-ESTIMATORS

The *R*-estimators have been proposed by Hodges and Lehmann (1963). The name *R*-estimator comes from the *rank* $R_i = r_i(x)$ of an observation x. Let

$$x_{(1)} \leq x_{(2)} \leq \cdots \leq x_{(n)} \tag{2.8.1}$$

be the order statistics of the observations $x_1, \ldots, x_n$. Let $r_i(x)$ be the number of x's that are less than or equal to x_i, i.e., the rank of x_i in the sequence (2.8.1):

$$x_i = x_{(r_i)}, \qquad 1 \leq i \leq n \tag{2.8.2}$$

The statistic $R_i = r_i(x)$ is called the *rank* of x_i [4]. The *R*-estimators of location $T_n = T(x_1, \ldots, x_n)$ have an implicit definition [5, p.111]:

Definition 1. An *R*-estimator of location T_n is chosen in such a way that the test statistic:

$$S_N = \frac{1}{n} \sum_{i=1}^{n} a_N(R_i) \tag{2.8.3}$$

becomes as close to zero as possible, when it is computed from the samples $x_1, \ldots, x_n$ and $2T_n - x_1, \ldots, 2T_n - x_n$.

R_i is the rank of the sample x_i in the observation set $\{ x_1, \ldots, x_n \}$ $\cup \{2T_n - x_1, \ldots, 2T_n - x_n\}$ which has size $N = 2n$. The coefficients $a_N(i)$ are given by:

$$a_N(i) = N \int_{(i-1)/N}^{i/N} J(u)du , \qquad N=2n \tag{2.8.4}$$

The function $J(u)$ is odd symmetric $J(1 - u) = -J(u)$ and satisfies $\int J(u) \, du = 0$. Thus the coefficients $a_N(i)$ satisfy $\sum_{i=1}^{n} a_N(i)=0$. The test statistic S_N is a rank test for location shift. If the observations $x_1, \ldots, x_n$ and their "mirror images" $2T_n - x_1, \ldots, 2T_n - x_n$ have the same location, the test statistic S_N detects no location shift and its values come close to zero. Based on the definition the R-estimator, T_n corresponds to the following functional $T(G)$:

$$\int J[\frac{1}{2}G(y)+\frac{1}{2}(1-G(2T(G)-y))]dG(y) = 0 \tag{2.8.5}$$

The influence function $IF(x;T,F)$ is obtained by substituting G by $F_{t,x} = (1-t)F + t\delta_x$ in (2.8.5) and differentiating. The result is

$$IF(x;T,F) = \frac{U(x)-\int U(x)f(x)dx}{\int U'(x)f(x)dx} \tag{2.8.6}$$

where $U(x)$ is given by:

$$U(x) = \int_0^x J'[\frac{1}{2}(F(y)+1-F(2T(F)-y))]f(2T(F)-y)d\lambda(y) \tag{2.8.7}$$

When F is symmetric, the following relations are valid: $T(F)=0$ and $U(x) = J(F(x))$. Therefore:

$$IF(x;T,F) = \frac{J(F(x))}{\int J'(F(x))f(x)^2 dx} \tag{2.8.8}$$

If J is monotone and integrable, the breakdown point ε^* of the R-estimator is given by [2, p. 67]:

$$\int_{\frac{1}{2}}^{1-\varepsilon^*/2} J(\lambda)d\lambda = \int_{1-\varepsilon^*/2}^{1} J(\lambda)d\lambda \tag{2.8.9}$$

According to (2.2.10) the R-estimators having maximal efficiency must satisfy:

$$J(F(x)) \quad \sim \quad -\frac{f'(x)}{f(x)} \tag{2.8.10}$$

for symmetric F. At $F = \Phi$, the *normal scores* estimator is found, having function $J(u)$:

$$J(u) = \Phi^{-1}(u) \tag{2.8.11}$$

Its influence function is given by (2.8.8):

$$IF(x;T,\Phi) = x \tag{2.8.12}$$

Therefore the normal scores estimator is not B-robust. Its breakdown point is given by $\varepsilon^* = 2\Phi(-\sqrt{\ln 4}) \approx 0.239$. For the logistic distribution $F(x) = [1 + exp(-x)]^{-1}$, the R-estimator having the maximal efficiency is the *Hodges-Lehmann* estimator:

$$T_n = med\{\frac{x_i + x_j}{2}, \quad i=1,..n, \quad j=1,..,n\} \tag{2.8.13}$$

Its function $J(u)$ is given by [2, p.65]:

$$J(u) = u - \frac{1}{2} \tag{2.8.14}$$

The Hodges-Lehmann estimator is related to the *Wilcoxon test* [4]. Its influence function is given by [2, p.65]:

$$IF(x;T,F) = \frac{F(x) - \frac{1}{2}}{\int f^2(y)dy} \tag{2.8.15}$$

Its breakdown point is $\varepsilon^* = 1 - 1/\sqrt{2} \approx 0.293$. The R-estimator having maximal efficiency for the Laplace distribution is the median having function $J(u)$:

$$J(u) = \begin{cases} -1 & u < 1/2 \\ 1 & u > 1/2 \end{cases} \tag{2.8.16}$$

The median is related to the *sign test* [4]. At the Cauchy distribution $f(x) = 1/[\pi(1+x^2)]$, the following R-estimator is obtained:

$$J(u) = -sin(2\pi u) \tag{2.8.17}$$

An alternative definition of the R-estimators is the following [3, p. 382]:

Definition 2. Assign the weights:

$$w_{jk} = \frac{d_{n-k+j}}{\sum\limits_{i=1}^{n} id_i} \tag{2.8.18}$$

to each of the $n(n+1)/2$ averages $(x_{(j)} + x_{(k)})/2$, $j \leq k$. The R-estimator T_n is the median of the discrete distribution that assigns probability w_{jk} to each of the averages $(x_{(j)} + x_{(k)})/2$.

This definition is due to Jaeckel [14]. It leads easily to some well known estimators. If we choose $d_1 = \cdots = d_{n-1} = 0$, $d_n = 1$, then

$$\sum_{i=1}^{n} id_i = n \qquad (2.8.19)$$

$$w_{jk} = \begin{cases} 1/n & j=k \\ 0 & otherwise \end{cases}$$

The discrete distribution assigns probability $1/n$ to each $x_{(1)}, \ldots, x_{(n)}$. Therefore, the R-estimator coincides with $med(x_i)$ in this case. If $d_1 = \cdots = d_n = 1$, the discrete distribution assigns equal probability to each of the values $(x_{(j)} + x_{(k)})/2$ and the R-estimator is the median of these values:

$$T_n = med\{ \frac{x_{(j)} + x_{(k)}}{2} , \ 1 \leq j \leq k \leq n \ \} \qquad (2.8.20)$$

(2.8.20) is equivalent to the Hodges-Lehmann estimator (2.8.13).

2.9 DISCUSSION

In many cases, the statistical estimation of certain parameters uses specific statistical models for the observation data. For example, a widespread noise model in digital signal processing applications is the Gaussian distribution. The corresponding estimators are chosen to be optimal or, at least, to have good performance in this data model. The robust estimation theory deals with the sensitivity of these estimators under mild or severe deviations from the data model. The estimators are robust if they have acceptable performance under such deviations. Three such classes of estimators, namely the M-, L-, and R-estimators have been described in the previous sections. Most of these estimators come from the pioneering work of Huber, Tukey, Lehmann, David and other researchers in the sixties. These estimators have found extensive applications in digital signal and image processing, as it will be seen in the subsequent chapters. Special emphasis will be given to the applications of order statistics, because they have played an important role in the development of various classes of nonlinear filters. Some other applications of robust statistics in signal processing, described in a different perspective, can be found in a very informative survey [24]. There exist also other classes of robust estimators, e.g. the W-estimators [19], A-estimators [15], and P-estimators [32]. The interested reader can find a brief description in [5, pp.113-116].

REFERENCES

[1] H. .A. David, *Order statistics,* John Wiley, 1980.

[2] P. S. Huber, *Robust statistics,* John Wiley, 1981.

[3] E. L. Lehmann, *Theory of point estimation,* John Wiley, 1983.

[4] J. Hajek, Z. Sidak, *Theory of rank tests,* Academic Press, 1967.

[5] F. Hampel, E. Ronchetti, P. Rousseeuw, W. Stahel, *Robust statistics,* John Wiley, 1986.

[6] B.J. Justusson, "Median filtering: Statistical properties", in *Topics in applied physics,* vol. 43, T.S. Huang editor, Springer Verlag, 1981.

[7] P. J. Huber, "Robust estimation of a location parameter", *Ann. Math. Statist.,* vol.35, pp. 73-101, 1964.

[8] D. F. Andrews, P. J. Bickel, F. R. Hampel, P. S. Huber, W. H. Rogers, J. W. Tukey, *Robust estimates of location: Survey and advances,* Princeton University Press, 1972.

[9] P. J. Bickel, "One step Huber estimates in the linear model", *J. Am. Statist. Assoc.,* vol. 70, pp. 428-434.

[10] R. B. Murphy, *On test for outlying observations,* Ph.D. Thesis, Princeton University, 1951.

[11] D. M. Hawkins, "Fractiles of an extented multiple outlier test", *J. Statist. Comput. Simulation,* vol.8, pp. 227-236, 1979.

[12] V. Barnett, T. Lewis, *Outliers in statistical data,* Wiley, 1978.

[13] J. L. Hodges Jr., E. L. Lehmann, "Estimates of location based on rank tests", *Ann. Math. Statist.,* vol.34, pp. 598-611, 1963.

[14] L. A. Jaeckel, *Robust estimates of location,* Ph.D. Thesis, University of California, Berkeley, 1969.

[15] D. A. Lax, *An interim report of a Monte Carlo study of robust estimators of width,* Technical report 93, series 2, Department of Statistics, Princeton University, 1975.

[16] J. Wolfowitz, "The minimum distance model", *Ann. Math. Statist.* vol.28, pp.75-88, 1957.

[17] M. V. Johns, "Robust Pitman-like estimators", in *Robustness in statistics,* R. L. Launer and G. N. Wilkinson editors, Academic Press, 1979.

[18] P. J. Rousseeuw, V. Yokai, "Robust regression by means of S-estimators", in *Robust and nonlinear time series analysis,* J. Franke, W. Hardle, R. D.

Martin editors, *Lecture notes in statistics*, vol.26, Springer, 1984.

[19] J. W. Tukey, *Exploratory data analysis*, Addison-Wesley, 1970, 1977.

[20] P. Papantoni-Kazakos, R.M. Gray, "Robustness of estimators on stationary observations", *Ann. Prob.*, vol.7, pp.989-1002, 1979.

[21] B. Kleiner, R.D. Martin, D.J. Thomson, "Robust estimation of power spectra", *J. Roy. Statist. Soc. Ser. B,* vol.41, pp.313-351, 1979.

[22] R.D. Martin, D.J. Thomson, "Robust resistant spectrum estimation", *Proceedings of IEEE*, vol. 70, pp. 1097-1115, Sept. 1982.

[23] B.T. Poljak, Y.Z.Tsypkin, "Robust identification", *Automatica*, vol. 16, pp. 53-63, 1980.

[24] S.A Kassam, H.V. Poor, "Robust techniques for signal processing: A survey", *Proceedings of IEEE*, vol. 73, no. 3, pp. 433-481, March 1985.

[25] H.V. Poor, "On robust Wiener filtering", *IEEE Transactions on Automatic Control*, vol. AC-25, pp. 531-536, June 1980.

[26] K.S. Vastola, H.V. Poor, "Robust Wiener-Kolmogoroff theory", *IEEE Transactions on Information Theory*, vol. IT-30, pp. 316-327, March 1984.

[27] S.A. Kassam, "Robust hypothesis testing and robust time series interpolation and regression", *Journal of Time Series Analysis*, vol. 3, pp. 185-194, 1982.

[28] K.M. Ahmed, R.J. Evans, "Robust signal and array processing", *Proceedings of IEE*, vol. 129, Pt. F, no. 4, pp. 297-302, Aug. 1982.

[29] A.H. El-Sawy, V.D. VandeLinde, "Robust detection of known signals", *IEEE Transactions on Information Theory*, vol. IT-23, pp. 722-727, Nov. 1977.

[30] J.W. Modestino, "Adaptive detection of signals in impulsive noise environments", *IEEE Transactions on Communications*, vol. COM-25, pp. 1022-1027, Sept. 1977.

[31] C.G. Boncelet, "Algorithms to compute order statistic distributions", *SIAM J. Stat. Comput.*, vol. 8, no. 5, pp. 868-876, Sept. 1987.

[32] M.V. Johns, "Robust Pitman-like estimators", in *Robustness in statistics*, R.L. Launer and G.N. Wilkinson editors, Academic Press, 1979.

CHAPTER 3

IMAGE FORMATION

3.1 INTRODUCTION

An image is a reproduction of a person or a thing, and image formation is the reproduction process. Thus, images are representations of objects, which are sensed through their radiant energy, e.g., light. Therefore, by its definition, image formation requires a *radiant source,* an *object,* and a *formation system.* Radiant sources can be of various kinds (e.g., white light sources, laser systems, X-ray tubes, thermal sources, even acoustic wave sources). Therefore, the physics of image formation can vary accordingly. The nature of the radiation also greatly influences the structure of the formation system. There exist formation systems which are biological (e.g., the vision system of the human and the animals), photochemical (e.g., photographic cameras) or photoelectronic (e.g., TV cameras). Thus it is very difficult to build an image formation model that can encompass this enormous variety of radiation sources and image formation systems. The model described in Figure 3.1.1 is quite general and can be used in various digital image processing and computer vision applications.

Figure 3.1.1: Model of a digital image processing system.

Its input can be radiant energy distribution **f**, which is reflected from an object. It passes from an optical system, which usually consists of lenses, and forms the image radiant energy distribution **b**. This light intensity is transformed to electric current **i** by the *photoelectric detector.* Finally, the electrical current, which is still an analog signal, is digitized by the *image digitizer (frame grabber).* The digital image can be stored in computer memory or it can be processed by an *image processor.* It can also be transmitted to another computer system. The stored or processed image can be displayed on an *image display device* and can be viewed by a human.

It is seen that a computer vision system is relatively complicated. Each of its blocks has its own transfer function, which transforms the input light energy distribution, sometimes in a rather complicated and nonlinear way. In the following sections, a brief analysis of the various steps in the computer vision system will be given. The aim of this analysis is to build a mathematical model for the vision system.

3.2 RADIANT SOURCES AND LIGHT REFLECTION

A narrow band of electromagnetic radiation can stimulate both biological photoreceptors and hardware image sensors. Generally the visible light has wavelength λ which lies between 380 nm (ultraviolet) and 700 nm (infrared). However, there exist image sensors whose range can be outside the visible light range. Such sensors are used for example in multispectral remote sensing. Visible radiant sources produce light either at a single wavelength λ or at a range of the visible spectrum. White light sources produce radiation in the entire visible spectrum. Sometimes the dimensions of the source are very small in comparison to its distance from the observer. In this case, a *point source* is assumed.

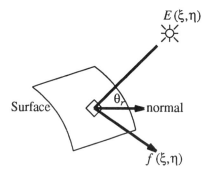

Figure 3.2.1: Reflection of the light from the surface of an object.

In image formation, a light source emits light which is reflected from objects in a scene. A simple geometry of light reflectance is shown in Figure 3.2.1. A point source of illumination emits radiant energy $E(\xi,\eta)$, and it is reflected from the surface of an object. If the surface behaves as an ideal diffuser of the light [1], the reflected radiant energy is given by:

$$f(\xi,\eta) = E(\xi,\eta)r(\xi,\eta)\cos\theta_r \qquad (3.2.1)$$

Thus, the reflected radiant energy depends on the input radiant energy, on the surface reflectance $r(\xi,\eta)$, and on the angle between the incident ray and the normal to the surface. If more complex objects are present in the scene, their parts contribute differently to the reflected radiant energy $f(\xi,\eta)$, according to their reflectances and to the angle between the incident light and the normal to the surface. The variation in the reflected radiant energy is exhibited as a spatial contrast variation in the gray levels of the image. This spatial gray-level contrast is used by humans or by the computer for object recognition. In the case of a complex object or a group of objects, the geometry of image formation is much more complicated because perspective transformations are required [3]. However, we shall not elaborate on the image formation geometry because it is outside of the scope of this book.

3.3 THE POINT-SPREAD FUNCTION OF THE IMAGING SYSTEM

The first part of an imaging system is the optical part, which usually consists of optical lenses and filters. Its input is light radiant energy $f(\xi,\eta)$ reflected from an object or light radiant energy transmitted through a photographic transparency. The input light intensity passes through the optical system whose output is the image $b(x,y)$. The plane (ξ,η) is the so-called *object plane,* whereas the coordinate system (x,y) is the so-called *image plane.* Their geometry is shown in Figure 3.3.1 [2,21].

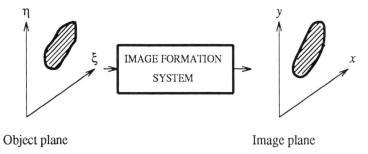

Object plane   Image plane

Figure 3.3.1: Simple geometrical model for image formation.

Since the image is formed by radiant energy, it cannot take negative values. Therefore:

$$f(\xi,\eta) \geq 0 \qquad\qquad (3.3.1)$$

$b(x,y) \geq 0$

The optical system has a transfer function, which relates the input light intensity $f(\xi,\eta)$ to its output image intensity $b(x,y)$. This transfer function is described by the *point-spread function h*:

$$b(x,y) = h(x,y,\xi,\eta,f(\xi,\eta)) \qquad (3.3.2)$$

The general image formation equation is found by summing all contributions of the object plane to the image plane:

$$b(x,y) = \int\limits_{-\infty}^{\infty} \int\limits_{-\infty}^{\infty} h(x,y,\xi,\eta,f(\xi,\eta))d\xi d\eta \qquad (3.3.3)$$

If the formation system is linear, its input-output relation is given by:

$$b(x,y) = \int\limits_{-\infty}^{\infty} \int\limits_{-\infty}^{\infty} h(x,y,\xi,\eta)f(\xi,\eta)d\xi d\eta \qquad (3.3.4)$$

If the formation system acts uniformly on both the image and the object planes, its point-spread function is position independent. Such a system is called *space-invariant system* and is described by the following equation:

$$b(x,y) = \int\limits_{-\infty}^{\infty} \int\limits_{-\infty}^{\infty} h(x-\xi,y-\eta,f(\xi,\eta))d\xi d\eta \qquad (3.3.5)$$

A formation system which is both linear and space-invariant is described by the well-known two-dimensional convolution:

$$b(x,y) = \int\limits_{-\infty}^{\infty} \int\limits_{-\infty}^{\infty} h(x-\xi,y-\eta)f(\xi,\eta)d\xi d\eta \qquad (3.3.6)$$

Since many practical imaging systems are assumed to be linear and space-invariant, (3.3.6) is of great importance in image formation.

3.4 IMAGE SENSORS AND RECORDERS

Image sensors can be divided in two families: *photochemical* and *photoelectronic* sensors. Positive and negative photographic films are photochemical image sensors. They have the advantage that they can detect and record the image at the same time. Their disadvantage is that the recorded image cannot be easily digitized. Image recording on a photographic film is based on the silver halides (AgCl and AgBr) contained in them. The incident light on the film causes precipitation of silver (Ag). The amount of silver precipitation is linearly proportional to the log of the total exposure of the film E. The characteristics of the film are measured by the *optical density D*:

$$D = \log_{10}(\frac{I_1}{I_2}) \qquad\qquad (3.4.1)$$

where I_2 is the intensity of a light which is transmitted through a film, if it is illuminated by a source having light intensity I_1. The $D-logE$ curve of a photographic film is shown in Figure 3.4.1.

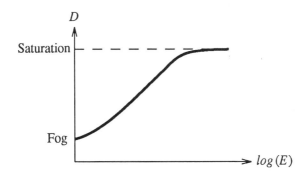

Figure 3.4.1: The D-logE curve of a photographic film.

It has a linear region and two nonlinear regions called *saturation* and *fog*, respectively. In saturation, all silver is deposited, whereas in fog an amount of silver is deposited, even in the absence of incident light. The following equation describes the film characteristics in the linear region:

$$D = \gamma \log_{10}E - D_0 \qquad\qquad (3.4.2)$$

The coefficient γ determines the contrast of the film. The negative films have positive γ, whereas the positive films have negative γ. If a film is exposed to the radiant energy $f_1(x,y)$, an image is formed having optical density:

$$D(x,y) = \gamma \log_{10}f_1(x,y) - D_0 \qquad\qquad (3.4.3)$$

The positive or negative film can be used as an image source, if the light is transmitted through it and then recorded again by a photoelectronic sensor. If I is the radiant energy of the light source, the radiant energy transmitted through the film is given by:

$$f_2(x,y) = Ie^{-D(x,y)} \qquad\qquad (3.4.4)$$

By combining (3.4.3-4), the following relation between $f_1(x,y)$, $f_2(x,y)$ holds:

$$f_2(x,y) = K\,(f_1(x,y))^{-\gamma} = K\,r(f_1(x,y)) \qquad\qquad (3.4.5)$$

where $r(.)$ is a nonlinearity of the form x^γ. Therefore, the radiant energy observed through the film is a highly nonlinear function of the radiant energy that originally exposed the film.

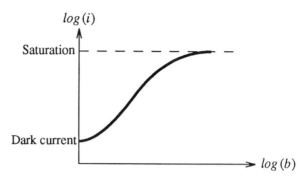

Figure 3.4.2: The current-illumination curve of a photoelectric sensor.

Photoelectronic sensors have the advantage over the film that they can be used to drive an image digitizer directly. There exist several photoelectronic sensors (e.g., standard vidicon tubes, silicon vidicon tubes, Charge Injection Devices (CID), and Charge Coupled Devices (CCD)) [4,5]. Each of them has its own physical characteristics. In vidicon tubes, an electron beam scans the whole image. Its output is current whose relation to the incident radiant energy is non-linearly related to the input radiant energy, as seen in Figure 3.4.2. The saturation comes when the maximum allowable output current is attained. The *dark current* is the output current that exists even in the absence of incident light. In the linear region of Figure 3.4.2, the relation between the input radiant energy and the output current is the following:

$$\log(i(x,y)) = \gamma \log(b(x,y)) + c_1 \qquad (3.4.6)$$

Thus, there always exist a nonlinear relation between the input radiant energy and the output current:

$$i(x,y) = c_2[b(x,y)]^{\gamma} \qquad (3.4.7)$$

For the vidicon tube $\gamma = 0.65$ and, therefore, the input-output nonlinearity is relatively high. For silicon vidicons, the nonlinearity is small ($0.95 < \gamma < 1.0$).

The new trend in image recording is to use solid-state photosensitive elements instead of tubes. The most widely used solid-state technology is the CCD (Charge Coupled Device). The CCD cameras consist of a two-dimensional array of solid state light sensing elements (cells), like the one shown in Figure 3.4.3. The incident light induces electric charge in each cell. Those charges are shifted to the right from cell to cell by using a two-phase clock [4,5] and they come to the readout register. The rows of cells are scanned sequentially during a vertical scan. Thus, the image is recorded and sampled simultaneously. Solid state cameras are very small in size and weight in comparison to the vidicon cameras. Thus they are very suitable for industrial and consumer applications.

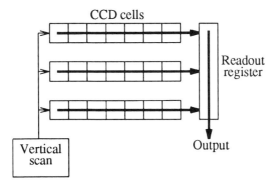

Figure 3.4.3: The structure of a CCD camera.

3.5 IMAGE SAMPLING AND DIGITIZATION

The image $i(x,y)$ produced by a photoelectronic detector is still an analog signal. It has to be sampled and digitized before it is stored on the computer. This is performed by the *image digitizer*. It contains an Analog to Digital (A/D) converter, which transforms the analog image $i_a(x,y)$ to its digital form $i(n_1, n_2)$, $n_1=1,..,N$, $n_2=1,..,M$. The size of the image is $N \times M$. Typical values of such digital image sizes are 256×256 or 512×512. Each digital image element $i(n_1, n_2)$ is called *pixel* . If a pixel is represented by b bits, the digitized image can have 2^b gray scale (intensity) levels. Usually each pixel is represented by 8 bits in image processing applications. Therefore, the discrete image can have 256 discrete intensity levels. Zero and 255 represent the black and white level, respectively.

The image is a two-dimensional signal and there are several ways to sample it. The most straightforward way is sampling along the rectangular coordinates (x,y) with sampling intervals T_1, T_2 along the dimensions x,y, respectively [8]. This sampling is called *rectangular sampling:*

$$i(n_1, n_2) = i_a(n_1 T_1, n_2 T_2) \tag{3.5.1}$$

Another way of sampling is *hexagonal sampling*, described by the following equation:

$$i(\mathbf{n}) = i_a(\mathbf{Vn}) \tag{3.5.2}$$

where $\mathbf{n} = [n_1, n_2]^T$ is the coordinate index vector and $\mathbf{V}$ is the sampling matrix [8, p. 44]:

$$\mathbf{V} = \begin{bmatrix} T_1 & T_1 \\ T_2 & -T_2 \end{bmatrix}$$

In the following, the relation between the spectrum $I_a(\Omega_x,\Omega_y)$ of the continuous image $i_a(x,y)$ and the spectrum $I(\Omega_x T_1,\Omega_y T_2)$ of the discrete image $i(n_1,n_2)$ will be given. The two-dimensional Fourier relations for the continuous image are the following:

$$I_a(\Omega_x,\Omega_y) = \int\limits_{-\infty}^{\infty} \int\limits_{-\infty}^{\infty} i_a(x,y)\, exp\,(-j\Omega_x x - j\Omega_y y)\,dx\,dy \tag{3.5.3}$$

$$i_a(x,y) = \frac{1}{4\pi^2} \int\limits_{-\infty}^{\infty} \int\limits_{-\infty}^{\infty} I_a(\Omega_x,\Omega_y)\, exp(j\Omega_x x + j\Omega_y y)\,d\Omega_x\,d\Omega_y$$

It can be proven [8, p. 37] that the spectrum $I(\Omega_x T_1,\Omega_y T_2)$ of the discrete image $i(n_1,n_2)$ is given by:

$$I(\Omega_x T_1,\Omega_y T_2) = \frac{1}{T_1 T_2} \sum_{k_1} \sum_{k_2} I_a(\Omega_x - \frac{2\pi k_1}{T_1},\ \Omega_y - \frac{2\pi k_2}{T_2}) \tag{3.5.4}$$

Relation (3.5.4) says that the spectrum of the discrete image is the periodic extension of the spectrum of the continuous image, as it is shown in Figure 3.5.1.

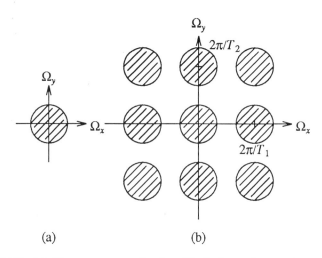

(a)   (b)

Figure 3.5.1: (a) The spectrum of a bandlimited continuous image; (b) The spectrum of the sampled image with sampling periods T_1, T_2.

If the continuous image is bandlimited, i.e.,

$$I_a(\Omega_x,\Omega_y) = 0 , \qquad |\Omega_x| \geq \frac{\pi}{T_1}, \qquad |\Omega_y| \geq \frac{\pi}{T_2} \qquad (3.5.5)$$

its periodic extensions after sampling do not overlap, and the continuous image can be reconstructed exactly from the sampled image [8]:

$$i_a(x,y) = \sum_{n_1} \sum_{n_2} i(n_1,n_2) \frac{sin\,[\pi(x-n_1T_1)/T_1]}{\pi(x-n_1T_1)/T_1} \frac{sin\,[\pi(y-n_2T_2)/T_2]}{\pi(y-n_2T_2)/T_2} \qquad (3.5.6)$$

Equation (3.5.6) is the extension of the Shannon sampling theorem to the two-dimensional case. It states that a bandlimited signal can be reconstructed from its samples, if the sampling intervals T_1, T_2 are small enough to ensure that the condition (3.5.5) is valid. If the sampling intervals are not sufficiently small, the periodic extension of the spectrum causes overlapping. In this case, the continuous image cannot be reconstructed exactly and the *aliasing* error occurs. Sometimes, images are not bandlimited. Therefore an analog lowpass filter precedes the A/D converter to force the continuous image to become bandlimited. This filter is called an *anti-aliasing* filter.

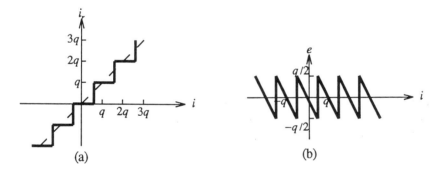

Figure 3.5.2: (a) Input-output curve of a quantizer; (b) Quantization error.

After sampling, the A/D converter performs quantization of the sampled image. The effect of quantization at signal levels iq, $i=1,2,3,..$ is shown in Figure 3.5.2. q is the quantization step. If b is the number of bits used for the representation of an image pixel, the quantization step is given by:

$$q = \frac{1}{2^b} \qquad (3.5.7)$$

Quantization introduces an error, as is seen in Figure 3.5.2 for a linear ramp input. The quantization error appears as noise which is added to the input signal:

$$i_r(n_1,n_2) = i(n_1,n_2) + e(n_1,n_2) \tag{3.5.8}$$

Its power P_e can be evaluated by integration over one period:

$$P_e = \frac{1}{q} \int_{-q/2}^{q/2} (-q)^2 dq = \frac{q^2}{12} = \frac{2^{-2b}}{12} \tag{3.5.9}$$

If P_i is the power of the image, the output SNR is given by:

$$SNR = 10 \log_{10} \frac{P_i}{P_e} = 10 \log_{10} P_i + 10.8 + 6b \tag{3.5.10}$$

This means that each additional bit used in the A/D increases the signal to noise ratio by 6 dB.

3.6 MATRIX REPRESENTATION OF DIGITIZED IMAGES

After digitization, the digital image $i(n_1,n_2)$ can be represented by an $N{\times}M$ matrix **i** or by a vector having NM elements:

$$\mathbf{i} = \begin{bmatrix} i(1,1) & i(1,2) & i(1,M) \\ i(2,1) & \cdots & i(2,M) \\ \cdots & \cdots & \cdots \\ i(N,1) & \cdots & i(N,M) \end{bmatrix} = \begin{bmatrix} i(1,1) \\ \cdots \\ i(1,M) \\ \cdots \\ i(N,M) \end{bmatrix} \tag{3.6.1}$$

It can be stored as a matrix in a file and it can be processed as a matrix on the computer. The vector notation is very convenient for the description of image processing operations. The addition of two images **i**, **f** can be described as follows:

$$\mathbf{s} = \mathbf{i} + \mathbf{f} \tag{3.6.2}$$

A pointwise operation r(.) on every image pixel can be described by:

$$\mathbf{s} = r(\mathbf{i}) \tag{3.6.3}$$

A two-dimensional convolution operator h can be described by an $NM{\times}NM$ matrix **H**. The two-dimensional convolution of an $N{\times}N$ image **i** by an operator **H** is described as a multiplication of a vector by a matrix:

$$\mathbf{s} = \mathbf{H}\mathbf{i} \tag{3.6.4}$$

Let **f** denote an original object image, **H** denote the transfer function of an imaging system, $r(.)$ denote the nonlinearity of the image sensor, and **i** denote the recorded and digitized image. If the noise is not taken into account, the image formation can be described as follows, in matrix notation:

$$\mathbf{i} = r(\mathbf{Hf})$$ (3.6.5)

The mean value $\mathbf{m}_i$ of a discrete image $\mathbf{i}$ is a vector of the form:

$$\mathbf{m}_i = E[\mathbf{i}] = \left[E[i(n_1, n_2)] \right]$$ (3.6.6)

The autocorrelation matrix $\mathbf{R}_i$ of an image $\mathbf{i}$ of dimensions $N \times M$ has dimensions $NM \times NM$:

$$\mathbf{R}_i = E[\mathbf{ii}^{*T}] = \left[E[i(n_1, n_2) i^*(n_3, n_4)] \right]$$ (3.6.7)

The covariance matrix $\mathbf{C}_i$ has dimensions $NM \times NM$ and it is given by:

$$\mathbf{C}_i = \mathbf{R}_i - \mathbf{m}_i \mathbf{m}_i^{*T}$$ (3.6.8)

A discrete image can be completely characterized by its probability density $p(\mathbf{i})$:

$$p(\mathbf{i}) = p\{i(1,1), .., i(1,M), .., i(N,M)\}$$ (3.6.9)

If the probability density function is joint Gaussian, it can be expressed as:

$$p(\mathbf{i}) = (2\pi)^{-NM/2} |\mathbf{C}_i|^{-1/2} \exp\{-\frac{1}{2}(\mathbf{i}-\mathbf{m}_i)^T \mathbf{C}_i^{-1}(\mathbf{i}-\mathbf{m}_i)\}$$ (3.6.10)

$|\mathbf{C}_i|$ denotes the determinant of the autocovariance matrix $\mathbf{C}_i$. If an image passes though a discrete linear system whose transfer function is $\mathbf{H}$, the mean $\mathbf{m}_s$ and the correlation matrix $\mathbf{R}_s$ of the output image $\mathbf{s}$ will be given by:

$$\mathbf{m}_s = E[\mathbf{s}] = \mathbf{H}\mathbf{m}_i$$ (3.6.11)

$$\mathbf{R}_s = E[\mathbf{ss}^{*T}] = \mathbf{H}\mathbf{R}_i\mathbf{H}^{*T}$$ (3.6.12)

3.7 NOISE IN IMAGE FORMATION

In photographic films, the recording noise is mainly due to the silver grains that precipitate during the film exposure. They behave randomly during both the film exposure and development. They are also randomly located on the film. This kind of noise which is due to the silver grains is called *film-grain noise*. It is a Poisson process and becomes a Gaussian process in its limit. Film-grain noise can be expressed in terms of the probability distribution of the optical density D, which is given by [6]:

$$p(D) = \frac{1}{\sigma_D \sqrt{2\pi}} \exp\left\{-\frac{1}{2}(\frac{D-\mu_D}{\sigma_D})^2\right\}$$ (3.7.1)

The noise mean and standard deviation are denoted by μ_D, σ_D, respectively. The standard deviation σ_D depends on the mean μ_D in a nonlinear way [7]:

$$\sigma_D = \alpha(\mu_D)^\beta \, , \qquad \beta = 1/2 \qquad\qquad (3.7.2)$$

The value $\beta=1/3$ has also been suggested as more realistic [2, p. 21]. If p is the ratio of the average grain area to the area of the microscopic region examined, coefficient α is given by:

$$\alpha = 0.66p^{1/2} \qquad\qquad (3.7.3)$$

The film-grain noise does not show statistical correlation for distances between samples greater than the grain size. Therefore, film-grain noise is a white-noise two-dimensional random process. Formulas (3.7.1-2) show that film-grain noise is *signal-dependent* noise. If $D(x,y)$ is the noise-free optical density of an image and $D_r(x,y)$ is the optical density of the noisy image, the following model, proposed by Huang, describes the film-grain noise [2]:

$$D_r(x,y) = D(x,y) + \alpha(D(x,y))^\beta n(x,y) \qquad\qquad (3.7.4)$$

where $n(x,y)$ is a two-dimensional random Gaussian process with zero mean and unit variance. Let $f_1(x,y)$ be the original image intensity which is recorded as optical density $D(x,y)$ and $f_2(x,y)$ be the observed image intensity, if the film is used as transparency. The relation of $f_1(x,y)$, $f_2(x,y)$ is given by (3.4.5). Let also $f_r(x,y)$ be the observed image, if the noisy film is used as transparency. The observed image has the following form, according to (3.4.4):

$$f_r(x,y) = f_2(x,y)e^{-\alpha(D(x,y))^\beta n(x,y)} = c\,(f_1(x,y))^{-\gamma}e^{-\alpha(D(x,y))^\beta n(x,y)} \quad (3.7.5)$$

It is seen that the form of the noise is highly nonlinear. However, for simplification, additive white Gaussian noise is assumed for the optical density:

$$D_r(x,y) = D(x,y) + \alpha m_D^\beta n(x,y) = D(x,y) + n_1(x,y) \qquad\qquad (3.7.6)$$

where m_D is mean optical density over the entire film. If this assumption is used the observation noise is *multiplicative* :

$$f_r(x,y) = c\,(f_1(x,y))^{-\gamma}e^{-n_1(x,y)} = c\,(f_1(x,y))^{-\gamma}n_2(x,y) \qquad\qquad (3.7.7)$$

The two-dimensional white noise process $n_2(x,y)$ has log-normal distribution [22, p. 99].

In photoelectronic detectors, two kinds of noise appear:
(a) *Thermal noise.* Its sources are the various electronic circuits. It is a two-dimensional additive white zero-mean Gaussian noise.
(b) *Photoelectron noise.* It is produced by random fluctuation of the number of photons on the light-sensitive surface of the detector. If its level is low, it has the Bose-Einstein statistics and it can be approximated by a Poisson-distributed noise [19]. In this case, its standard deviation is equal to the square root of its mean. When its level is high, its distribution is Gaussian, having standard deviation equal to the square root of the mean. Therefore, in both cases, it is signal-

dependent. It can be modeled as follows:

$$i_r(x,y) - c_2(b(x,y))^{\gamma} + (c_2(b(x,y))^{\gamma})^{\frac{1}{2}} n(x,y) \qquad (3.7.8)$$

where $n(x,y)$ is a Gaussian two-dimensional random process, having zero mean and unit variance. If the thermal noise is taken into account, the model takes the following form:

$$i_r(x,y) = c_2(b(x,y))^{\gamma} + (c_2(b(x,y))^{\gamma})^{\frac{1}{2}} n(x,y) + n_t(x,y) \qquad (3.7.9)$$

If the imaging system point-spread function is taken into account, the final image formation model is derived:

$$g(x,y) = c_2(f(x,y)**h(x,y))^{\gamma} \qquad (3.7.10)$$

$$+ (c_2(f(x,y)**h(x,y))^{\gamma})^{\frac{1}{2}} n(x,y) + n_t(x,y)$$

The final recorded image is denoted by $g(x,y)$. The two-dimensional convolution is denoted by $**$. The block diagram of this model is shown in Figure 3.7.1. The nonlinearities $r(x)$, $s(x)$ in this model are of the form x^{γ}.

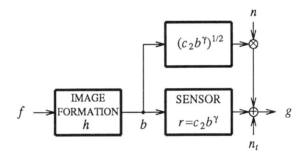

Figure 3.7.1: The complete image formation model.

Another kind of noise that is present during the image transmission is the *salt-pepper noise*. It appears as black and/or white impulses on the image. Its source is usually man-made or atmospheric noise which appears as impulsive noise. It has the following form:

$$n(k,j) = \begin{cases} z(k,j) & \textit{with probability } p \\ i(k,j) & \textit{with probability } 1-p \end{cases} \qquad (3.7.11)$$

where $z(k,j)$ denotes an impulse and $i(k,j)$ denotes the original image intensity at the pixel (k,j). The impulsive noise $z(k,j)$ probability distribution $P(z)$ is usually long-tailed, i.e., it favors very large or very small values.

(a)   (b)

(c)   (d)

Figure 3.7.2: (a) Noise-free image; (b) Image corrupted by additive Gaussian noise; (c) Image corrupted by signal-dependent noise; (d) Image corrupted by salt-pepper noise.

An example of an image corrupted by additive white Gaussian noise having variance 100 is shown in Figure 3.7.2b. The noise is distributed homogeneously in the whole image. The same image corrupted by signal-dependent noise is shown in Figure 3.7.2c. The dynamic range of the image is compressed due to the nonlinearity $r(.)$ in the image formation model. The noise is more evident in the bright image regions because the mean image intensity is greater in these regions. Finally, the same image corrupted by salt-pepper noise having probability of occurrence 10% is shown in Figure 3.7.2d. The black and white impulses are characteristic of this type of noise.

In the case of the CCD cameras, the main form of noise is the *transfer-loss* noise. In CCD technology, charges are transferred from one cell to the other. However, in practice, this transfer is not complete. A fraction ε of the charge is not transferred and it represents the transfer-loss noise. This noise occurs along the rows of cells and therefore has strong horizontal correlation. It usually appears as white smear located on one side of a bright image spot. If this bright image spot induces charge Q_0 at a CCD cell and this charge undergoes n transfers before reaching the CCD output, the final charge corresponding to this spot is given by:

$$Q_n = (1-\varepsilon)^n Q_0 \simeq (1-n\varepsilon)Q_0 \qquad\qquad (3.7.12)$$

In most practical cases the leaking factor ε is small ($\sim 10^{-5}$). Thus the inefficiency product $n\varepsilon$ is about 1% for a 512×512 CCD camera. Other types of noise due to capacitance coupling of clock lines and output lines or due to noisy cell recharging are present in the CCD cameras [4]. Finally thermal noise in the photocells can give rise to fixed pattern noise.

3.8 ELEMENTS OF HUMAN VISUAL PERCEPTION

In many cases, the purpose of digital image processing is to process images, so that their quality is improved or so that they are better perceived by the human eye. In such cases, humans evaluate the result of a digital image processing algorithm. Therefore, the structure of the human vision must be studied so that it is well understood. The ultimate purpose, from the digital image processing point of view, is to construct a mathematical model of human vision which will be incorporated in several image processing applications. The research into human vision is an interdisciplinary area. Opticians, neurophysiologists, psychologists, and, recently, computer vision scientists are working in this area. However, the human visual system is very complex and still relatively poorly understood.

The first part of the human visual system is the eye. Its vertical cross-section is shown in Figure 3.8.1. Its form is nearly spherical and its diameter is approximately 20 mm. Its outer cover consists of the *cornea* and *sclera*. The cornea is a tough transparent tissue in the front part of the eye. The sclera is an opaque membrane, which is continuous with cornea and covers the remainder of the eye. Directly below the sclera lies the *choroid*, which has many blood vessels. At its anterior extreme lies the *iris diaphragm*. The light enters in the eye through the central opening of the iris, whose diameter varies from 2 mm to 8 mm, according to the illumination conditions. Behind the iris is the *lens*, which consists of concentric layers of fibrous cells and contains up to 60-70% of water. Its operation is similar to that of the man-made optical lenses. It focuses the light on the *retina*, which is the innermost membrane of the eye.

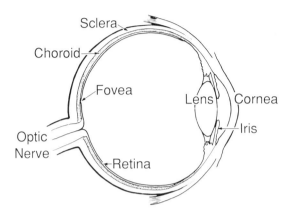

Figure 3.8.1: The human eye.

Retina has two kinds of photoreceptors: *cones* and *rods*. The cones are highly sensitive to color. Their number is 6-7 million and they are mainly located at the central part of the retina. Each cone is connected to one nerve end. Cone vision is the *photopic* or *bright-light* vision. Rods serve to view the general picture of the vision field. They are sensitive to low-levels of illumination and cannot discriminate colors. This is the *scotopic* or *dim-light vision*. Their number is 75-150 million and they are distributed over the retinal surface. Several rods are connected to a single nerve end. This fact and their large spatial distribution explain their low resolution. Both cones and rods transform light to electric stimulus, which is carried through the optical nerve to the human brain for the high-level image processing and perception.

Figure 3.8.2: A model of the human eye.

Based on the anatomy of the eye, a model can be constructed, as shown in Figure 3.8.2. Its first part is a simple optical system consisting of the cornea, the opening of the iris, the lens, and the fluids inside the eye. Its second part consists of the retina, which performs the photoelectrical transduction, and the visual nerve, which performs simple image processing operations and carries the information to the brain.

3.9 IMAGE FORMATION IN THE EYE

The image formation in the human eye is not a simple phenomenon. It is only partially understood, and research continues in this area. Only some of the visual phenomena have been measured and understood. And most of them are proven to have nonlinear characteristics. An example of such a phenomenon is *contrast sensitivity* .

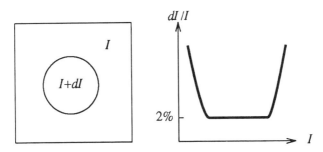

Figure 3.9.1: The Weber ratio without background.

Let us consider a spot of intensity $I+dI$ in a background having intensity I, as is shown in Figure 3.9.1. dI is increased from 0 until it becomes noticeable. The ratio dI/I, called *Weber ratio*, is nearly constant at about 2% over a wide range of illumination levels, except for very low or very high illuminations, as it is seen in Figure 3.9.1. The range over which the Weber ratio remains constant is reduced considerably, when the experiment of Figure 3.9.2 is considered. In this case, the background has intensity I_0 and two adjacent spots have intensities I and $I+dI$, respectively. The Weber ratio is plotted as a function of the background intensity in Figure 3.9.2. The envelope of the lower limits is the same with that of Figure 3.9.1. The derivative of the logarithm of the intensity I is the Weber ratio:

$$d[log(I)] = \frac{dI}{I} \tag{3.9.1}$$

Thus equal changes in the logarithm of the intensity result in equal noticeable changes in the intensity for a wide range of intensities. This fact suggests that the human eye performs a pointwise logarithm operation on the input image.

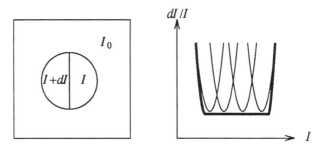

Figure 3.9.2: The Weber ratio with background.

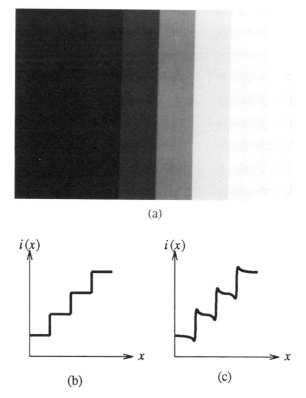

Figure 3.9.3: The Mach-band effect: (a) Vertical stripes having constant illumination; (b) Actual image intensity profile; (c) Perceived image intensity profile.

Another characteristic of the human visual system is that it tends to "overshoot" around image edges (boundaries of regions having different intensity). As a result, regions of constant intensity, which are close to edges, appear to have varying intensity. Such an example is shown in Figure 3.9.3. The stripes appear to have varying intensity along the horizontal dimension, whereas their intensity is constant. This effect is called *Mach band effect*. It indicates that the human eye is sensitive to edge information and that it has high-pass characteristics.

3.10 A SYSTEM MODEL FOR HUMAN VISION

A model of the human eye has been given in section 3.8. However, this model is not sufficient for the description of the various visual phenomena, which have been analyzed in the previous section. Therefore, the objective is to give a mathematical system model that fits to the experimental results described in the previous section. This model must explain both the nonlinear (logarithmic) characteristics in contrast sensitivity and the high-pass characteristics observed in the Mach-band effect. Furthermore, it must have as much biological evidence as possible. This means that it has to match with the biological model 3.8.2 and to the characteristics of the visual receptors and the visual path.

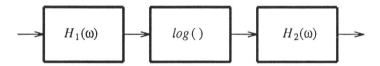

Figure 3.10.1: Mathematical model of the human vision.

The first block of the mathematical model of Figure 3.10.1 corresponds to the optical system of the eye. This system is passive and its characteristics are low-pass. Therefore, it can be represented by a low-pass two-dimensional isotropic filter. A proposed line-spread function of the system is the following [10],[1, p.142]:

$$h_1(x) = exp\left(-\alpha |x|\right) \tag{3.10.1}$$

where $\alpha = 2\pi u$ is the spatial angular frequency. A typical value of α is 0.7 for a 3mm diameter of the iris opening. The frequency response of the optical system is given by:

$$H_1(\omega) = \frac{2\alpha}{\alpha^2 + \omega^2} \tag{3.10.2}$$

Typical lens systems have similar impulse responses [11].

The second subsystem of the model is a pointwise nonlinearity. The logarithmic nonlinearity can be used. However other nonlinearities of the form x^s have been proposed [12]. A typical value for s suggested in [13] is $s=0.33$. The pointwise nonlinearity explains the logarithmic characteristics of the eye.

The third subsystem in the vision model is a high-pass isotropic two-dimensional filter. Its transfer function, given by Hall and Hall, has been represented by a second order expression [14],[1,p. 145]:

$$H_2(\omega) = \frac{\alpha^2+\omega^2}{2\alpha_0\alpha+(1-\alpha_0)(\alpha^2+\omega^2)} \tag{3.10.3}$$

The typical values for α_0 (called *distance factor*) and α (called *strength factor*) are $\alpha_0=0.2$, $\alpha=0.01$. In this case, the transfer function becomes:

$$H_2(\omega) = \frac{10^{-4}+\omega^2}{4\times10^{-3}+0.8\omega^2} \tag{3.10.4}$$

The two-dimensional high-pass filter explains the Mach-band effect and other related visual phenomena. It also has strong biological evidence. A model of human neurons is shown in Figure 3.10.2.

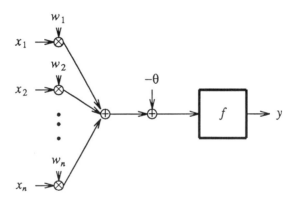

Figure 3.10.2: Mathematical model of a neuron.

Their mathematical model is given by:

$$y = f(\sum_{i=1}^{n} w_i x_i - \theta) \tag{3.10.5}$$

where f is a monotonic function with saturation at both negative and positive values. If f is the sign function, the model (3.10.5) is the classical McCulloch-

Pitts neuron model [15]. Some of the synaptic weights w_i are positive and some are negative. Visual neurons have synapses to each other and form a network. The negative synapses are *inhibitory* ones and contribute to the lateral inhibition process which explains the high-pass characteristics of the human eye. Such a backward inhibitor model for the optic nerve is shown in Figure 3.10.3 [14].

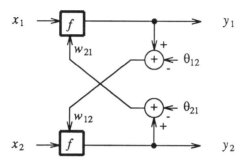

Figure 3.10.3: The backward inhibition process in the optical nerve.

The output of the model is given by:

$$y_1 = \begin{cases} x_1 - w_{21}(y_2 - \theta_{21}) & y_2 \geq \theta_{21} \\ x_1 & y_2 < \theta_{21} \end{cases} \qquad (3.10.6)$$

$$y_2 = \begin{cases} x_2 - w_{12}(y_1 - \theta_{12}) & y_1 \geq \theta_{12} \\ x_2 & y_1 < \theta_{12} \end{cases} \qquad (3.10.7)$$

This model can be generalized to an n input- n output system:

$$y_i = \begin{cases} x_i - \sum_{j=1}^{n} w_{ji}(y_j - \theta_{ji}) & y_j \geq \theta_{ji} \\ x_i & y_j < \theta_{ji} \end{cases} \qquad (3.10.8)$$

Symmetry can be assumed ($\theta_{ij} = \theta_{ji}$, $w_{ij} = w_{ji}$). The coefficients w_{ij} can be chosen to be an exponentially decreasing function of the distance between the receptors:

$$w_{ij} = \begin{cases} 0 & i=j \\ \alpha_0 \exp(-\alpha |i-j|) & i \neq j \end{cases} \qquad (3.10.9)$$

This backward inhibition model has been used to derive the high-pass filter $H_2(\omega)$ in (3.10.3).

3.11 THE IMPLICATIONS OF THE MODEL OF HUMAN VISION TO IMAGE PROCESSING

The visual characteristics of the human eye have been extensively used in digital image processing. The sensitivity of the eye to edge information has forced the development of filters having edge preservation properties.

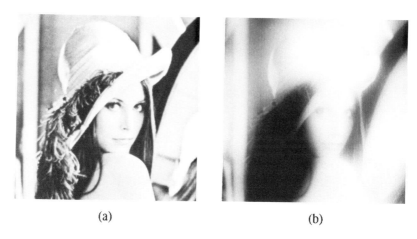

(a)   (b)

Figure 3.11.1: (a) Original image; (b) Perception of the image (a), if the high-pass filter H_2 is compensated by the insertion of its inverse filter.

Most of them are nonlinear filters. The logarithmic nonlinearity has led to the so-called *homomorphic image processing*, to be discussed in chapter 7. Several researchers have used visual image models for the evaluation of existing image filters or for the development of new ones. A striking example of the application of the visual model is shown in Figure 3.11.1, where an inverse filter has been inserted in a vision model to cancel the effect of the high-pass filter $H_2(\omega)$. The elimination of the high-pass characteristics leads to a complete blurring of the image and to the destruction of all the perceivable details.

One criterion used in the evaluation of the performance of a filter is the *normalized mean square error* :

$$NMSE = \frac{\sum\limits_{k=1}^{N}\sum\limits_{j=1}^{M}(i(k,j)-\hat{i}(k,j))^2}{\sum\limits_{k=1}^{N}\sum\limits_{j=1}^{M}(i(k,j))^2} \qquad (3.11.1)$$

where $i(k,j)$ is the input image to the filter and $\hat{i}(k,j)$ is the filter output. It has been proven that the NMSE does not necessarily correlate to the subjective notion of image quality. In fact, it cannot take into account the loss of information close to image edges. Thus a filter reducing the noise in the homogeneous image regions and simultaneously blurring the edges has low NMSE, but its subjective performance is poor. Therefore, other performance criteria have been used, which match better the human visual characteristics. Such a measure is the *Perceptual Mean Square Error* [17]:

$$PMSE = \frac{\sum\limits_{k=1}^{N}\sum\limits_{j=1}^{M}(z(k,j)-\hat{z}(k,j))^2}{\sum\limits_{k=1}^{N}\sum\limits_{j=1}^{M}(z(k,j))^2} \tag{3.11.2}$$

where z is the perceived image:

$$z = H_2 r (H_1 i) \tag{3.11.3}$$

where H_1, H_2 are the transfer functions of the human vision model. The effects of the first stage are not so important and thus, the model (3.11.3) can be simplified:

$$z = H_2 r (i) \tag{3.11.4}$$

The nonlinear function $r(.)$ can be either the logarithm or a function of the form $r(x) = x^{0.33}$ [13]. It has been shown that the PMSE has strong correlation to the subjective image evaluation. However, the measurement of the correlation between PMSE and subjective image quality is based only on one image [17,20]. Therefore, there is no concrete proof that such a correlation exists for a wide range of images.

Another subjective quality criterion is the following:

$$E = \left[\frac{1}{L}\sum\limits_{l=1}^{L}|D_l|^p\right]^{\frac{1}{p}} \tag{3.11.5}$$

where

$$D_l = a_l g (i_l - \hat{i_l}) \tag{3.11.6}$$

The weights a_l and the function g are chosen appropriately. The coefficient p is chosen large enough to increase the importance of large errors. Since such errors occur close to edges or at impulses that have not been filtered, the filters that fail to filter salt-pepper noise or the filters that do not preserve edges tend to have high E. The opposite happens with the filters which remove impulses and preserve edges. It has been found that the actual value of p and the form of D_i depend on the input image [18]. Thus, the following parameters have been found for different images:

$$D_l = i_l - \hat{i}_l\,, \qquad p = 3.0 \tag{3.11.7}$$

$$D_l = m_l - \hat{m}_l\,, \qquad p = 2.0 \ \ or \ \ p = 5.0 \tag{3.11.8}$$

where m_l, $\hat{m}_l$ denote the local means of the images $\mathbf{i}$, $\hat{\mathbf{i}}$ in a 2×2 neighborhood.

As shown, the human visual properties have had a large impact on digital image filtering techniques and image quality criteria. However, most of the problems in this area are still open to research.

REFERENCES

[1]   M.D. Levine, *Vision in man and machine*, McGraw Hill, 1985.

[2]   H.C Andrews, B.R. Hunt, *Digital image restoration*, Prentice-Hall, 1977.

[3]   J. Foley, A. Van Dam, *Fundamentals of interactive computer graphics*, Addison-Wesley, 1981.

[4]   T. E. Jenkins, *Optical sensing techniques and signal processing*, Prentice-Hall, 1987.

[5]   D.F. Barbe, "Imaging devices using the charge coupled concept", *Proceeding of the IEEE*, vol. 63, pp.38-67, 1975.

[6]   C.E.K. Mees, *The theory of the photographic process*, McMillan Company, 1954.

[7]   D.G. Falconer, "Image enhancement and film-grain noise", *Optica Acta*, vol. 17, pp. 693-705, 1970.

[8]   D.E. Dudgeon, R.M. Mersereau, *Multidimensional digital signal processing*, Prentice-Hall, 1984.

[9]   W.K. Pratt, *Digital image processing*, J. Wiley, 1978.

[10] G. Westheimer, F.W. Campel, "Light distribution in the image formed by the living human eye", *Journal of the Optical Society of America*, vol. 52, No. 9, pp. 1040-1045, 1962.

[11] G. Wizecki, W.S. Stiles, *Color science, concepts and methods, quantitative data and formulas*, Wiley, 1967.

[12] W.F. Schreiber, "Image processing for quality improvement", *Proc. of IEEE*, vol. 66, no. 12, pp. 1640-1651, Dec. 1978.

[13] J.L. Mannos, D. Sacrison, "The effects of the visual fidelity criterion on the encoding of images" *IEEE Transactions on Information Theory* , vol. IT-20, pp. 525-536, 1974.

[14] C.F. Hall, E.L. Hall, "A nonlinear model for the spatial characteristics of the human visual system", *IEEE Transactions on Systems, Man and Cybernetics* , vol. SMC-7, no.3, pp.161-170, March 1977.

[15] J. Metzler (editor), *Systems neuroscience* , Academic Press, 1977.

[16] T.G. Stockham Jr., "Image processing in the context of a visual model", *Proc. of IEEE* , vol. 60, no. 7, pp.828-842, July 1972.

[17] C.F Hall, "Subjective evaluation of a perceptual quality metric", *Image Quality, Proc. Soc. Photo Opt. Instr. Eng.*, vol. 310, pp. 200-204, 1981.

[18] H. Marmolin, "Subjective MSE measures", *IEEE Transactions on Systems, Man, Machine and Cybernetics,* vol. SMC-16, no. 3, pp. 486-489, 1986.

[19] L. Mandel, "Fluctuations of photon beams, the distribution of the photoelectrons", *Proc. Phys. Soc. of London*, vol. 74, pp. 233-243, 1959.

[20] C.F. Hall, "The application of human visual system models to digital color image compression", *Proc. IEEE Int. Comm. Conf.*, pp. 436-441, Boston, 1983.

[21] B.R. Hunt, "Digital image processing", *Proc. IEEE*, vol. 63. no. 4, pp. 693-708, April 1975.

[22] A. Papoulis, *Probability, random variables and stochastic processes*, McGraw-Hill, 1984.

CHAPTER 4

MEDIAN FILTERS

4.1 INTRODUCTION

A major approach to nonlinear filtering is based on *robust estimation* and especially on local *L*-estimators, i.e., on order statistics. The main advantage of this approach is its computational simplicity and speed. Filters based on order statistics usually have good behavior in the presence of additive white Gaussian noise and long-tailed additive noise. They have good edge preservation properties and they can become adaptive. Thus, they are suitable in a variety of applications where classical linear filters fail, notably in digital image filtering. The best known and most widely used filter based on order statistics is the *median* filter. Originally, the median was widely used in statistics. It was introduced by Tukey in time series analysis in 1970. Later on, the median filter and its modifications have found numerous applications in digital image processing [2,3,13], in digital image analysis [15,46], in digital TV applications [44,47], in speech processing and coding [20,23], in cepstral analysis [45], and in various other applications. The reason for its success is its good performance and computational simplicity. The theoretical analysis of its deterministic and statistical properties has started at the end of the seventies. A description of the early theoretical results can be found in three very good review chapters in edited books, namely in [4,13,21]. The material of this chapter is based on the recently published results, as well as in the classical work described in [4, 13,21].

4.2 DEFINITION OF THE MEDIAN FILTER

The median of *n* observations x_i, $i=1,..n$ is denoted by $med(x_i)$ and it is given by:

$$med(x_i) = \begin{cases} x_{(v+1)} & n=2v+1 \\ \frac{1}{2}(x_{(v)}+x_{(v+1)}) & n=2v \end{cases} \qquad (4.2.1)$$

where $x_{(i)}$ denotes the *i*-th order statistic. In the following, mainly the definition

(4.2.1) for an odd n will be used.

A one-dimensional *median filter* of size $n=2v+1$ is defined by the following input-output relation:

$$y_i = med(x_{i-v},..,x_i,..,x_{i+v}) \qquad i \in \mathbf{Z} \qquad (4.2.2)$$

Its input is the sequence x_i, $i \in \mathbf{Z}$ and its output is the sequence y_i, $i \in \mathbf{Z}$. Definition (4.2.2) is also called *moving median* or *running median*.

A two-dimensional median filter has the following definition:

$$y_{ij} = med\{ x_{i+r,j+s} ; (r,s) \in A \} \qquad (i,j) \in \mathbf{Z}^2 \qquad (4.2.3)$$

The set $A \subseteq \mathbf{Z}^2$ defines a *neighborhood* of the central pixel (i,j) and it is called the *filter window*.

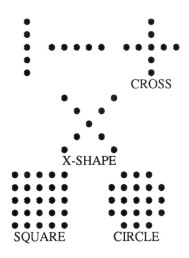

CROSS

X-SHAPE

SQUARE   CIRCLE

Figure 4.2.1: Two-dimensional windows used in median filtering.

Such windows are shown in Figure 4.2.1. If the input image is of finite extent $N \times M$, $1 \le i \le N$, $1 \le j \le M$, definition (4.2.3) is valid only for the interior of the output image, i.e., for those (i,j) for which:

$$1 \le i+r \le N, \quad 1 \le j+s \le M \qquad (r,s) \in A \qquad (4.2.4)$$

At the border of the image, (4.2.4) is not valid. There are two approaches to solve this problem. In the first one, the filter window A is truncated in such a way so that (4.2.4) becomes valid and definition (4.2.3) is used again. In the second approach, the input sequence is appended with sufficient samples and the definition (4.2.3) is applied for $1 \le i \le N$, $1 \le j \le M$.

An example of one-dimensional median filtering of size $n=3$ is shown in Figure 4.2.2. The input sequence x_i takes 3 values: 0, 1, 2. It has also been appended by 0 at the beginning and at the end to solve the boundary problems. As can be seen in Figure 4.2.2, the output sequence is 3-valued and contains no impulses.

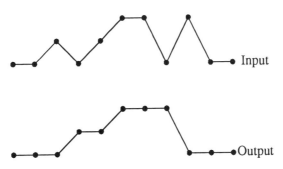

Figure 4.2.2: Median filtering of size $n=3$.

4.3 PROBABILITY DISTRIBUTIONS OF THE MEDIAN FILTERS

It is well-known that the median of $n=2v+1$ independent identically distributed (iid) variables X_i, $i=1,..,n$ having distribution $F(x)$, has probability distribution $F_{v+1}(x)$ given by (2.6.5):

$$F_{v+1}(x) = \sum_{i=v+1}^{n} \binom{n}{i} F(x)^i [1-F(x)]^{n-i} \qquad (4.3.1)$$

Its probability density function is given by:

$$f_{v+1}(x) = n\binom{n-1}{v} F^v(x)[1-F(x)]^v f(x) \qquad (4.3.2)$$

If X_i, $i=1,..,n$ are independent random variables, each having probability distribution $F_i(x)$, the distribution of the median is given by:

$$F_{v+1}(x) = \sum_{i=v+1}^{n} \sum_{S_i} \prod_{l=1}^{i} F_{j_l}(x) \prod_{l=i+1}^{n} [1-F_{j_l}(x)] \qquad (4.3.3)$$

where the summation extends over all permutations $(j_1,..,j_n)$ of $1,2,..,n$ for which $j_1<..<j_i$ and $j_{i+1}<..<j_n$. If k random variables X_i, $i=1,..,k$ are distributed according to $F_1(x)$ and $n-k$ random variables X_i, $i=k+1,..,n$ are distributed according to $F_2(x)$, the median x_{v+1} has the following pdf, according to (2.6.15-17):

$$f_{v+1}(x) = p_1(x) + p_2(x) \qquad (4.3.4)$$

$$p_1(x) = \sum_j k\binom{k-1}{j}\binom{n-k}{v-j} f_1(x) F_1(x)^j F_2(x)^{v-j} \qquad (4.3.5)$$

$$[1-F_1(x)]^{k-j-1}[1-F_2(x)]^{n-k-v+j}$$

$$p_2(x) = \sum_j (n-k)\binom{k}{j}\binom{n-k-1}{v-j} f_2(x) F_1(x)^j F_2(x)^{v-j} \qquad (4.3.6)$$

$$[1-F_1(x)]^{k-j}[1-F_2(x)]^{n-k-v+j-1}$$

where the summations are carried over all j for which all involved binomials $\binom{p}{q}$ satisfy the condition $p \geq q \geq 0$.

Let X_i, $i \in \mathbf{Z}$ be a time series, whose samples are iid random variables having distribution $F(x)$. Let Y_i, Y_j, $i \geq j$ be two output samples of the median filtered sequence. If $i-j \geq n$, the variables Y_i, Y_j are independent because their windows do not overlap and they are generated by iid random variables. If $i-j < n$ their windows overlap by k samples ($0 \leq k \leq n$). Let:

$$F_{ij}(y_i, y_j) = Pr\{Y_i \leq y_i , Y_j \leq y_j\} \qquad (4.3.7)$$

be the bivariate distribution of Y_i, Y_j. If $y_i \leq y_j$, the joint distribution function is given by the following formula:

$$F_{ij}(y_i,y_j) = \sum_{\mu=0}^{k} S_{\mu,k}(y_j) \sum_{l=v+1-\mu}^{n-k} S_{l,n-k}(y_j) \sum_{l=v+1}^{n-k+\mu} \sum_{r=0}^{\mu} S_{l-r,n-k}(y_i) \qquad (4.3.8a)$$

$$S_{r,\mu}(y_i \mid y_i \leq y_j)$$

In the case where $y_i > y_j$, the joint distribution function is given by:

$$F_{ij}(y_i,y_j) = \sum_{\mu=0}^{k} S_{\mu,k}(y_j) \sum_{l=v+1-\mu}^{n-k} S_{l,n-k}(y_j) \sum_{l=v+1-\mu}^{n} \sum_{r=0}^{k-\mu} S_{l-r,n-k}(y_i) \qquad (4.3.8b)$$

$$S_{r,k-\mu}(y_i \mid y_i > y_j)$$

where

$$S_{k,j}(x) = \binom{j}{k}^* F(x)^k [1-F(x)]^{j-k} \qquad (4.3.9)$$

$$S_{k,j}(x \mid x \leq y) = \binom{j}{k}^* F_{X \mid X \leq y}(x)^k [1-F_{X \mid X \leq y}(x)]^{j-k} \qquad (4.3.10)$$

$$S_{k,j}(x \mid x > y) = \binom{j}{k}^* F_{X \mid X > y}(x)^k [1-F_{X \mid X > y}(x)]^{j-k} \qquad (4.3.11)$$

$$\binom{j}{k}^* = \begin{cases} \dfrac{k!}{k!(j-k)!} & \text{if } j \geq k \geq 0 \\ 1 & \text{if } j > 0, k = 0 \\ 0 & \text{otherwise} \end{cases} \qquad (4.3.12)$$

A proof of (4.3.8-12) can be found in [7].

A similar formula can be found for the two-dimensional case [8]. Expression (4.3.8), although computationally tractable, has limited practical significance because it is valid for iid variables, which is not the case for most of the real world images.

4.4 THE MEDIAN AS AN ESTIMATOR OF LOCATION

The median is a well-known robust estimator of location. Therefore, it has been thoroughly studied by the statisticians, as has already been described in chapter 2. In the following, the results of these studies will be summarized.

Let us suppose for simplicity reasons that the model distribution $F(x)$ satisfies $F(0)=1/2$. In this case the influence function of the median filter is given by (2.2.20):

$$IF(x;T,F) = \frac{1}{2f(0)} sign(x) \qquad (4.4.1)$$

Its asymptotic variance is given by:

$$V(T,F) = \int IF(x;T,F)^2 dF(x) = \frac{1}{4f^2(0)} \qquad (4.4.2)$$

The functional $T(F)$ corresponding to the median filter is the following:

$$T(F) = F^{-1}(\frac{1}{2}) \qquad (4.4.3)$$

Relation (4.4.3) is consistent with the definition of the median of a distribution F at the point x that satisfies $F(x)=1/2$. If the median is used as an estimator of the location of the model distribution $F(x-\theta)$, $T(F)$ is given by:

$$T(F) = F^{-1}(\frac{1}{2}) = \theta \qquad (4.4.4)$$

The median is asymptotically normal [10, p.352]. Its distribution, for large n and for iid variables having zero mean and unit variance, tends to the normal distribution having zero mean and variance:

$$\sigma^2_{med} = \frac{V(T,F)}{n} = \frac{1}{4f^2(0)n} \qquad (4.4.5)$$

The variance σ^2_{med} depends on the input probability density function $F(x)$. For small values of n, which is usually the case in image filtering, (4.4.5) is only approximately valid. A better approximation is the following [13, p. 165]:

$$\sigma^2_{med} = \frac{1}{4f^2(0)(n+b)} \qquad (4.4.6)$$

$$b = \frac{1}{4f^2(0)} - 1 \qquad (4.4.7)$$

Coefficient b is chosen in such a way that σ^2_{med} becomes equal to 1 for $n=1$. The output median variance (4.4.5-6) is valid both for one- and two-dimensional median filtering. In both cases n is the number of samples in the filter window.

To obtain an idea about the performance of the median, it will be compared to the arithmetic mean $\bar{x}$. It is well known that $\bar{x}$ also has asymptotic normal behavior [10, p. 356] and that its variance is given by:

$$\sigma_{\bar{x}}^2 = \sigma_x^2/n \qquad (4.4.8)$$

The asymptotic relative efficiency $ARE(med(x_i),\bar{x})$ will be used as a measure of the performance for various distributions:

$$ARE(med(x_i),\bar{x}) = \frac{V(\bar{x},F)}{V(med(x_i),F)} = 4f^2(0) \qquad (4.4.9)$$

When $ARE(med(x_i),\bar{x})$ is greater than one, the median has better performance, i.e. lower output variance than the arithmetic mean. It is interesting to note that the ARE (4.4.9) is not affected by scale changes in F.

The most significant distribution is naturally the Gaussian distribution $\Phi(x)$ having unit variance $\sigma^2_x = 1$ and $\phi(0) = \frac{1}{\sqrt{2\pi}}$. In this case ARE becomes:

$$ARE(med(x_i),\bar{x}) = \frac{2}{\pi} \simeq 0.637 \qquad (4.4.10)$$

Therefore, the arithmetic mean has better performance than the median for the Gaussian distribution. The modified formula (4.4.6) gives the following output variance of the median for the Gaussian distribution:

$$\sigma^2_{med} \simeq \frac{1}{n+\pi/2-1} \frac{\pi}{2} \qquad (4.4.11)$$

and the ARE is given approximately by (4.4.10). Another interesting distribution is the *logistic distribution* $L(0,1)$ having cdf and pdf respectively:

$$F(x) = \frac{1}{1+e^{-x}} \qquad (4.4.12)$$

$$f(x) = \frac{e^{-x}}{(1+e^{-x})^2} \qquad (4.4.13)$$

It can be easily proven that the *ARE* for the logistic distribution is less than one [10, p.356]:

$$ARE\,(med\,(x_i),\bar{x}) = \frac{\pi^2}{12} \simeq 0.82 \qquad (4.4.14)$$

Thus, the median has better performance for the logistic distribution than for the Gaussian distribution. This comes from the fact that logistic distribution is more long-tailed than the Gaussian distribution. However, its performance is still worse than that of the arithmetic mean.

A distribution with even longer tails is the Laplacian distribution:

$$f\,(x) = \frac{1}{2}exp\,(-|x|) \qquad (4.4.15)$$

Since $f\,(0) = 1/2$, the asymptotic variance $V\,(med,F)$ and σ^2_{med} are given by:

$$V\,(med,F) = J^{-1}(F) = 1 \qquad (4.4.16)$$

$$\sigma^2_{med} = \frac{1}{n} \qquad (4.4.17)$$

Therefore:

$$ARE\,(med\,(x_i),\bar{x}) = 2 \qquad (4.4.18)$$

because $\sigma_{\bar{x}}^2 = 2$. Thus the median has better performance than the mean $\bar{x}$ for the Laplacian distribution. If the median is used as a location estimator with the model $f_\theta(x) = \frac{1}{2}exp\,(-|x-\theta|)$, the equality (4.4.16) shows that the Cramer-Rao bound has been reached and that median is the asymptotically efficient estimator. This is verified by the fact that (2.2.11) is valid, as can be seen by comparing the expression:

$$J^{-1}(F)\frac{\partial}{\partial\theta}[lnf_\theta(x)] = \frac{\partial f_\theta(x)/\partial\theta}{f_\theta(x)} = sign\,(x) \qquad \text{for } \theta=0 \qquad (4.4.19)$$

and (4.4.1). The median is the maximum linear estimate (MLE) of location for the Laplacian distribution because it minimizes:

$$\sum_{i=1}^{n}[-lnf_{T_n}(x_i)] = \sum_{i=1}^{n}|x_i-T_n| = min \qquad (4.4.20)$$

(4.4.20) shows that the median is the best estimator in the L_1 sense. By choosing $\psi(x) = sign(x)$, (4.4.20) is equivalent to the solution of the following equation:

$$\sum_{i=1}^{n}\psi(x_i,T_n) = \sum_{i=1}^{n}sign\,(x_i-T_n) = 0 \qquad (4.4.21)$$

(4.4.21) is an implicit definition of the median. It states that the median lies exactly at the "middle" of the observation data x_i, $i=1,..,n$.

An excellent tool for the study of the dependence of the performance of the median to the tail behavior of the model distribution $F(x)$ are the $t-distributions$ with m *degrees of freedom* [10, p. 356]:

$$f_m(x) = \frac{\Gamma[(m+1)/2]}{\sqrt{m\pi}\ \Gamma(m/2)} \frac{1}{(1+x^2/m)^{(m+1)/2}}$$ (4.4.22)

The Cauchy distribution is obtained for $m=1$. It is a long-tailed distribution and has infinite expectation $E[x]=\infty$. The expectation exists, but the variance is infinite for $m=2$. For $m\geq3$ the variance is finite and it is given by:

$$\sigma_m^2 = \frac{m}{m-2}$$ (4.4.23)

As m tends to infinity, the t-distribution tends to the normal distribution and its tail diminishes. The $ARE(med(x_i),\bar{x})$ for the t-distributions is given in Table 4.4.1. It is clearly seen that the performance of the median deteriorates when the distribution tail becomes shorter.

Table 4.4.1: Asymptotic relative efficiency of the median vs. the arithmetic mean for t-distributions (adapted from [10], p. 357).

m	3	4	5	8	∞
ARE	1.62	1.12	0.96	0.8	0.64

Another important family of distributions is the *contaminated normal family* introduced by Tukey [11]:

$$F(x) = (1-\varepsilon)\Phi(x)+\varepsilon\Phi(\frac{x}{\tau})$$ (4.4.24)

ε governs the percentage of the contamination (outliers) and τ^2 is the variance of the outliers. In this case $f(0)$ and variance are given by [12, p.357]:

$$f(0) = \frac{1}{\sqrt{2\pi}}(1-\varepsilon+\frac{\varepsilon}{\tau})$$ (4.4.25)

$$\sigma^2 = (1-\varepsilon)+\varepsilon\tau^2$$ (4.4.26)

The values of the asymptotic relative efficiency are tabulated in Table 4.4.2. It is easily seen that the performance of the median increases with the percentage of the outliers and with their variance, i.e., with the length of the tail of the distribution.

Table 4.4.2: Asymptotic relative efficiency of the median vs. the arithmetic mean for contaminated Gaussian distributions (adapted from [10], p. 357).

τ	ε	0.01	0.05	0.1
2		0.65	0.70	0.75
3		0.68	0.83	1.0
4		0.72	1.03	1.36

The median has its worst performance for the uniform distribution, which is a very short-tailed distribution. In this case, the asymptotic relative efficiency attains its lowest value:

$$ARE\,(med\,(x_i),\bar{x}) = \frac{1}{3} \qquad (4.4.27)$$

The output variance of the median for the uniform distribution in $[0,1]$ is given by:

$$\sigma^2_{med} = \frac{3\sigma_x^2}{n} \qquad (4.4.28)$$

A more accurate formula for the output variance for small n is the following [13, p. 163]:

$$\sigma^2_{med} = \frac{3\sigma_x^2}{n+2} \qquad (4.4.29)$$

As a conclusion, we can say that median filtering is preferred when the observation data have long-tailed distributions. Its performance is worse than that of the arithmetic mean for distributions, which have shorter tails than the Laplacian distribution.

All the previously mentioned theoretical results have an important application in the filtering of additive white noise in image regions having constant intensity s (called *homogeneous image regions*):

$$x_{ij} = s + n_{ij} \qquad (4.4.30)$$

$$E\,(n_{ij}) = 0$$

$$E\,(n_{ij}^2) = \sigma_n^2$$

In this case, the observed image intensity x_{ij} is distributed according to $F\,(x-s)$, where F is the noise distribution. The median filter can be used as an estimator of the location s.

(a)   (b)

(c)   (d)

Figure 4.4.1: (a) Original image; (b) Image corrupted by additive white Gaussian noise; (c) Output of a 7×7 moving average filter; (d) Output of a 7×7 median filter.

An example of median and moving average filtering of images corrupted by additive white Gaussian noise having standard deviation 50 is shown in Figure 4.4.1. The window dimensions of both filters is 7×7. As expected, the moving average filter performs better than the median in homogeneous image regions. However, the moving average filter tends to blur the edges, as it will be proven in section 4.8. Therefore, its output is less pleasant to the human eye.

In this section it has been assumed that the input data are independent identically distributed random variables. This assumption is valid only in homogeneous image regions and in the presence of additive white noise. However, in many image processing applications the variables are dependent on each other. Thus the median filtering of dependent variables will be examined.

4.5 MEDIAN FILTERING OF NONWHITE NOISE

In most practical cases, images are random processes with statistically dependent random variables. In this case, it is very difficult to obtain exact formulas for the probability distribution and for the output variance of the median filter. However, there exist limit theorems which give approximations to the output variance [13]. The conditions needed for the validity of these theorems are that the input process x_i are stationary and mixing. The mixing condition means that variables lying far apart are almost independent. If the input process x_i is stationary, normal, and mixing with covariance function:

$$C(x_i, x_{i+\tau}) = \sigma^2_x r_x(\tau) , \qquad \tau \in \mathbf{Z} \tag{4.5.1}$$

the following approximate expression gives the variance of the output y_i of the median filter [13]:

$$\sigma^2_{med} \simeq \frac{\sigma^2_x}{n + \pi/2 - 1} \sum_{j=-n+1}^{n-1} (1 - \frac{|j|}{n}) arcsin[r_x(j)] \tag{4.5.2}$$

In the same case, the moving average filter has the following output variance:

$$\sigma^2_{\bar{x}} = \frac{\sigma^2_x}{n} \sum_{j=-n+1}^{n-1} (1 - \frac{|j|}{n}) r_x(j) \tag{4.5.3}$$

Therefore, the output variances of a median and a moving average filter are given by very similar formulas. If the normal process x_i has a nonnegative correlation function:

$$r_x(\tau) \geq 0 , \qquad \tau \in \mathbf{Z} \tag{4.5.4}$$

the following inequalities hold:

$$r_x(\tau) \leq arcsin[r_x(\tau)] \leq r_x(\tau) \frac{\pi}{2} \tag{4.5.5}$$

Therefore, the ratio $\sigma^2_{med}/\sigma^2_{\bar{x}}$ is bounded:

$$1 \leq \frac{\sigma^2_{med}}{\sigma^2_{\bar{x}}} \leq \frac{\pi}{2} \tag{4.5.6}$$

In this case, the performance of the moving average filter is always better than that of the median filter. However, if the input process has both positive and negative correlation values, the upper bound (4.5.6) is not valid any more. This happens in the case of the first order autoregressive process AR(1):

$$x_i = ax_{i-1} + n_i \tag{4.5.7}$$

having correlation coefficient:

$$r_x(\tau) = a^{|\tau|} \tag{4.5.8}$$

The ratio $\sigma^2_{med}/\sigma^2_{\bar{x}}$ has much greater values than $\pi/2 \simeq 1.57$ for negative values of a [13]. Similar results hold also for two-dimensional processes [13].

4.6 AUTOCORRELATION FUNCTIONS OF THE MEDIAN FILTER OUTPUT

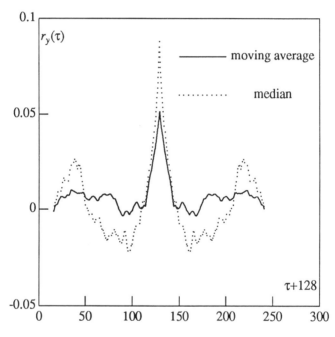

Figure 4.6.1: Autocorrelation functions of the moving average filter output (solid line) and of the median filter output (dotted line) for normal white noise and a window of size $n=15$.

It has been proven in section 4.3 that the output y_i of the median filter is not independent from the output y_j, $i \neq j$, if their respective windows overlap at $k \neq 0$ points ($0 \leq k \leq n$). Furthermore, the bivariate probability distribution has been derived in the same section. In this section the properties of the autocorrelation function of the output of the median filter will be discussed. The autocorrelation function of the output y of a median filter, when its input is white noise, is given by:

$$R_y(\tau) = E[y_i y_{i+\tau}] = \int\limits_{-\infty}^{\infty} \int\limits_{-\infty}^{\infty} y_i y_{i+\tau} f(y_i, y_{i+\tau}) dy_i dy_{i+\tau} \qquad (4.6.1)$$

where f is the joint probability density function corresponding to the bivariate distribution (4.3.8). The corresponding formula for a moving average filter is given by [38, p.273]:

$$R_{\bar{x}}(\tau) = \frac{\sigma^2_x}{n}(1 - \frac{|\tau|}{n}) \qquad |\tau| \leq n \qquad (4.6.2)$$

The autocorrelation function of the median filter output $r_y(\tau)$ can also be calculated by integrating (4.6.1) numerically [7]. Both autocorrelation functions (4.6.1-2) have been calculated by simulation and are plotted in Figure 4.6.1 for $n=15$. Their form is very similar. This similarity will be explained later in this section.

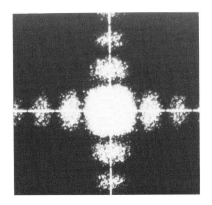

Figure 4.6.2: Power spectrum of the median filter output.

The power spectrum $S_y(\omega)$ of the median filter output:

$$S_y(\omega) = \sum_{\tau=-\infty}^{\infty} R_y(\tau) e^{-j\tau\omega} \qquad (4.6.3)$$

If $R_y(\tau_1, \tau_2)$ is the autocorrelation function of the output of a two-dimensional filter, its power spectrum is given by:

$$S_y(\omega_1, \omega_2) = \sum_{\tau_1=-\infty}^{\infty} \sum_{\tau_2=-\infty}^{\infty} R_y(\tau_1, \tau_2) e^{-j\tau_1\omega_1} e^{-j\tau_2\omega_2} \qquad (4.6.4)$$

It can also be calculated numerically and it is plotted in Figure 4.6.2 for a white Gaussian input process. The low-pass characteristics of the median filter are evident.

If the input of the median is a nonwhite process, exact formulas for the covariance functions cannot be found. However, asymptotic results have been derived by Justusson for large filter windows $(n \to \infty)$. These results work fairly well for small window sizes [13, p.180].

An approximate formula for the median filter output at a distribution $F(x)$ can be found by a Taylor series expansion of its influence function [12, p.85]:

$$T_n(F_n) \simeq T(F) + \frac{1}{n} \sum_{i=1}^{n} IF(x_i; T, F) \tag{4.6.5}$$

The influence function and the functional $T(F)$ of the median filter are given by (4.4.1), (4.4.4), respectively. Therefore, the approximate formula (4.6.5) takes the following form for the median filter:

$$y_i \simeq \tilde{m} + \frac{1}{2f(\tilde{m})n} \sum_{j=-v}^{v} sign(x_{i+j} - \tilde{m}) \tag{4.6.6}$$

$$\tilde{m} = F^{-1}(\frac{1}{2})$$

Formula (4.6.6) is the so-called *Bahadur representation* of the median. According to this representation, the median behaves asymptotically as a moving average of sign variables. This fact explains the low-pass characteristics of the median filter.

4.7 IMPULSIVE NOISE FILTERING BY MEDIAN FILTERS

Impulsive noise appears as black and/or white spots in an image. Therefore, it is also called salt-pepper noise. It is usually caused by errors during the image acquisition or transmission through communication channels. During this process some image pixels are destroyed and they take high positive (white spots) or low values (black spots). In the statistical terminology, impulses outlie from the distribution of the rest of the data and, thus, they are called outliers. Two models have been proposed for the description of the impulsive noise [13]. The first model produces one-sided positive or negative impulses having probability of occurrence p:

$$x_{ij} = \begin{cases} d & \text{with probability } p \\ s_{ij} & \text{with probability } 1-p \end{cases} \tag{4.7.1}$$

The noiseless and the noisy images are denoted by s_{ij}, x_{ij}, respectively. The value of the outlier is denoted by d. The second model of impulsive noise can produce mixed impulses. It assumes that the image data follow a distribution $F(s)$, whereas the impulses follow a distribution $H(z)$, which has usually much longer tails than the data distribution. The mixed noise with probability of occurrence p is given by:

$$x_{ij} = \begin{cases} z_{ij} & \text{with probability } p \\ s_{ij} & \text{with probability } 1-p \end{cases} \qquad (4.7.2)$$

The probability distribution of the corrupted data x_{ij} is given by:

$$G(x) = (1-p)F(x) + pH(x) \qquad (4.7.3)$$

The distribution $G(x)$ lies in the neighborhood of the distribution $F(x)$. In many cases $F(x), H(x)$ are symmetric. The outlier distribution $H(x)$ can vary from $F(x)$ in location $H(x)=F(x-a)$ and/or in variance $H(x)=F(x/b)$ or it can be completely different from the data distribution $F(x)$.

The median is a B-robust estimator and has gross-error sensitivity $\gamma^* \simeq 1.253$ for the Gaussian distribution. One impulse, however large it may be, has no effect on the median output. Furthermore, the median has a breakdown point $\varepsilon^* \simeq 1/2$. Thus, a median having window dimension $n=2v+1$ can reject up to v impulses. Therefore, the median filter has good impulsive noise filtering capabilities. In the following, the performance of the median will be investigated in the presence of impulsive noise.

The performance of the median in one-sided impulsive noise (4.7.1) in a constant value image neighborhood A:

$$s_{i+r,j+s} = s_{ij} = c \neq d , \qquad (r,s) \in A \qquad (4.7.4)$$

is measured by the probability of correct reconstruction, i.e., by the probability $Pr\{y_{ij}=c\}$ that the output y_{ij} of the median filter is equal to the image background level c. The number of the impulses in A follow a binomial distribution. If the number of impulses is less than or equal to v, the median gives a correct reconstruction. Thus:

$$Pr\{y_{ij} = c\} = \sum_{k=0}^{v} \binom{n}{k} p^k (1-p)^{n-k} \qquad (4.7.5)$$

The probability of erroneous reconstruction $1-Pr\{y_{ij} = c\}$ is tabulated in Table 4.7.1 (adapted from [13]). If the probability of impulse occurrence is less or equal to 30%, medians of lengths 3×3 or 5×5 reduce the impulse occurrence rate to less than 10%. It is also observed that, if the outliers have probability of occurrence 50%, the probability of correct reconstruction is 50%, no matter how large the window size is. However, generally speaking, larger filter windows improve the performance of the median filter in impulsive noise removal.

Table 4.7.1: Rate of failure of the median filter in impulsive noise.

Impulse rate	Window size n			
	3	5	9	25
0.1	0.028	0.0086	0.00089	0.0000002
0.2	0.104	0.058	0.0196	0.00037
0.3	0.216	0.163	0.099	0.017
0.4	0.352	0.317	0.267	0.154
0.5	0.5	0.5	0.5	0.5

If the impulse noise follows the model (4.7.2) and the outliers are uni-formly distributed in the range $[0,d]$, it can be proven that (4.7.5) still gives the probability of correct reconstruction. It can also be proven that for background level $c=0$, the expected value of the median filter output is given by [13, p.171]:

$$E[y_{ij}] = d \sum_{k=(n+1)/2}^{n} \frac{k-v}{k+1} \binom{n}{k} p^k (1-p)^k \qquad (4.7.6)$$

The expected value of the remaining impulses is the following:

$$E[y_{ij} \mid erroneous\ reconstruction\ at\ (i,j)] = \frac{E[y_{ij}]}{1-Pr\{y_{ij}=c\}} \qquad (4.7.7)$$

Figure 4.7.1 illustrates the performance of the median filter in the presence of mixed impulses having probability of occurrence 10%. The impulses have been removed almost completely by using a 3×3 median filter, as can be shown in Figure 4.7.1c. The moving average of the same size has very unsatisfactory results, as can be seen in Figure 4.7.1d. The poor results of the moving average filter in the presence of impulses can be explained by some simulations of the performance of the median filters in the presence of outliers having distribution $F(x-A)$ [16]. The performance of a 3-point median filter is compared to the per-formance of a 3-point moving average filter. The input samples x_{i-1}, x_i, x_{i+1} in the filter window are assumed to be independent and to follow the distributions $F_i(x) = N(s_i, \sigma^2)$. Two of the samples are assumed to have expected values $s_1 = s_2$, whereas the third one is supposed to be an outlier having $s_3 = A$.

The expected values of the median filter output and of the moving average filter output are shown in Figure 4.7.2 as a function of A.

(a)   (b)

(c)   (d)

Figure 4.7.1: (a) Original image; (b) Image corrupted by mixed impulses; (c) Output of a 3×3 median filter; (d) Output of a 3×3 moving average filter.

The effect of the outlier on the median filter output is limited because the median is a B-robust estimator. In contrast, the expected value of the output of the moving average filter tends to infinity as A increases. This is explained by the fact that the influence function of the moving average filter is not bounded. Thus, even one outlier can destroy its performance.

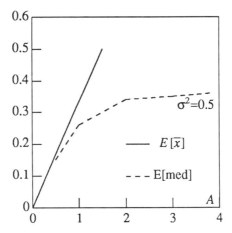

Figure 4.7.2: The effect of one outlier in the expected value of the output of a 3-point median filter and moving average filter (adapted from [16]).

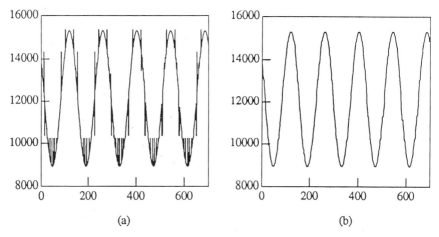

(a)   (b)

Figure 4.7.3: (a) Sinusoidal signal corrupted due to A/D converter malfunction; (b) The same signal filtered by a 5-point median filter (courtesy of Dr. A. Hadjopoulos, University of Thessaloniki, Greece).

Another example of the performance of the median filter in impulsive noise removal is shown in Figure 4.7.3. A corrupted sinusoidal signal is shown in Figure 4.7.3a. The impulsive noise is due to A/D converter malfunction. The same signal filtered by a 5-point median filter is shown in Figure 4.7.3b. The impulses have been removed without any significant distortion of the signal.

4.8 EDGE PRESERVATION PROPERTIES OF THE MEDIAN FILTER

Edge information is very important for human perception. Therefore, its preservation and, possibly, its enhancement is a very important subjective feature for a good image filter. However, the notion of an image *edge* is ill-defined and it has no generally accepted mathematical definition. The following definition will be used in this section [18]:

Definition An edge is the border between two image regions having different illumination intensities.

This definition implies that an edge is a local variation of illumination (but not vice versa). If the image pixels have value a, $a+h$ on the two edge sides, respectively, a *step edge* of height h is present. If the image intensity increases linearly from the level a to the level $a+h$, a *ramp edge* is present.

The median filter has very good edge preservation properties. It can be easily proven that, if the filter window A is symmetric about the origin and includes the origin:

$$(r,s) \in A \quad \rightarrow \quad (-r,-s) \in A \tag{4.8.1}$$

$$(0,0) \in A \tag{4.8.2}$$

the corresponding median filter preserves any step edge [13]. Edges have high frequency content and are destroyed by linear low-pass filters. Such a filter is the moving average filter. It tends to decrease the high frequency content of the edges and to blur them. It transforms the step edges to ramp edges and makes the ramp edges wider. The wider its window is, the more profound the edge blurring becomes.

In the following, the edge preservation properties of the median filter in the presence of additive Gaussian noise $N(0,\sigma^2)$ will be investigated. The following one-dimensional model will be used:

$$x_i = s_i + z_i \tag{4.8.3}$$

$$s_i = \begin{cases} 0 & i \leq 0 \\ h & i > 0 \end{cases}$$

where z_i is white Gaussian noise having distribution $N(0,\sigma^2)$. Such an edge is shown in Figure 4.8.1a. The output of a 15-point moving average filter and of a 15-point median filter shown in Figures 4.8.1b and 4.8.1c, respectively. The moving average filter clearly transforms the step edge to a ramp edge, whereas the median filter has better edge preservation properties. When the median window covers both sides of the edge, its input pixels are independent from each other and some are distributed according to the distribution $N(0,\sigma^2)$, whereas the rest of them are distributed according to $N(h,\sigma^2)$.

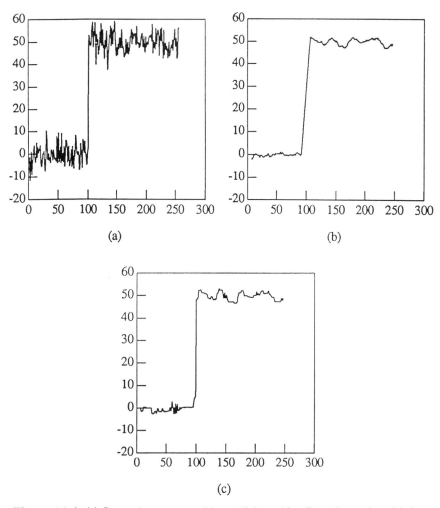

(a)

(b)

(c)

Figure 4.8.1: (a) Step edge corrupted by additive white Gaussian noise; (b) Output of the moving average filter ($n=15$); (c) Output of the median filter ($n=15$).

Therefore, (4.3.4) can be used to find the distribution $f_{v+1}(x)$ of the median output. The expected values $E(y_i)$ and the variances of the output points can be found by numerical integration of (4.3.4). The expected values of the median filter output and of the moving average filter output are given in Figure 4.8.2.

The distribution of the moving average output is $N(hk/n, \sigma^2/n)$, where k is the number of the pixels in the filter window having $s_i = h$. It can be easily seen in Figure 4.8.2 that the moving average filter blurs edges much more than the median filter.

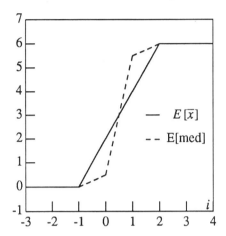

Figure 4.8.2: Expected values of the output of the median and the moving average filters (adapted from [13]).

Let us suppose that we want to investigate the effect of the edge height on the performance of the median filter. The filter window size is chosen to be $n=3$. A good measure of this performance is the expected value $E[y_0]$ of the median filter output y_0. If it stays close to 0 (as it should be) for large edge heights, the median filter output can be considered to be immune to the edge height. This is in fact the case. The two points x_{-1}, x_0, which contribute to y_0, are distributed according to $N(0, \sigma^2)$. The third point x_1 is distributed according to $N(h, \sigma^2)$ and can be considered as outlier in the computation of y_0. Therefore, its influence on $E[y_0]$ is bounded, since the median is a B-robust estimator, as it has already been described in the previous section. Thus the expected value of y_0 remains close to 0 and is not influenced greatly by the edge height.

In the following, the properties of the two-dimensional median filters will be investigated in the presence of two-dimensional noisy edges. The median filter employing an $n \times n$ square window has been used in the simulations. The idealized noisy edges used are the vertical edge:

$$x^v_i = \begin{cases} z_{ij} & j \leq 0 \\ h + z_{ij} & j > 0 \end{cases} \qquad (4.8.4)$$

and the diagonal edge:

$$x^d_i = \begin{cases} z_{ij} & i+j \leq 0 \\ h+z_{ij} & i+j > 0 \end{cases} \qquad (4.8.5)$$

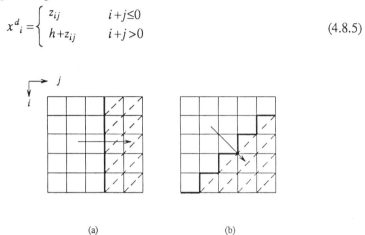

(a)   (b)

Figure 4.8.3: Models for a vertical and a diagonal edge.

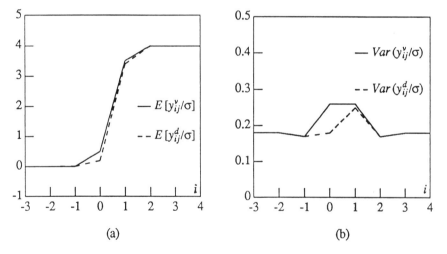

(a)   (b)

Figure 4.8.4: (a) Scaled expected value of the output of the 3×3 square median filter in the presence of a vertical or a diagonal edge; (b) Scaled variance of the output of the 3×3 square median filters in the presence of a vertical or a diagonal edge (adapted from [17]).

Both edge models are shown in Figure 4.8.3. The noise process z_{ij} is Gaussian zero mean process having variance σ^2. The expected values and the variance of the filter output can be found by integrating (4.3.4) numerically. The expected values $E[y^v_{ij}]$ and $E[y^d_{ij}]$, scaled by σ, are plotted in Figure 4.8.4a for vertical and diagonal edges, respectively. Similarly, the scaled variances $var(y^v_{ij}/\sigma)$ and $var(y^d_{ij}/\sigma)$ are plotted in Figure 4.8.4b. It can be easily seen that the median filter having square window preserves fairly well both the vertical and the diagonal edges. Simulation experiments using square-shaped, cross-shaped, and x-shaped median filters are given in [17]. The windows of those filters are shown in Figure 4.2.1. As expected from the shapes of the windows, cross-shaped median filters perform better than square or x-shaped median filters close to vertical or horizontal edges. X-shaped median filters perform better than square or cross-shaped median filters close to the diagonal edges. The square-shaped median filter has no preference to the edge direction. Therefore, it has relatively good performance for both the vertical and the diagonal edges. Since most images have edges along all directions, the use of the square-shaped median filter is recommended.

4.9 STREAKING IN MEDIAN FILTERING

The median filter not only smooths the noise in homogeneous image regions but it also tends to produce regions of constant or nearly constant intensity. The shape of these regions depends on the geometry of the filter window A. They are usually linear patches (streaks) or amorphous blotches. These side-effects of the median filter are highly undesirable, because they are perceived as either lines or contours (between blotches) that do not exist in the original image.

The probability that two successive outputs of the median filter y_i, y_{i+1} have the same value is quite high [13, p.186]:

$$Pr\{y_i = y_{i+1}\} = 0.5(1 - \frac{1}{n}) \tag{4.9.1}$$

when the input x_i is a stationary random process. When n tends to infinity, this probability tends to 0.5. More generally, let us suppose that two median filter windows of length $n = 2v+1$ include the samples $x_1, ..., x_n$ and $x_{n+1-k}, ..., x_{2n-k}$ and overlap in k samples. The probability that their respective outputs y_1, y_2 are equal to each other is given by [19]:

$$Pr\{y_1 = y_2\} = \frac{k}{2n-k} \sum_{s=s_1}^{s_2} \binom{2v+1-k}{s-1}^2 \frac{\binom{k-1}{v-s+1}}{\binom{4v+1-k}{v+s-1}} \tag{4.9.2}$$

where $s_1 = \max(1, v+2-k)$, $s_2 = \min(v+1, 2v+2-k)$. The probability given by

(4.9.2) is valid both for one- and two-dimensional filtering. Therefore, it can be used for the comparison of the streaking effect of the square-shaped, cross-shaped, and x-shaped median filters, shown in Figure 4.2.1. The number of pixels in the filter windows is $(2v+1)^2$ for the square-shaped filter of size $n \times n$ and $4v+1$ for the cross-shaped and x-shaped filters of size n. If the median filter window is centered at (i,j) and it is shifted afterwards by d pixels horizontally to the pixel $(i,j+d)$, the number k of the overlapping pixels is given by:

$$k = \begin{cases} n(n-d) & \text{for square-shaped} \\ n-d & \text{for cross-shaped} \\ 0 & \text{for x-shaped and } k \text{ odd} \\ 2 & \text{for x-shaped and } k \text{ even} \end{cases} \qquad (4.9.3)$$

The probabilities that the output y_{ij} of the median filter is equal to the output $y_{i,j+d}$ $P(d,v)=Pr\{y_{ij}=y_{i,j+d}\}$ is tabulated in Table 4.9.1 for $n=9$.

Table 4.9.1: Probabilities that the outputs of the median filter of length $n=9$ lying d pixels apart (horizontally or vertically) are equal (adapted from [19]).

d	1	2	3	4	5	6	7	8
square	0.1649	0.1028	0.0721	0.0521	0.0373	0.0256	0.0158	0.0074
cross	0.0879	0.0732	0.0599	0.0479	0.0369	0.0267	0.0172	0.0084

The adjacent x-shaped filter windows do not overlap for k odd and they overlap at only 2 pixels for k even. Thus x-shaped filters have very small probability to produce horizontal streaks. The probability of small streaks $(d \rightarrow 1)$ is larger for the square filter than the cross filter. The opposite is true for large horizontal streaks $(d \rightarrow 2v)$. Therefore, the cross filter is more likely to produce long horizontal (and vertical) streaks. If two filter windows are placed d pixels apart along the diagonal direction, the number k of the overlapping pixels is the following:

$$k = \begin{cases} (n-d)^2 & \text{for square-shaped} \\ 2 & \text{for cross-shaped} \\ n-d & \text{for x-shaped} \end{cases} \qquad (4.9.4)$$

The streaking probabilities can by computed in a similar way, by using (4.9.2), and they are shown in Table 4.9.2 for $n=9$.

Table 4.9.2: Probabilities that the outputs of the median filter of length $n=9$ lying d pixels apart diagonally are equal (adapted from [19]).

d	1	2	3	4	5	6	7	8
square	0.1074	0.0602	0.0373	0.0233	0.0138	0.0074	0.0032	0.0008
cross	0.0172	0.0172	0.0172	0.0172	0.0172	0.0172	0.0172	0.0172
x-shaped	0.0879	0.0732	0.0599	0.0479	0.0369	0.0267	0.0172	0.0084

It is found that the x-shaped filter has a high probability of producing diagonal streaks. The square filter has higher probability of producing diagonal streaks than the cross filter. Since the square median filter produces both horizontal, vertical, and diagonal streaks, it performs almost homogeneously along all directions. Thus, the end result is that it produces blotches rather than streaks. These results are expected because of the geometry of the filter windows. The result of a 15×15 square median filter is shown in Figure 4.9.1. The blotching is evident. Simulations with cross-shaped and x-shaped median filters agree with the previous analysis and show strong horizontal/vertical and diagonal streaks, respectively [19].

A streak of length L produced by an one-dimensional median filter is an event of the form:

$$y_{i-1} \neq y_i = y_{i+1} = \cdots = y_{i+L-1} \neq y_{i+L} \qquad (4.9.5)$$

The streak length L is a discrete random variable having a probability density function $P_L(\lambda)$ [19]:

$$P_L(\lambda) = Pr\{L=\lambda\} = \qquad (4.9.6)$$

$$Pr\{\, y_i = y_{i+1} = \cdots = y_{i+\lambda-1} \neq y_{i+\lambda} \mid y_{i-1} \neq y_i \,\}$$

It is proven that for large filter lengths n [19]:

$$\lim_{n \to \infty} P_L(\lambda) = 2^{-\lambda} \qquad (4.9.7)$$

(4.9.7) coincides with (4.9.1) for $\lambda = 1$. The mean value of a streak is given by:

$$\mu_L = \sum_{\lambda=1}^{\infty} \lambda P_L(\lambda) = \frac{2\nu+1}{\nu+1} \;\; \to 2 \qquad (4.9.8)$$

The expected value of the streak length for a large filter window tends to 2 rather fast when n increases. This suggests that filters having large windows tend to produce successive streaks of moderate length, rather than long streaks [19].

(a)   (b)

(c)

Figure 4.9.1: (a) Original image; (b) Image corrupted by impulsive noise; (c) Output of a square 15×15 median filter.

Streaking and blotching are undesirable effects. Post-processing of the median filter output by a linear smoother having a small window has been proposed in [20]. A better solution is to use other nonlinear filters based on order statistics, which have performance similar to that of the median filter.

4.10 DETERMINISTIC PROPERTIES OF THE MEDIAN FILTER

A very important tool in the analysis of the linear systems is their steady state behavior with sinusoidal inputs. It is known that the frequency of a sinusoid is not changed, when it passes through a linear system. Only its phase and amplitude are changed. This fact is, of course, not valid for median filters, because they are highly nonlinear systems. A great effort has been made by many researchers to find signals that are invariant under median filtering. Such signals are called *roots* or *fixed points* of the median filter. Their meaning for median filtering is analogous to the meaning of the sinusoids in the passband of linear filters. An example of a root of a median filter of length $n=3$ is shown in Figure 4.10.1. The root is obtained by filtering twice the input sequence.

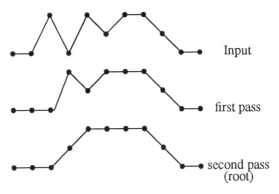

Figure 4.10.1: Root of a one-dimensional median filter of length $n=3$ obtained in two median passes.

There are several problems, which are related to the median roots. The first problem is the determination of the shape of a median root. The second is the construction of all possible roots of a certain length. The third problem is the rate of convergence of a non-root signal to a root after consecutive median filtering. These three problems are the subject of the deterministic analysis of the median filters.

The clarification of some notions is important for the deterministic analysis of the median filters. Therefore, some definitions follow which are useful for median filtering of length $n=2v+1$ [4,21,22]:

A sequence x_i is *monotonic*, if $x_i \leq x_j$ or $x_i \geq x_j$ for every $i < j$.

A sequence x_i is *locally monotonic* of length m, denoted by $LOMO(m)$, if the subsequence $(x_i,..,x_{i+m-1})$ is monotonic for every i.

A *constant neighborhood* consists of at least $v+1$ consecutive identically valued points.

An *edge* is a monotonic region between two constant neighborhoods having different values.

An *impulse* is a constant neighborhood followed by at least one but no more than n points, which are then followed by another constant neighborhood having the same value as the first constant neighborhood. The two boundary points of these at most n points do not have the same value as the two constant neighborhoods.

Oscillation is a sequence of points that is not part of a constant neighborhood, edge, or an impulse.

4.11 SHAPE OF THE ONE-DIMENSIONAL ROOT SIGNALS

The shape of the signals that are invariant to median filtering generally depends on the window size $n=2v+1$. Signals that are roots of a median filter of size n may not be roots of a median filter of greater size m ($m>n$). Therefore, a root will be associated with the length n of the median filter. In the following, the shapes of the root signals will be described by theorems [21]. Some of the proofs of these theorems are lengthy and therefore they are omitted.

Theorem 4.11.1 If a sequence x_i is monotonic, it is a root of median filters of any size.

Furthermore, monotonic functions do not have any impact on median filtering, as is stated by the following theorem:

Theorem 4.11.2 If $g(x)$ is a monotonic function, then:

$$med(g(x_1),..,g(x_n)) = g(med(x_1,..,x_n)) \qquad (4.11.1)$$

An implication of Theorem 4.11.2 is that scale is irrelevant to median filtering. A sufficient condition for a signal to be root of the median filter is given by the following theorem [21, p.199]:

Theorem 4.11.3 A $LOMO(m)$ sequence is invariant to the median filtering of length $n=2v+1$ for all $v \leq m-2$.

This theorem can also be restated as follows: a sufficient condition for a signal to be root of a median filter is that it consists only of constant neighborhoods and edges [22]. The following two theorems give the necessary conditions for a signal to be a root of the median filter [21].

Theorem 4.11.4 If the sequence x_i is a root of the median filter of extent $n=2v+1$ and if there exists a monotonic segment $(x_p, x_{p+1}, .., x_{p+v})$ of length $v+1$, then x_i is $LOMO$ $(v+2)$.

Theorem 4.11.5 If the sequence x_i is root of a median of extent $n=2v+1$ and it is nowhere $LOMO$ $(v+1)$, then x_i is a bi-valued sequence.

Theorem 4.11.5 suggests, rather surprisingly, that median filters cannot remove binary oscillations. However, a similar property has been already found in the statistical analysis of the medians, where it has been stated that the medians do not suppress effectively first order autoregressive processes of the form (4.5.7) having a negative coefficient a. Such processes have highly oscillatory behavior.

Median filters of length $n=3$ do not possess oscillatory roots because any segment of two samples is monotonic and, therefore, any sequence is $LOMO$ (3). By using mathematical induction this property is generalized as follows [21]:

Theorem 4.11.6 If a sequence x_i is invariant to median filters of extent $2p+1$, for every $p=1,2,..,v$, then it is $LOMO$ $(v+2)$.

In most cases, sequences of finite extent are inputs to median filters. If the sequence x_i is of finite extent $1 \leq i \leq L$, it is extended at its beginning and at its end by v points before filtering to account for the start-up and end effects. Usually, the appended points x_i, $i=-v+1,..,0$ are chosen to be equal to x_1 and the points x_i, $i=L+1,..,L+v$ are chosen to be equal to x_L. These extended subsequences have by definition two monotonic subsequences of length $v+1$, namely $x_{-v+1},..,x_1$ and $x_L,..,x_{L+v}$. If such a sequence is root of the median filter of extent $n=2v+1$, it is $LOMO$ $(v+2)$ by virtue of Theorem 4.11.4. Thus, for most practical cases median roots consist only of constant neighborhoods and edges.

4.12 SHAPE OF TWO-DIMENSIONAL ROOT SIGNALS

The deterministic analysis of the two-dimensional root signals is much more difficult than the corresponding analysis of the one-dimensional signals. The first reason for this difficulty is the variety of the shapes of the filter window A. The second reason is the variety of the patterns of the two-dimensional signals. Thus, several notions, e.g., monotonicity, have to be restated for the two-dimensional case. A two-dimensional sequence x_{ij} is called *locally monotonic* with respect to a window A, if $x_{i+r,j+s}$ is monotonic for $(r,s) \in L \cap A$ and for all lines L passing through (0,0) [21, p.206].

In the following, some theorems will be given that describe the shape of two-dimensional root signals [21]. It is always assumed that the window A is symmetric about and includes the origin (0,0).

Theorem 4.12.1 If an image is locally monotonic with respect to a window A, then it is a root of the median filter with a window A or subset of A.

The requirement of local monotonicity can be further relaxed, if the so-called *p-symmetric* windows are used. A window is *p*-symmetric, if it is symmetric about (0,0), contains (0,0) and all points of the intersection of $\mathbf{Z}^2$ and the finite line segment $\theta(i,j)$, $0 \le \theta \le 1$, which links (0,0) and (i,j) for every $(i,j) \in A$. Let $\{A+(r,s)\}$ be a *p*-symmetric window centered at (r,s), L be an arbitrary line in $\mathbf{Z}^2$ and $N_{L,A}$ be the number of points contained in $L \cap \{A+(r,s)\}$. $N_{L,A}$ is independent of (r,s) due to the periodicity of $\mathbf{Z}^2$. It is also an odd number, because A is symmetric. All lines in $\mathbf{Z}^2$ parallel to L have the same $N_{L,A}$. Based on these definitions, the following theorem gives a sufficient condition for the shape of two-dimensional root signals [21]:

Theorem 4.12.2 Let A be a *p*-symmetric window and x_{ij} be a two-dimensional sequence. If, for every line L, the samples x_{ij} on it are locally monotonic of length $(N_{L,A}+3)/2$, then x_{ij} is a root of a two-dimensional median filter, whose window is A or a subset of A.

A necessary condition for a sequence to be a two-dimensional root is given by the following theorem [21]:

Theorem 4.12.3 If a two-dimensional sequence x_{ij} is a root of all medians having a *p*-symmetric window which is equal to A or to a subset of A (A is *p*-symmetric), then for any line L, the samples x_{ij} on it are locally monotonic of length $(N_{L,A}+3)/2$.

The two-dimensional sequence described by this theorem is monotonic on any line in $\mathbf{Z}^2$ and it is called *monotonic in all directions*. The sufficient conditions described by Theorem 4.12.2 are usually difficult to meet in natural images, especially for large filter windows A. Some binary patterns which satisfy the conditions of the Theorem 4.12.2 and are roots of two-dimensional medians are shown in Figure 4.12.1. These patterns are the smallest convex finite objects, which are preserved under median filter of the respective window. Although the restriction that each line L is locally monotonic and invariant to median filtering of length $N_{L,A}$ is required, it appears that, for finite convex objects, the sufficient condition of Theorem 4.12.2 is also necessary. However, if the roots are infinite patterns, this is not valid any more. Consider the patterns of Figure 4.12.2. All three patterns are invariant to two-dimensional 3×3 square median filtering. Patterns (a) and (b) are nowhere locally monotonic with respect to the median window Λ. Pattern (c) is locally monotonic everywhere except at the so-called *saddle points a, b, c, d*. This example shows that the structure of the roots of two-dimensional median filters is very complicated, especially if the roots are infinite patterns.

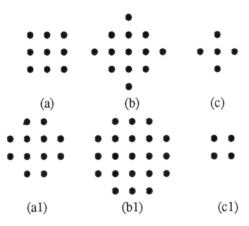

Figure 4.12.1: Two-dimensional windows (a)-(c) and their corresponding two-dimensional root patterns (a1)-(c1) (adapted from [13]).

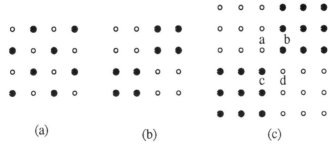

Figure 4.12.2: Two-dimensional roots of a square 3×3 window (adapted from [13]).

In certain cases two-dimensional median roots must be constructed. An easy way to construct such roots is described in [37]. Let $x_i, y_i, i \in \mathbf{Z}$ be two one-dimensional binary roots of a median filter of length $n=2v+1$, such that there exist always ν or $\nu+1$ ones in the filter window. Two-dimensional binary roots of a median filter having $n \times n$ square window can be constructed as follows:

$$x_{ij} = \begin{cases} 1 & \text{if } x_i = y_j \\ 0 & \text{if } x_i \neq y_j \end{cases} \qquad (4.12.1)$$

Oscillatory binary two-dimensional roots can be easily constructed by using (4.12.1). Such a root is the periodic extension of the signal shown in Figure 4.12.2b.

4.13 CONSTRUCTION OF ROOT SIGNALS

It is well known from section 4.11 that the one-dimensional root signals x_i of finite extent $1 \leq i \leq L$ are *LOMO* $(\nu+2)$ sequences, if the median filter has window length n. If the signal x_i is k-valued the total number of roots of length L is finite. In the following, algorithms for the construction of such roots will be given. We shall start from the construction of binary $(k=2)$ root signals because their construction is easier than that of the k-valued roots.

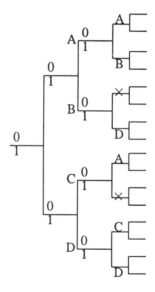

Figure 4.13.1: Tree structure for the construction of binary root signals for a median filter of length 3 (adapted from [23]).

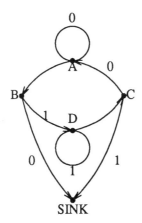

Figure 4.13.2: State diagram for the root construction for n=3 (adapted from [23]).

The algorithm for the construction of binary root signals is based on the so-called *discrete state model* [23]. To start, it is assumed that the window size is $n=3$ ($v=1$). Such a median filter has root signals with minimal constant neighborhood length 2. The root signal is constructed as follows. The first sample x_1 is chosen to be 0, without loss of generality. A sample $x_0=0$ is appended to the left to accommodate the start effect. The subsequence (x_0,x_1) is already a constant neighborhood (00) of length 2. The second sample x_2 can be either 0 or 1. If $x_2=0$ is chosen, the constant neighborhood is extended to $(x_0,x_1,x_2)=(000)$. If $x_2=1$ is chosen, an edge (001) is formed. In this case, the third sample x_3 cannot be 0 because the sequence $(x_0,x_1,x_2,x_3)=(0010)$ is not *LOMO* (3). In contrast, the value $x_3=1$ is allowed because the subsequence $(x_0,x_1,x_2,x_3)=(0011)$ is *LOMO* (3). This process is repeated and results in the tree structure in Figure 4.13.1. The nonallowable paths in this tree are denoted by ×. Each allowable path in the tree corresponds to a root signal. Thus the number of all possible root signals is finite, although it increases with the sequence length. Figure 4.13.1 indicates that four discrete states are encountered in the construction of a root signal [23]:

State A: Two successive samples have zero value (00). The next digit can take any value 0 or 1.

State B: Two successive samples have values (01). The next digit must be 1.

State C: Two successive samples have values (10). The next digit must be 0.

State D: Two successive samples have zero value (11). The next digit can take any value 0 or 1.

Table 4.13.1: Number of roots of a median filter of length $n=3$ for k-leveled signals (adapted from [23,25]).

Length L	Number of levels	
	$k=2$	$k=8$
2	4	64
3	6	232
4	10	932
5	16	3704
6	26	14932
7	42	60112
8	68	241718

Figure 4.13.2 shows the *state transition diagram*. It is seen that state A generates itself and state B. State B generates only state D. State C generates state A. State D generates itself and state C. The nonallowable path is denoted by a sink in Figure 4.13.2. Let us denote by $A(i)$, $B(i)$, $C(i)$, $D(i)$ the number of the states when the sample x_i is appended to the root. From the state transition diagram it follows that:

$$A(i+1) = A(i)+C(i) \tag{4.13.1}$$

$$B(i+1) = A(i)$$

$$C(i+1) = D(i)$$

$$D(i+1) = D(i)+B(i)$$

with the initial conditions: $A(2)=B(2)=C(2)=D(2)=1$. The number of roots at a given root signal length i is given by:

$$R(i) = A(i)+B(i)+C(i)+D(i) \tag{4.13.2}$$

By combining (4.13.1-2) the following recursive relation is found for R(i):

$$R(i+1) = 2A(i)+2D(i)+B(i)+C(i) = R(i)+R(i-1) \tag{4.13.3}$$

with initial conditions $R(1)=2$ and $R(2)=4$. The solution of this difference equation is the following:

$$R(L) = 2f(L+1) \tag{4.13.4}$$

$$f(L) = \frac{1}{\sqrt{5}}\left[(\frac{1+\sqrt{5}}{2})^L-(\frac{1-\sqrt{5}}{2})^L\right]$$

The number of the median roots having length L is tabulated in Table 4.13.1 for filter length 3 and $k=2$. It is clearly seen that the number of the roots increases tremendously with L.

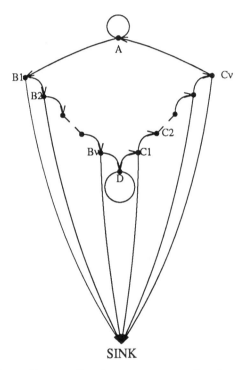

SINK

Figure 4.13.3: State diagram for the root construction for arbitrary $n=2v+1$ (adapted from [23]).

The state diagram approach can be generalized for binary roots of a median filter of arbitrary length $n=2v+1$ [23]. The corresponding state diagram is shown in Figure 4.13.3. It is odd symmetric, and the number of Bi states is equal to the number of the Ci states. States Bi, Ci are transitional, i.e., each of them is transformed into only another state, as we move along the diagram. States A,D generate one state of their own kind plus one transitional state. Therefore:

$$R(i) = A(i) + B1(i) + B2(i) +...+ B[v](i) + \qquad (4.13.5)$$

$$D(i) + C1(i) + C2(i) +...+ C[v](i)$$

where:

$$A(i) = D(i) \qquad (4.13.6)$$

$$B\,1(i) = A\,(i-1)$$

$$B\,2(i) = A\,(i-2)\ ....$$

$$B\,[v](i) = A\,(i-v)$$

$$C\,1(i) = B\,1(i)$$

$$C\,2(i) = B\,2(i)\ ....$$

$$C\,[v](i) = B\,[v](i)$$

Based on (4.13.5), $R\,(i)$ can be expressed as a function of $A\,(i)$:

$$R\,(i) = 2 \sum_{j=0}^{v} A\,(i-j) \tag{4.13.7}$$

The state A at step i is generated by the state A at step $i-1$ and the state D at step $i-v-1$, as it can be seen in Figure 4.13.3. Thus the following recursive relation gives $A\,(i)$:

$$A\,(i) = A\,(i-1)+D\,(i-v-1) = A\,(i-1)+A\,(i-v-1) \tag{4.13.8}$$

By combining (4.13.7-8), the following recursive equation is found, which gives the number of roots at step i:

$$R\,(i) = R\,(i-1)+R\,(i-v-1) \tag{4.13.9a}$$

or equivalently:

$$R\,(i+v+1) = R\,(i+v)+R\,(i) \tag{4.13.9b}$$

The number of roots $R\,(i)$ is found by an inverse z transform of the z transform of $R\,(i)$ obtained from the difference equation (4.13.9b) [23]:

$$R\,(i) = Z^{-1}\{\frac{(z-1)}{z^{v+1}-z^{v}-1}(z^{v}R\,(0)+..+zR\,(v-1)+\frac{z}{z-1}R\,(v))\} \tag{4.13.10}$$

A more descriptive solution for $R\,(i)$ can be found by using a matrix state variable approach [24].

The same methodology can also be used for the construction of k-valued root signals. However, in this case the number of states and the number of root signals is much larger. An example of the increase of the number of roots with their length L for $k=2$ and $k=8$ is shown in Table 4.13.1 for a 3-point median filter [25].

4.14 RATES OF CONVERGENCE OF MEDIAN FILTERS AND APPLICATIONS OF MEDIAN ROOTS

It is known from previous sections that signals consisting of constant neighborhoods and edges, i.e., being *LOMO* (v+2), are roots of median filters of length n=2v+1. It can also be proven that every signal of length L converges to a root after consecutive median filtering.

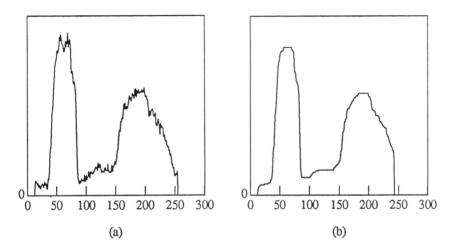

(a)   (b)

Figure 4.14.1: (a) Original signal; (b) Root of the median filter of length 15.

An example of a non-root signal converging to a root after twenty passes is shown in Figure 4.14.1. The rate of convergence is governed by the length of the non-root sections of the input signals. Such sections in a binary input signal are oscillations of the type ..010101.. After each median pass, the length of the oscillations is reduced by 2, until the oscillation vanishes. Therefore, the longer the original length of the oscillations, the slower the rate of convergence. There exists an upper bound on convergence given by the following theorem [22]:

Theorem 4.14.1 After successive median filter passes, any non-root signal of length L will become a root after a maximum of $(L-2)/2$ successive filterings.

An even tighter bound of the number of passes is given by the following theorem [26]:

Theorem 4.14.2 A nonroot signal of length L will converge to a root in at most

$$3 \left\lceil \frac{(L-2)}{2(v+2)} \right\rceil$$ passes of a median of length $n=2v+1$.

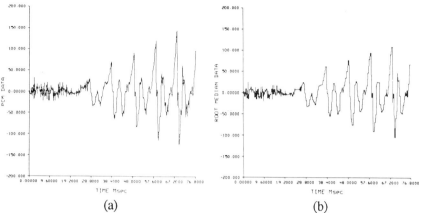

(a)   (b)

Figure 4.14.2: (a) Original speech signal; (b) Modified BTC reconstructed signal (from G. R. Arce, N. C. Gallagher, "State description of the root-signal set of median filters", IEEE Trans. on ASSP, vol.30, pp. 894-902, Dec. 1982, C 1982 IEEE).

In the following an example of the use of the median roots in speech processing is given [23]. The speech signal is sampled at 10 KHz and 10 bits per sample. Such a speech signal is shown in Figure 4.14.2a. With standard PCM, 100Kbits/s are needed for speech coding. Data compression can be achieved by a method similar to block truncation coding (BTC), originally proposed in [27]. This method consists of two steps. First, the sampled data are divided into smaller samples (e.g., 6 samples) and their average and standard deviation are calculated. Next the samples are truncated to 1 or to 0, if they lie above or below the average. For instance, the segment [200 191 162 159 165 205] has average 180, standard deviation 21 and binary representation [1 1 0 0 0 1]. The original signal is reconstructed as follows. The samples truncated to 1 are represented by the average plus the standard deviation, whereas the ones truncated to zero are represented by the average minus the standard deviation. The signal reconstruction in the previous example is [201 201 159 159 159 201]. If 11 bits are required for the transmission of the mean and of the standard deviation, 17 bits are required for the transmission of the whole packet, instead of 60 bits required in the original PCM coding. Using the standard BTC, the bit rate is reduced to 28.4 Kbits/s. A further data compression can be achieved as follows. The binary signal [1 1 0 0 0 1] can be reduced to its root signal by repetitive median filtering. The number of roots of length $L=6$ is 26, if the median filter of size 3 is considered. For median filters of size 5, the number of roots of length $L=6$ is 18. A root number can be transmitted instead for the BTC binary segment, as it is shown in Figure 4.14.3. This modified BTC scheme achieves 26 Kbits/s bit rate if the median filter of length n=3 is used, and 25 Kbits/s, if the median filter of length n=5 is used. The reconstructed speech signal is shown in Figure 4.14.2b and it is very close to the original signal of Figure 4.14.2a. This fact is explained easily. Nonroot sequences, e.g., [1 0 1 0 1 0], are very unlikely to occur. Furthermore, the binary combinations with higher probability of occurrence are very close to the root of the median filter. However, the bit rate of the modified BTC scheme is not much lower than that of the standard BTC. This fact is explained by the same overhead, which is used in both schemes.

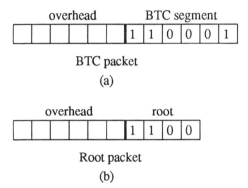

Figure 4.14.3: (a) BTC packet; (b) Modified BTC packet.

4.15 THRESHOLD DECOMPOSITION OF MEDIAN FILTERS. STACK FILTERS

The properties of the binary signals in median filtering are well understood, as has been described in the previous sections. Furthermore, the computation of the median filter for binary signals x_i is reduced to counting the ones in the filter window. If the number of ones is greater or equal to $v+1$, the output of the median filter is 1, otherwise it is 0. Therefore, it is attractive both for theoretical and practical reasons to reduce median filtering of k-valued sequences to the median filtering of binary sequences. This is obtained by the so-called *threshold decomposition* of the median filter [28].

Let x_i be the k-valued input signal taking values in $[0,k-1]$. A threshold decomposition function $g_j(x_i)$ at level j is defined as follows:

$$g_j(x_i) = \begin{cases} 1 & \text{if } x_i \geq j \\ 0 & \text{if } x_i < j \end{cases} \qquad (4.15.1)$$

By using $k-1$ such decomposition functions $g_j(x_i)$, $1 \leq j \leq k-1$ the sample x_i is decomposed to $k-1$ binary values $x_i^{(j)}$, $1 \leq j \leq k-1$:

$$x_i^{(j)} = g_j(x_i) \qquad (4.15.2)$$

The k-valued sample x_i can be reconstructed from the binary samples $x_i^{(j)}$, $1 \leq j \leq k-1$ as follows:

$$x_i = \sum_{j=1}^{k-1} x_i^{(j)}$$ (4.15.3)

The output y_i of the median filter:

$$y_i = med(x_{i-v}, .., x_i, x_{i+v})$$ (4.15.4)

can also be decomposed to $k-1$ binary samples $y_i^{(j)}$, $1 \leq j \leq k-1$. Each of them can be computed as follows:

$$y_i^{(j)} = med(x_{i-v}^{(j)}, .., x_i^{(j)}, x_{i+v}^{(j)})$$ (4.15.5)

according to Theorem 4.11.2, because the functions $g_j(x_i)$ are monotonic. The median filter output is reconstructed as follows:

$$y_i = \sum_{j=1}^{k-1} y_i^{(j)}$$ (4.15.6)

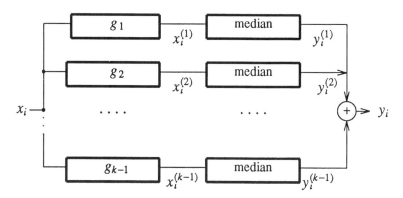

Figure 4.15.1: Threshold decomposition of the median filter.

Thus, the following procedure can be used for the computation of the median. The sequence x_i of length L is decomposed in $k-1$ binary sequences of length L. Each of them is filtered by a median filter of length n. The median filter k-valued output is reconstructed from these binary median filtered sequences. This procedure is illustrated in Figure 4.15.1. A demonstration example of the threshold decomposition for a three-valued sequence is shown in Figure 4.15.2. The input sequence is decomposed in two binary sequences which are filtered by a median filter of length $n=3$ and are recombined to produce the k-valued median output.

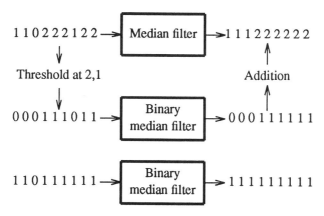

Figure 4.15.2: Median filtering of a three-valued sequence by threshold decomposition.

Threshold decomposition is a very important tool for the theoretical analysis of the median filters. For example, it can be used to find the statistical distribution of the output of the median filter [29]. The distribution $F_{v+1}(y)$ is given by:

$$F_{v+1}(j) = Pr\{y_i \leq j\} \tag{4.15.7}$$

By using the decomposition property (4.15.3), this relation reduces to:

$$Pr\{y_i \leq j\} = Pr\{y_i^{(j+1)} = 0\} = Pr\{\sum_{l=i-v}^{i+v} x_l^{(j+1)} < v\} = \tag{4.15.8}$$

$$Pr\{\sum_{l=i-v}^{i+v} [1-x_l^{(j+1)}] \geq v+1\}$$

Let

$$z_i = \sum_{l=i-v}^{i+v} [1-x_i^{(j+1)}] \tag{4.15.9}$$

be the sum of $2v+1$ Bernoulli trials. The probability density function of z_i is given by:

$$B(2v+1, F(j)) = \binom{2v+1}{k} F(j)^k [1-F(j)]^{2v+1-k}, \qquad k=0,1,..,2v+1 \tag{4.15.10}$$

where $F(j)$ is the cdf of the input data x_j Therefore (4.15.8) becomes:

$$F_{v+1}(j) = Pr\{z_i \geq v+1\} = \sum_{l=v+1}^{2v+1} Pr\{z_i = l\} \tag{4.15.11}$$

$$= \sum_{l=v+1}^{n} B(2v+1, F(j)) = \sum_{l=v+1}^{n} \binom{n}{l} F^{l}(j)[1-F(j)]^{n-l}$$

As expected, this formula is the same with (4.3.1).

Let $y_q(i)$ denote the root of a median filter obtained after a sufficiently high number q of median passes. Threshold decomposition can be used to find the probability distribution $F_{y_q}(j)$ of the root signal $y_q(i)$ in the following way:

$$F_{y_q}(j) = Pr\{y_q(i) \le j\} = Pr\{y_q^{(j+1)}(i) = 0\} \tag{4.15.12}$$

where $y_q^{(j)}(i)$, $j=1,...,k-1$ denotes the thresholded values of $y_q(i)$. It can be proven that [29]:

$$Pr\{y_q^{(j+1)}(i)=0\} = a_{j+1}^{2}(2-a_{j+1}) + \tag{4.15.13}$$

$$\frac{a_{j+1}^{4}[1-a_{j+1}^{2}(2-a_{j+1})]+2(a_{j+1}(1-a_{j+1}))^3}{[1-a_{j+1}(1-a_{j+1})]^2}$$

where:

$$a_{j+1} = Pr\{x_0^{(j+1)}(i)=0\} = Pr\{x_i \le j\} = F(j) \tag{4.15.14}$$

By combining (4.15.12-14) the root probability distribution can be found:

$$F_{y_q}(j) = \frac{F^4(j)[1-F^2(j)(2-F(j))]+2F^3(j)(1-F(j))^3}{[1-(1-F(j))F(j)]^2} \tag{4.15.15}$$

$$+ F^2(j)[2-F(j)]$$

It can be seen from (4.15.15) that the cdf of the root signal does not depend on the number of passes q.

The binary median filter operating on a binary signal x_i is essentially a Boolean function of the binary variables $x_{i-v},..,x_i,..,x_{i+v}$. Its form is given by the following relation for the filter length $n=3$ [39]:

$$y_i = x_{i-1}x_i + x_i x_{i+1} + x_{i+1} x_{i-1} \tag{4.15.16}$$

where .,+ denote Boolean operators. Such a binary function operates at each binary signal $x_i^{(j)}$, $1 \le j \le k-1$ obtained by the threshold decomposition. The outputs of those binary median filters possess the so-called *stacking property*: if the binary output signals $y_i^{(j)}$, $1 \le j \le k-1$ are piled, the pile at time i consists of a column of 1's having a column of 0's on top. The desired multilevel output y_i is the level just before the (10) transition. This remark greatly facilitates the VLSI implementation of the median filter based on threshold decomposition [39]. An interesting generalization is to find a class of filters, called *stack filters*, which possess both the threshold decomposition structure and the stacking property. This problem reduces to the problem of finding *stackable* Boolean functions, because it is the Boolean functions that determine the properties of the decomposed filter. It is proven that a Boolean function is stackable if and only if it contains no complements of the input variables of the form $\bar{x}_i$. Thus the stackable functions are also called *positive Boolean functions*. The binary median (4.15.16) is stackable because it does not contain complements of any of the inputs x_{i-1}, x_i, x_{i+1}. The number of the positive Boolean functions is finite for finite filter length n. There exist 20, 7581 such functions for $n=3,5$, respectively. Examples of stackable functions for $n=3$ are the following:

$$y_i = x_{i-1} x_i x_{i+1} \qquad \text{minimum} \tag{4.15.17}$$

$$y_i = x_{i-1} + x_i + x_{i+1} \qquad \text{maximum} \tag{4.15.18}$$

$$y_i = x_i + x_{i-1} x_{i+1} \tag{4.15.19}$$

$$y_i = x_{i-1} x_i + x_i x_{i+1} \tag{4.15.20}$$

The stack filters based on (4.15.19-20) are called *asymmetric medians* because they eliminate negative and positive impulses, respectively. Any composition of the stack filters can be expressed by a single stack filter, due to a similar property of the positive Boolean functions. Any appended signal having constant neighborhoods at its start and at its end converges to a root after consecutive applications of a stack filter [39]. The use of stack filters in noise filtering by employing the mean absolute error criterion is described in [40].

4.16 SEPARABLE TWO-DIMENSIONAL MEDIAN FILTERS

In the recent years there has been a burst in the literature of median filters. Several modifications of the standard one-dimensional and two-dimensional median filters have been proposed. Such a modification is the *separable median filter* [31], which will be analyzed in this section.

The separable two-dimensional filter of size n results from two successive applications of an one-dimensional filter of length n along rows and then along columns of an image (or vice versa):

$$z_{ij} = med(x_{i,j-v}, .., x_{ij}, x_{i,j+v}) \tag{4.16.1}$$

$$y_{ij} = med(z_{i-v,j}, .., z_{ij}, z_{i+v,j}) \tag{4.16.2}$$

v points are appended to the beginning and to the end of each row, equal to the first and to the last sample, respectively, before row-wise filtering. The same procedure is repeated along columns before the vertical pass. The whole procedure is illustrated in Figure 4.16.1.

The output distribution $F_z(z)$ of the row-wise filtering is given by:

$$F_z(z) = \sum_{j=v+1}^{2v+1} \binom{n}{j} F^j(z) [1 - F(z)]^{n-j} \tag{4.16.3}$$

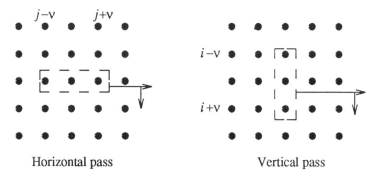

Horizontal pass   Vertical pass

Figure 4.16.1: The separable median filter.

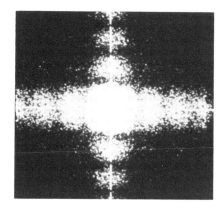

Figure 4.16.2: Output power spectral density of a 3×3 separable median filter.

If the data x_{ij} are independent identically distributed (iid), the output z_{ij} of the row-wise filtering remains independent along columns. Therefore, the probability distribution of the output of the separable median filter is given by:

$$F_y(y) = \sum_{j=v+1}^{2v+1} \binom{n}{j} F_z^{\,j}(y)[1-F_z(y)]^{n-j} \tag{4.16.4}$$

The probability density function of the output of the rowwise filtering is given by:

$$f_z(z) = \frac{n!}{(v!)^2} F^v(z)[1-F(z)]^v f(z) \tag{4.16.5}$$

Therefore, the variance of the output of the separable median filter is given by:

$$\sigma_y^2 = \frac{1}{4f_z^2(0)(n+b)} = \frac{(\nu!)^4 2^{2n}}{16(n+b)(n!)^2} \frac{1}{f^2(0)} \tag{4.16.6}$$

The value $b=2$ is suggested in [31]. For comparison, the variance σ^2_{med} of an $n \times n$ nonseparable two-dimensional median filter is given:

$$\sigma_{med}^2 = \frac{1}{4f^2(0)(n^2+b)} \tag{4.16.7}$$

(a)   (b)

(c)   (d)

Figure 4.16.3: (a) Original image; (b) Image corrupted by 10% impulsive noise; (c) Output of the 3×3 separable median filter; (d) Output of the 3×3 standard median filter.

Monte Carlo simulations have been performed to verify the relations (4.16.7-8) [31]. It has been proven that the separable median filter has always a slightly greater variance than the $n \times n$ standard two-dimensional filter. The joint probability distribution $F_y(y_m, y_n)$ has been derived in [8]. Its formula is lengthy and cumbersome. However, it is numerically tractable and has been used for the numerical calculation of the autocorrelation function and of the power spectrum of the separable median filter. The output power spectral density of a 3×3 separable median filter with uniform input distribution is shown in Figure 4.16.2. The low-pass characteristics of the separable median filter are evident. The interested reader can find the deterministic analysis of the separable median filters in [34]. An example of separable two-dimensional median filtering of impulsive noise is shown in Figure 4.16.3. The separable and the standard median filters remove the spikes, as it is seen in Figures 4.16.3a and 4.16.3b, respectively. The separable median filter is inferior but has comparable noise filtering properties to the standard median filter. Its great advantage is that it is much faster than the nonseparable median filter because it requires only two calculations of the median of n points per output pixel. The nonseparable median filter requires the calculation of the median of n^2 points per output pixel. Therefore, it is much slower.

4.17 RECURSIVE MEDIAN FILTERS

One intuitive modification of the median filters is to use the already computed samples $y_{i-v}, .., y_{i-1}$ for the calculation of y_i:

$$y_i = med(y_{i-v}, .., y_{i-1}, x_i, .., x_{i+v}) \tag{4.17.1}$$

The filter (4.17.1) is the so-called *recursive median filter*. Its properties are different of that of the standard median filter. In the two-dimensional case, the properties also depend on the direction of scanning of the image. The output of the recursive median tends to be more correlated than the output of the standard median. This is easily explained by the fact that the output samples $y_{i-v}, .., y_{i-1}$ contribute directly to the computation of y_i.

The recursive median filter possesses some nice deterministic properties. A signal is invariant to recursive median filtering if and only if it is invariant to standard median filtering [34]. This is easily explained by the definition of the root: $x_i = y_i$, $i \in \mathbf{Z}$. Therefore, definitions (4.17.1) and (4.2.2) are equivalent for root signals. However, this does not mean that every signal converges to the same root after standard or recursive median filtering. The root of the recursive median filter is different from the root of the standard median filter. Any one-dimensional signal is reduced to a root after just one pass by a recursive median filter. This property can easily be proven by mathematical induction. It is a very important property because it shows that every one-dimensional signal converges to its root very quickly, i.e., in just one recursive median pass. Such a recursive median root is shown in Figure 4.17.1. However, no similar property holds in the general case of the recursive median filtering of two dimensional signals [34,42]. Fast recursive algorithms for the calculation of binary and multilevel two-dimensional median roots can be found in [42].

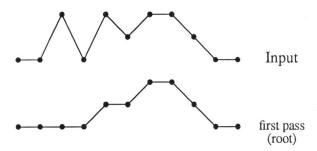

Figure 4.17.1: Root of a recursive median filter of length $n=3$.

The threshold decomposition can also be used for recursive median filters as follows:

$$y_i = \sum_{j=1}^{k-1} y_i^{(j)} \tag{4.17.2}$$

$$y_i^{(j)} = med(y_{i-v}^{(j)},..,y_{i-1}^{(j)},x_i^{(j)},..,x_{i+v}^{(j)}) \tag{4.17.3}$$

where:

$$x_i^{(j)} = g_j(x_i) , \quad y_i^{(j)} = g_j(y_i)$$

and $g_j(x)$ is given by (4.15.1).

A variation of the recursive two-dimensional median filter is the *separable recursive median filter*:

$$z_{ij} = med(z_{i,j-v},..,x_{ij},x_{i,j+v}) \tag{4.17.4}$$

$$y_{ij} = med(y_{i-v,j},...,z_{ij},z_{i+v,j}) \tag{4.17.5}$$

The statistical analysis of the recursive median filters can be found in [43]. Recursive separable filters have been proposed for positive impulse noise filtering of the form:

$$x_{ij} = \begin{cases} s=0 & \text{\textit{with probability} } q=1-p \\ d=1 & \text{\textit{with probability} } p \end{cases} \tag{4.17.6}$$

In this model the image background is zero and the impulse amplitude is one. The filtering in one dimension does not introduce correlation along the second dimension. Therefore, the probability of correct reconstruction by using a nonrecursive separable median filter is given by (4.7.5) both for rowwise and columnwise filtering. Let $P_h(1)$, $P_v(1)$ denote the probability of incorrect reconstruction along rows and columns, respectively. They are given by:

$$P_h(1) = \sum_{v+1}^{n} \binom{n}{k} p^k q^{n-k} \tag{4.17.7}$$

$$P_v(1) = \sum_{v+1}^{n} \binom{n}{k} P_h{}^k(1)(1-P_h(1))^{n-k} \tag{4.17.8}$$

If recursive filter is used for the impulse filtering, the probability of correct or incorrect reconstruction is given by [36]:

$$p(0) = \frac{q^{v+1} + v q^{v+1} p^{v+1}}{p^{v+1} + 2v q^{v+1} p^{v+1} + q^{v+1}} \tag{4.17.9}$$

$$p(1) = \frac{p^{v+1} + v q^{v+1} p^{v+1}}{p^{v+1} + 2v q^{v+1} p^{v+1} + q^{v+1}} \tag{4.17.10}$$

respectively.

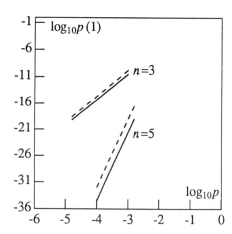

Figure 4.17.2: Probability of incorrect reconstruction for positive impulsive noise filtering as a function of the impulse probability p (the solid and dashed lines denote the recursive and the nonrecursive median, respectively).

These relations must be applied consecutively for impulse noise filtering by a separable recursive median filter. The probabilities of incorrect reconstruction for recursive and nonrecursive separable median filters are shown in Figure 4.17.2 as a function of p. It is seen that the recursive separable median filtering rejects the positive impulses slightly better than the nonrecursive median filter. A similar analysis has also been performed for mixed impulsive noise [36] and it has been proven that recursive separable median filters perform better than the nonrecursive ones. It has also been proven that the recursive median has lower breakdown probabilities and higher immunity to impulsive noise than the nonrecursive median filters [43]. Finally, a deterministic analysis of the recursive separable filters can be found in [33].

4.18 WEIGHTED MEDIAN FILTERS

It has been proven in section 4.4 that the median is the best location estimator in the L_1 sense, i.e., it minimizes:

$$\sum_{i=1}^{n} |x_i - T_n| \rightarrow \min \tag{4.18.1}$$

The *weighted median* is the estimator that minimizes the weighted L_1 norm of the form:

$$\sum_{i=1}^{n} w_i |x_i - T_n| \rightarrow \min \tag{4.18.2}$$

It can be proven that the minimization of (4.18.2) leads to the following explicit form of the weighted median [48]:

$$T_n = med\{w_1 \square x_1, .., w_n \square x_n\} \tag{4.18.3}$$

where $w \square x$ denotes w times duplication of x:

$$w \square x = x, .., x \qquad\qquad w \text{ times} \tag{4.18.4}$$

The sum of the weights w_i, $i=1,..,n$ must be an odd number. The weighted L_2 norm:

$$\sum_{i=1}^{n} w_i (x_i - T_n)^2 \rightarrow \min \tag{4.18.5}$$

is minimized by the weighted arithmetic mean:

$$T_n = \frac{\sum_{i=1}^{n} w_i x_i}{\sum_{i=1}^{n} w_i} \tag{4.18.6}$$

Therefore, there exists a close relation between the weighted median and the weighted arithmetic mean that uses the same weights. The weighted median filter was first introduced by Justusson [13]:

$$y_i = med\{w_{-v}\square x_{i-v},..,w_v\square x_{i+v}\} \tag{4.18.7}$$

It is closely related to the FIR filter having the form:

$$y_i = \frac{\sum\limits_{j=-v}^{v} w_j x_{i+j}}{\sum\limits_{j=-v}^{v} w_j} \tag{4.18.8}$$

The analysis of the performance of the weighted median filter can be found in [49,50]. In particular, the connection between the stack filters and the weighted median filters is described in [50]. This connection is used to derive the statistical and deterministic properties of the weighted median filters. It is proven that their output distribution $F_y(y)$ for iid input data is given by:

$$F_y(y) = \sum_{i=1}^{n} c_i F^i(y) \tag{4.18.9}$$

$$\sum_{i=1}^{n} c_i = 1$$

where $F(x)$ is the cdf of the input data. The coefficients c_i depend on the weights w_i ,$i=1,..,n$. For example, the output distribution of the weighted median filter having coefficients 2,1,1,1 is given by:

$$F_y(y) = -2F^3(y) + 3F^2(y) \tag{4.18.10}$$

It has been proven that several weighted median filters having different coefficient sets may have the same output cdf [50]. It has also been proven that in the case of the iid input data, the best noise attenuation is obtained by a weighted median having all coefficients equal to 1, i.e., with the standard median filter.

4.19 DISCUSSION

The median filter has been described in this chapter. It has been analyzed both from the statistical and the deterministic point of view. Its statistical analysis revealed that the median filter outperforms the moving average filter in the case of additive long-tailed noise. Thus it is very suitable for the removal of impulsive noise. It has also good edge preservation properties. These two facts make it very attractive for digital image filtering applications. However, it has some disadvantages: it destroys fine details and produces streaks and blotches, especially for large filter windows. The deterministic analysis of the median filter aims to find signals that remain invariant under one- and two-dimensional median filtering, called roots. The shape of such roots has been described in the one- and in the two-dimensional case. As expected, the analysis of the two-dimensional root signals is much more difficult. It has also been proven that every signal converges to a root after successive median filtering. The rate of convergence depends both on the signal length and on the filter length.

Some important modifications and extensions of the standard median filter have also been presented. The threshold decomposition property of the median has been proven to be an important tool for the theoretical analysis of the properties of the median filters as well as for their VLSI implementation. It is also the basis for the definition of the stack filters, which are an important extension of median filters. Another useful modification is the two-dimensional separable filters. Although they have small computational complexity, their statistical performance is similar to that of the standard median filter. Finally, recursive versions of the median have been presented. Their main advantage is that they converge to their roots much faster than the standard median.

REFERENCES

[1]   M.D. Levine, *Vision in man and machine* , Mc Graw-Hill, 1985.

[2]   W.K. Pratt, *Digital image processing,* Wiley, 1978.

[3]   H.C. Andrews, B.R. Hunt, *Digital image restoration,* Prentice-Hall 1977.

[4]   G.R. Arce, N.C. Gallagher, T.A. Nodes, "Median filters: theory for one- and two-dimensional filters", in *Advances in Computer Vision and Image Processing*, T.S. Huang editor, JAI Press, 1986.

[5]   H.A. David, *Order statistics,* ,Wiley, 1981.

[6]   J.W. Tukey, *Exploratory data analysis*, Addison-Wesley 1977.

[7]   F. Kuhlmann, G.L. Wise, "On the second moment properties of median filtered sequences of independent data", *IEEE Transactions on Communications*, vol. COM-29, no.9, pp. 1374-1379, Sept. 1981.

[8]   G.Y. Liao, T.A. Nodes, N.C. Gallagher, "Output distributions of two-dimensional median filters", *IEEE Transactions on Acoustics, Speech and Signal Processing*, vol. ASSP-33, no.5, pp. 1280-1295, Oct. 1985.

[9]   T.A. Nodes, N.C. Gallagher, "The output distribution of the median type filters", *IEEE Transactions on Communications*, vol. COM-32, pp. 532-541, May 1984.

[10]   E.L. Lehmann, *Theory of point estimation*, J.Wiley, 1983.

[11]   J.W. Tukey, "A survey of sampling for contaminated distributions", in *Contributions to probability and statistics*, Olkin editor, Stanford University Press, 1960.

[12]   F. Hampel, E. Ronchetti, P. Rousseevw, W. Stahel, *Robust statistics: An approach based on influence functions* , Wiley, 1986.

[13]   B.I. Justusson, "Median filtering: statistical properties" in *Two-dimensional digital signal processing II* , T.S. Huang editor, Springer Verlag, 1981.

[14]   P.J. Huber, "Robust estimation of a location parameter", *Ann. Math. Statist.*, vol. 35, pp. 73-101, 1964.

[15]   G.J. Yong, T.S. Huang, "The effect of median filtering in edge location estimation", *Computer Vision, Graphics and Image Processing*, vol. 15, pp. 224-245, 1981.

[16]   E. Ataman, V.K. Aatre, K.M. Wong, "Some statistical properties of median filters", *IEEE Transactions on Acoustics, Speech and Signal Processing,* vol. ASSP-29, no. 5, pp. 1073-1075, Oct. 1981.

[17]   A.C. Bovik, T.S. Huang, D.C. Munson, "The effect of median filtering on edge estimation and detection", *IEEE Transactions on Pattern Analysis and Machine Intelligence*, vol. PAMI-9, no. 2, pp. 181-194, March 1987.

[18]   D.H. Ballard, C.M. Brown, *Computer vision*, Prentice-Hall, 1983.

[19]   A.C. Bovik, "Streaking in median filtered images", *IEEE Transactions on Acoustics, Speech and signal Processing,* vol. ASSP-35, pp.493-503, April 1987.

[20]   L.R. Rabiner, M.R. Sambur, C.E. Schmidt, "Applications of a nonlinear smoothing algorithm to speech processing", *IEEE Transactions on Acoustics, Speech and Signal Processing,* vol. ASSP-23, pp. 552-557, Dec. 1975.

[21]   S.G Tyan, "Median filtering: Deterministic properties", *Two- dimensional Signal Processing II* , T.S. Huang editor, Springer Verlag, 1981.

[22]   N.C. Gallagher, G.L. Wise, "A theoretical analysis of the properties of the median filter", *IEEE Transactions on Acoustics, Speech and Signal Processing,* vol. ASSP-29, no. 6, pp. 1135-1141, Dec. 1981.

[23]   G. Arce, N.C. Gallagher, "State description for the root-signal set of median filters", *IEEE Transactions on Acoustics, Speech and Signal Processing,* vol. ASSP-30, no. 6, pp. 894-902, Dec. 1982.

[24]   G.R. Arce, N.C. Gallagher, "Root-signal set analysis for median filters", *Proc. Allerton Conference Commun. Contr. Comp.*, Oct. 1980.

[25] J.P. Fitch, E.J. Coyle, N.C. Gallagher, "Root properties and convergence rates for median filters", *IEEE Transactions on Acoustics, Speech and Signal Processing,* vol. ASSP-33, no. 1, pp. 230-240, Feb. 1985.

[26] P. D. Wendt, E.J. Coyle, N.C. Gallagher, "Some convergence properties of median filters", *IEEE Transactions on Circuits and Systems,* vol. CAS-33, no. 3, pp. 276-286, March 1986.

[27] O.R. Mitchell, E. Delp, "Multilevel graphics representation using block truncation coding", *Proc. IEEE,* vol. 68, July 1980.

[28] J.P. Fitch, E.J. Coyle, N.C. Gallagher, "Median filtering by threshold decomposition", *IEEE Transactions on Acoustics, Speech and Signal Processing,* vol. ASSP-32, no. 6, pp. 1183-1188, Dec. 1984.

[29] G.R. Arce, R.L. Stevenson, "On the synthesis of median filter systems", *IEEE Transactions on Circuits and Systems,* vol. CAS-34, no. 4, pp. 420-429, April 1987.

[30] G.R. Arce, "Statistical threshold decomposition for recursive and nonrecursive median filters", *IEEE Transactions on Information Theory,* vol. IT-29, March 1986.

[31] P.M. Narendra, "A separable median filter for image noise smoothing", *IEEE Transactions on Pattern Analysis and Machine Intelligence,* vol. PAMI-3, no. 1, pp. 20-29, Jan. 1981.

[32] T.A. Nodes, N.C. Gallagher, "Two-dimensional root structures and convergence properties of the separable median filter", *IEEE Transactions on Acoustics, Speech and Signal Processing,* vol. ASSP-31, no. 6, pp. 1350-1365, Dec. 1983.

[33] M.P. Loughlin, G.R. Arce, "Deterministic properties of the recursive separable median filter", *IEEE Transactions on Acoustics, Speech and Signal Processing,* vol. ASSP-35, no. 1, pp. 98-106, Jan. 1987.

[34] T.A. Nodes, N.C. Gallagher, "Median filters: some modifications and their properties", *IEEE Transactions on Acoustics, Speech and Signal Processing,* vol. ASSP-30, no. 5, pp. 739-746, Oct. 1982.

[35] G.R. Arce, R.J. Crinon, "Median filters: analysis of two-dimensional recursively filtered signals" *IEEE Int. Conf. on Acoustics, Speech and Signal Processing,* 1984.

[36] R.J. Crinon, G.R. Arce, "Median filters: analysis for signals with additive impulsive noise", *Proc. 21st Allerton Conference,* Oct. 1983.

[37] J. Astola, P. Heinonen, Y. Neuvo, "On root structure of median and median-type filters", *IEEE Transactions on Acoustics, Speech and Signal Processing,* vol. ASSP-35, no. 8, pp. 1199-1201, Aug. 1987.

[38] A. Papoulis, *Probability, random variables and stochastic processes*, McGraw-Hill, 1984.

[39] P.D. Wendt, E.J. Coyle, N.C. Gallagher Jr., "Stack filters", *IEEE Transactions on Acoustics, Speech and Signal Processing,* vol. ASSP-34, no. 4, pp. 898-911, Aug. 1986.

[40] E.J. Coyle, J.H. Lin, "Stack filters and the mean absolute error criterion", *IEEE Transactions on Acoustics, Speech and Signal Processing,* vol. ASSP-36, no. 8, pp. 1244-1254, Aug. 1988.

[41] E.N. Gilbert, "Lattice-theoretic properties of frontal switching functions", *Journal of Mathematical Physics*, vol. 33, pp. 57-67, Apr. 1954.

[42] W.W. Boles, M. Kanewski, M. Simaan, "Recursive two-dimensional median filtering algorithms for fast image root extraction", *IEEE Transactions on Circuits and Systems,* vol. CAS-35, no. 10, pp. 1323-1326, Oct. 1988.

[43] G. Arce, N.C. Gallagher, "Stochastic analysis for the recursive median filter process", *IEEE Transactions on Information Theory,* vol. IT-34, no. 4, pp. 669-679, July 1988.

[44] S.S. Perlman, S. Eisenhandler, P.W. Lyons, M.J. Shumila, "Adaptive median filtering for impulse noise elimination in real-time TV signals", *IEEE Transactions on Communications,* vol. COM-35, no. 6, pp. 646-652, June 1987.

[45] K.J. Hahn, K.M. Wong, "Median filtering of cepstra", *Proc. 1983 International Electrical and Electronics Conference*, pp. 352-355, Toronto, Canada, 1983.

[46] P. Zamperoni, "Feature extraction by rank-order filtering for image segmentation", *International Journal of Pattern Analysis and Artificial Intelligence*, vol. 2, no. 2, pp. 301-319, 1988.

[47] S.S.H. Naqvi, N.C. Gallagher, E.J. Coyle, "An application of median filter to digital TV", *Proc. 1986 IEEE Int. Conf. on Acoustics, Speech and Signal Processing*, Tokyo, Japan, 1986.

[48] V.J. Gebski, "Some properties of splicing when applied to nonlinear smoothers", *Comput. Statist. Data Analysis*, no. 3, pp. 151-157, 1985.

[49] D.R.K. Brownrigg, "Weighted median filters", *Commun. Assoc. Comput. Machinery*, vol. 27, pp. 807-818, Aug. 1984.

[50] O. Yli-Harja, "Median filters: extensions, analysis and design", *Technical Report*, Lappeenranta University of Technology, Finland, 1989.

CHAPTER 5

DIGITAL FILTERS BASED ON ORDER STATISTICS

5.1 INTRODUCTION

The median filter and its modifications, described in chapter 4, are a special case of a large class of nonlinear filters that is based on order statistics. This class includes a variety of nonlinear filters, e.g., L-filters, α-trimmed mean filters, max/median filters and median hybrid filters. Most of them are based on the well-known L-estimators. Therefore, they have excellent robustness properties. Others are modifications or extensions of the median filter. Nonlinear filters based on order statistics have been designed to meet various criteria, e.g., robustness, adaptivity to noise probability distributions, preservation of edge information, preservation of image details. Thus, each of them has optimal performance for specific figures of merit and specific noise characteristics.

In the following sections several nonlinear filters will be analyzed. The extensions of the median filter (max/median filter, median hybrid filter) are treated first, and the filters based on L estimators follow. At the end of this chapter some filters are presented that are based on M estimators and R estimators. Thus, most of the filters based on robust estimation theory are covered in this chapter.

5.2 MAX/MEDIAN FILTERS AND MULTISTAGE MEDIAN FILTERS

In the theoretical analysis of median filters several rather unrealistic assumptions were made about the image signal. It was assumed that the signal was of constant value, or that it was an ideal step edge. However, real images have a much more complex structure. This structure (e.g., lines, sharp corners) is destroyed by median filters having relatively large windows (e.g., more than 5×5). Fine details, sharp corners and lines are removed in this case. This is natural for median filters because the ordering process destroys any structural and spatial neighborhood information. Several efforts have been made to modify the median filter so that it takes into account structural information. Such a modification is the *max/median filter*:

$$y_{ij} = \max(z_1, z_2, z_3, z_4) \tag{5.2.1}$$

$$z_1 = \text{med}(x_{i,j-v}, \dots, x_{ij}, \dots, x_{i,j+v}) \tag{5.2.2}$$

$$z_2 = \text{med}(x_{i-v,j}, \dots, x_{ij}, \dots, x_{i+v,j}) \tag{5.2.3}$$

$$z_3 = \text{med}(x_{i+v,j-v}, \dots, x_{ij}, \dots, x_{i-v,j+v}) \tag{5.2.4}$$

$$z_4 = \text{med}(x_{i-v,j-v}, \dots, x_{ij}, \dots, x_{i+v,j+v}) \tag{5.2.5}$$

The subsections z_1, z_2, z_3, z_4 of the max/median filter span the horizontal line, the vertical line, and the two diagonal lines that pass through the point (i,j). Hence, a max/median filter of size $n = 2v+1$ spans $8v+1$ sample points, as it can be seen from (5.2.2-5). The recursive max/median filter has a definition entirely similar to (5.2.1-5).

The output distribution of the max/median filter can be found as follows. Define the sets M_l, $l=1,\dots,4$ as follows:

$$M_l = \{x_{k,l} \text{ contributing to } z_l \text{ and } (k,l) \neq (i,j)\} \tag{5.2.6}$$

Define also the event $B(k,M,y)$ as follows:

$$B(k,M,y) = \{\text{at least } k \text{ samples in the set } M \text{ are } \leq y\} \tag{5.2.7}$$

The output distribution $F_y(y)$ is given by [5]:

$$F_y(y) = Pr\{y_{ij} \leq y\} = Pr\{z_1 \leq y, z_2 \leq y, z_3 \leq y, z_4 \leq y\} = \tag{5.2.8}$$

$$Pr\{B(v+1, M_1 \cup x_{ij}, y), B(v+1, M_2 \cup x_{ij}, y), B(v+1, M_3 \cup x_{ij}, y), B(v+1, M_4 \cup x_{ij}, y)\}$$

Using the law of total probability, $F_y(y)$ becomes:

$$F_y(y) = Pr\{B(v, M_1, y), B(v, M_2, y), B(v, M_3, y), B(v, M_4, y)\} Pr\{x_{ij} \leq y\} + \tag{5.2.9}$$

$$Pr\{B(v+1, M_1, y), B(v+1, M_2, y), B(v+1, M_3, y), B(v+1, M_4, y)\} Pr\{x_{ij} > y\}$$

The events $B(v, M_i, y)$, $i = 1, \dots, 4$ are independent from each other and possess the following probability, according to (2.6.5):

$$Pr\{B(v, M_i, y)\} = \sum_{j=v}^{2v} \binom{2v}{j} F^j(y)[1-F(y)]^{2v-j} \tag{5.2.10}$$

where $F(x)$ is the cdf of x_{ij}. Therefore, $F_y(y)$ becomes:

$$F_y(y) = \left[\sum_{j=v}^{2v} \binom{2v}{j} F^j(y)[1-F(y)]^{2v-j} \right]^4 F(y) + \tag{5.2.11}$$

$$+ \left[\sum_{j=\nu+1}^{2\nu} \binom{2\nu}{j} F^j(y)[1-F(y)]^{2\nu-j} \right]^4 [1-F(y)]$$

A plot of the pdf $f_y(y)$ is shown in Figure 5.2.1 for Gaussian input distribution having mean 100 and standard deviation 50. Clearly the max/median filter produces an output whose mean is shifted to 120. Therefore, it is a biased estimator of location.

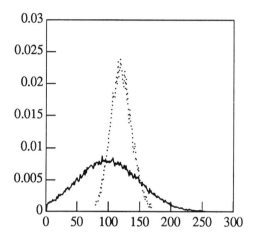

Figure 5.2.1: Output probability distribution (dotted curve) of the max/median filter whose input distribution (solid curve) is Gaussian having mean 100 and standard deviation 50. The output mean is shifted to 120.

The max/median filter consists of the median and max operators. Both of them satisfy the monotonicity property (4.11.1). Therefore, the threshold decomposition is valid for the max/median and it can be used in the deterministic analysis of the max/median filter [5].

Examples of max/median filters are shown in Figures 5.2.2 and 5.2.3 for impulsive and additive white noise, respectively. The size n of the max/median filter was $n=7$ in both cases. It is evident in both cases that the max operator in (5.2.1) introduces a very strong bias toward large output values. These values appear along vertical, horizontal, or diagonal lines in the output image because of the geometry of the max/median window. Such distortions of the output image are less evident for smaller window sizes. It is also seen that max/median preserves the image details better than the median filter of 7×7 extent.

(a)   (b)

(c)   (d)

Figure 5.2.2: (a) Original image; (b) Image corrupted by 10% mixed impulses; (c) Output of the $n=7$ max/median filter; (d) Output of the 7×7 median filter.

The performance of this filter can be improved if the median operator is used in (5.2.1) in the place of the max operator because the resulting filter will be less biased toward large values [20]. The resulting filters are the so-called *multistage median filters*. Such a filter is described by the following relation:

$$y_{ij} = \text{med}(\text{med}(z_1,z_2,x_{ij}),\ \text{med}(z_3,z_4,x_{ij}),\ x_{ij}) \qquad (5.2.12)$$

It is called multistage because it involves three stages of median filtering. This filter is unidirectional because it involves median operation along one direction in each of the medians z_1,z_2,z_3,z_4. Another multistage filter, called *bidirectional multistage median filter*, has the following form:

$$y_{ij} = \text{med}(\, z_{12}, z_{34}, x_{ij})$$ (5.2.13)

$$z_{12} = \text{med}(\, \{x_{i,j-v}, .., x_{ij}, .., x_{i,j+v}\} \cup \{x_{i-v,j}, .., x_{ij}, .., x_{i+v,j}\})$$ (5.2.14)

$$z_{34} = \text{med}(\, \{x_{i+v,j-v}, .., x_{ij}, .., x_{i-v,j+v}\} \cup \{x_{i-v,j-v}, .., x_{ij}, .., x_{i+v,j+v}\})$$ (5.2.15)

(a)   (b)

(c)   (d)

Figure 5.2.3: (a) Original image; (b) Image corrupted by additive white Gaussian noise having standard deviation 50; (c) Output of the $n=7$ max/median filter; (d) Output of the 7×7 median filter.

Filter (5.2.13) is a two-stage filter. It can preserve details in horizontal, diagonal, and vertical directions because it involves two median filters (5.2.14-15) that are sensitive to those directions. The derivation of the probability distributions of the multistage median filters by using the threshold decomposition property can be found in [34]. The cdf of the output of the filter (5.2.12) is given by the following relation:

$$F_y(y) = Pr\{y_{ij} \leq y\} = F(y)\left[1 - \left[\sum_{j=v+1}^{2v}\binom{2v}{j}(1-F(y))^j F(y)^{2v-j}\right]^4\right] + \qquad (5.2.16)$$

$$(1-F(y))\left[\sum_{j=v+1}^{2v}\binom{2v}{j}(1-F(y))^{2v-j}F(y)^j\right]^4$$

Multistage median filters have been proven to preserve two-dimensional structures and to be good smoothers in heavy-tailed noise [34].

5.3 MEDIAN HYBRID FILTERS

The structure of the input signal can be taken into account in median filtering if the filter subsections operate on different regions of the filter window A. Thus, the max/median filter possesses four sections z_1, z_2, z_3, z_4 which operate horizontally, vertically, and diagonally. These filter sections, however, can be linear.

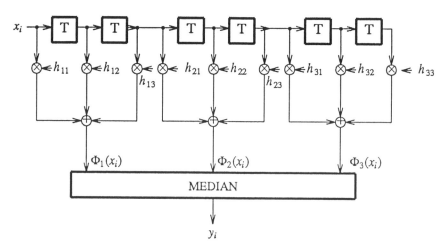

Figure 5.3.1: Structure of the median hybrid FIR filter.

This combination of the median filter with linear finite impulse response (FIR) or infinite impulse response (IIR) filter produces the *median hybrid* filters:

$$y_i = \text{med}(\Phi_1(x_i),...,\Phi_M(x_i)) \tag{5.3.1}$$

where the filters $\Phi_j(x_i)$, $j=1,...,M$ are linear FIR or IIR filters. The median hybrid filter structure employing FIR filters is shown in Figure 5.3.1. In this section, attention will be focused on the performance of median hybrid filters of the form:

$$y_i = \text{med}(\Phi_1(x_i),\Phi_2(x_i),\Phi_3(x_i)) \tag{5.3.2}$$

where $\Phi_1(x_i)$, $\Phi_3(x_i)$ are linear lowpass filters following the slower trends of the input signal x_i and $\Phi_2(x_i)$ is a linear filter reacting to fast signal changes.

A simple choice of FIR filters $\Phi_1(x_i)$, $\Phi_2(x_i)$, $\Phi_3(x_i)$ is the following:

$$\Phi_1(z) = \frac{1}{v}(z^v + z^{v-1} + ... + z) \tag{5.3.3}$$

$$\Phi_2(z) = 1 \tag{5.3.4}$$

$$\Phi_3(z) = \frac{1}{v}(z^{-1} + ... + z^{-v}) \tag{5.3.5}$$

$\Phi_1(x_i)$, $\Phi_3(x_i)$ are simple averaging filters. Therefore, the median hybrid filter is described by the following equation in the time domain:

$$y_i = \text{med}\left[\frac{1}{v}\sum_{j=1}^{v} x_{i-j}, x_i, \frac{1}{v}\sum_{j=1}^{v} x_{i+j}\right] \tag{5.3.6}$$

The lowpass filters $\Phi_1(x_i)$, $\Phi_3(x_i)$ can also be chosen in such a way, so that they are forward and backward signal predictors, respectively. In this case the median hybrid filter takes the median of the two signal predictions $\hat{x}_{1ij}, \hat{x}_{3ij}$ and of x_{ij}. The use of the median facilitates the performance of the filter in signal transitions, i.e., close to edges. The calculation of the coefficients of the two FIR predictors is described in [35].

The distributions $F_1(x)$, $F_2(x)$, $F_3(x)$ of the outputs of $\Phi_1(x_i)$, $\Phi_2(x_i)$, $\Phi_3(x_i)$ are different from each other. The output probability density function $f_y(y)$ is given by [7]:

$$f_y(y) = f_1(y)[F_2(y)+F_3(y)]+f_2(y)[F_1(y)+F_3(y)]+f_3(y)[F_1(y)+F_2(y)] \tag{5.3.7}$$
$$-2[f_1(y)F_2(y)F_3(y)+f_2(y)F_1(y)F_3(y)+f_3(y)F_1(y)F_2(y)]$$

The distributions $F_1(x)$, $F_2(x)$, $F_3(x)$ can be obtained by numerical convolution of the input distribution $F(x)$. If the mean of the input signal is zero, so is the mean of the output of the filters $\Phi_j(x_i)$, $j=1,2,3$ and of the output y_i of the median hybrid filter. The output variance σ_y^2 can be found by direct integration of (5.3.7). If the output samples of Φ_1, Φ_2, Φ_3 are random variables having Gaussian distribution and variances $\sigma_1^2, \sigma_2^2, \sigma_3^2$, the direct integration of (5.3.7) gives the following output variance σ_y^2 [8]:

$$\sigma_y^2 = \frac{1}{\pi}\left[\sum_{j=1}^{3}\sigma_j^2\arctan\frac{P}{\sigma_j^2}-P\right] \qquad (5.3.8)$$

where

$$P = (\sigma_1^2\sigma_2^2+\sigma_2^2\sigma_3^2+\sigma_3^2\sigma_1^2)^{1/2}$$

The output of the sections Φ_1, Φ_2, Φ_3 of the filter (5.3.6) are independent Gaussian variables, if the input x_i is Gaussian. Their variances are $\sigma_1^2=\sigma_3^2=1/v$, $\sigma_2^2=1$, if the input variance is one. Therefore, (5.3.8) becomes:

$$\sigma_y^2 = \frac{1}{\pi}\left[\frac{2}{v}\arctan\sqrt{2v+1}+\arctan\frac{\sqrt{2v+1}}{v}-\frac{\sqrt{2v+1}}{v}\right] \qquad (5.3.9)$$

An approximation of (5.3.9) for large v is the following:

$$\sigma_y^2 = \frac{1}{v}+O(v^{-3/2}) \qquad (5.3.10)$$

Therefore, the output variance of a median hybrid filter is the same as that of a moving average of half extent. The output variances of the median filter (5.3.6) and of its recursive counterpart:

$$y_i = \text{med}\left[\frac{1}{v}\sum_{j=1}^{v}y_{i-j},x_i,\sum_{j=1}^{v}x_{i+j}\right] \qquad (5.3.11)$$

have been determined by simulation [8]. The output variance of the median filter (5.3.6) is plotted in Figure 5.3.2. It is seen that it follows approximately (5.3.10).

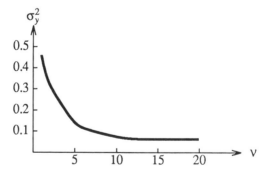

Figure 5.3.2: Output variance of the median hybrid filter, if the input noise is Gaussian.

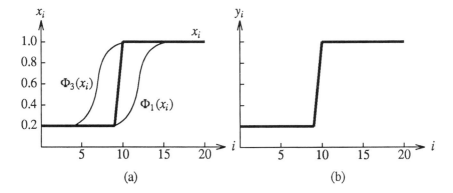

Figure 5.3.3: (a) The output of the filter sections $\Phi_1(x_i)$ and $\Phi_3(x_i)$ of the median hybrid filter (5.3.6) in the presence of a step edge; (b) Output of the median hybrid filter.

The performance of the median hybrid filter (5.3.6) in a step edge is shown in Figure 5.3.3. The sections Φ_1, Φ_3 lag and lead the real edge, respectively, whereas Φ_2 reacts at the edge. Therefore, the median hybrid filter preserves the edge. The performance figures of merit used are the absolute bias $|E[y_i-s_i]|$ and the output variance σ_y^2. They are plotted in Figure 5.3.4 for input noise variance 1, edge height 4, and $v=50$ [8]. For $i>0$ the behavior is symmetrical for the variance and antisymmetrical for the bias.

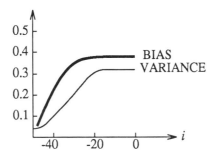

Figure 5.3.4: Absolute bias and output variance of the median hybrid filter in the presence of a step edge of height 4 corrupted by Gaussian noise of unit variance (adapted from [8]).

If the edge height h is large, the influence of $\Phi_1(x_i)$ is small, since it lags the edge. Therefore, the output y_i of the median hybrid filter has the form:

$$y_i \simeq \min(x_i, \frac{1}{v} \sum_{j=1}^{v} x_{i+j}) \qquad (5.3.12)$$

If x_i are iid Gaussian variables and have variance 1, the maximum bias and variance of the median hybrid filter near the edge is given by:

$$E(y_i - s_i)|_{max} = \left[\frac{v+1}{2\pi v}\right]^{1/2} \qquad (5.3.13)$$

$$\sigma_y^2|_{max} = \frac{(\pi-1)(v+1)}{2\pi v} \qquad (5.3.14)$$

It is known that a moving average of length $n=2v+1$ introduces maximum bias $hv/(2v+1)$ and variance $1/n$. It can also be shown [8] that the median filter of length $n=2v+1$ introduces bias error of the order $\sqrt{\log v}$ for large v. Therefore, median hybrid filters have less bias error and more output noise variance than either the moving average or the standard median filter.

Median hybrid filters can be easily extended to two dimensions. There exist several varieties of such filters, because of the variety of the regions of support of the linear sections. An example of a two-dimensional filter is the so-called $1LH+$ filter [6]:

$$y_{1LM+}(i,j) = med(y_N(i,j), y_E(i,j), y_S(i,j), y_W(i,j), x_{ij}) \qquad (5.3.15)$$

and its rotated version, called $R1LM+$ filter:

$$y_{R1LM+}(i,j) = med(y_{NE}(i,j), y_{SE}(i,j), y_{SW}(i,j), y_{NW}(i,j), x_{ij}) \qquad (5.3.16)$$

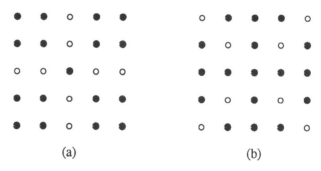

(a)   (b)

Figure 5.3.5: (a) N, E, W, S regions of support of the $1LH+$ filter (b) NE, SE, SW, NW regions of support of the $R\,1LM\,1+$ filter.

The sections $y_k(i,j)$, $k=N,E,S,W,NE,SE,SW,NW$ are FIR filters whose regions of support are shown in Figure 5.3.5. The 1LM+ filter preserves horizontal and vertical lines, whereas the R1LM+ filter preserves diagonal lines. A combination of 1LM+ and R1LM+ filters, called 2LM+ filter:

$$y_{2LM+}(i,j) = \mathrm{med}(y_{1LM+}(i,j), y_{R1LM+}(i,j), x_{ij}) \qquad (5.3.17)$$

is not direction-sensitive and preserves any line. An example of Gaussian noise filtering by such a filter is shown in Figure 5.3.6. It is seen that a 3×3 2LM+ filter preserves the fine detail better than the 3×3 median filter. The median hybrid filters of the form (5.3.15-17) have structure similar to that of the multistage median filters. However, their statistical analysis is more difficult because they do not possess the threshold decomposition property due to the existence of the linear filters.

(a) (b)

(c) (d)

Figure 5.3.6: (a) Original image; (b) Image corrupted by additive white Gaussian noise having standard deviation 50; (c) Output of the 3×3 $2LM+$ median hybrid filter; (d) Output of the 3×3 median filter.

5.4 RANKED-ORDER FILTERS

An *r-th ranked-order filter* is the r-th order statistic of the samples x_i in the filter window:

$$y_i = r\text{-}th \text{ order statistic of } \{x_{i-v},...,x_i,...,x_{i+v}\} \tag{5.4.1}$$

The two-dimensional r-th ranked-order filters are defined as

$$y_{ij} = r\text{-}th \text{ order statistics of } \{x_{i+r,j+s} \quad (r,s) \in A\} \tag{5.4.2}$$

where A is the filter window. Special cases of the r-th ranked-order filter are the median ($r=v+1$), the max operator ($r=n$):

$$y_i = \max\{x_{i-v},...,x_i,...,x_{i+v}\} \tag{5.4.3}$$

and the min operator ($r=1$):

$$y_i = \min\{x_{i-v},...,x_i,...,x_{i+v}\} \tag{5.4.4}$$

Definition of ranked-order filters using pixel neighborhoods different from the ones used in (5.4.2) can be found in [33]. If the input data x_i are iid having distribution $F(x)$, the output cdf $F_y(y)$ is given by (2.6.5):

$$F_y(y) = \sum_{j=r}^{n} \binom{n}{j} F^j(y)[1-F(y)]^{n-j} \tag{5.4.5}$$

and the output pdf $f_y(y)$ is given by (2.6.8):

$$f_y(y) = n\binom{n-1}{r-1} F^{r-1}(y)[1-F(y)]^{n-r} f(y) \tag{5.4.6}$$

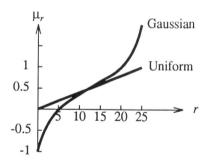

Figure 5.4.1: Output mean of an r-th ranked-order filter of 5×5 extent for uniform and Gaussian input distributions (adapted from [10]).

The mean and the variance of the output of the r-th ranked-order filter can be found by a direct integration of (5.4.6). However, such an integration is difficult, and usually the means and variances are tabulated. If the input samples are iid uniformly distributed with mean 0.5, the output mean μ_r of y_i is given by [10]:

$$\mu_r = \frac{r}{n+1} \qquad (5.4.7)$$

(a)   (b)

(c)

Figure 5.4.2: (a) Original image; (b) Output of the 2nd ranked-order filter having window size 3×3; (c) Output of the 8th ranked-order filter having window size 3×3.

Therefore, as expected, the r-th ranked-order filter introduces a strong bias toward small values (if $r < v+1$) or toward larger values (if $r > v+1$). This bias is even stronger for heavily-tailed distributions, as is seen in Figure 5.4.1. Generally speaking, the r-th ranked-order filter must be used with care, especially for large or small r, because its performance varies considerably for different noise distributions.

r-th ranked-order filters can be used for the elimination of positive impulses of width less than $2v+2-r$ samples and for the elimination of negative impulses of width less than r samples [14]. Any constant region of $2v+2-r$ or more points, surrounded by constant neighborhoods of lesser values, will be changed in width by $2(r-v-1)$ points after being passed through an r-th ranked-order filter. Any constant region of r or more points surrounded by constant neighborhoods of greater values will, after being operated on, be changed by $2(v+1-r)$ points. Thus, if $r > v+1$, r-th ranked-order filters tend to increase the high intensity image constant regions. If $r < v+1$, r-th ranked-order filters tend to increase the low intensity image plateaus. This behavior of the r-th ranked-order filters can be seen in Figure 5.4.2. The low r-th ranked-order filters ($r < v+1$) reproduce the dark region even darker and less structured. They thin, or even entirely suppress, the thin bright lines, e.g., the diagonals. In contrast, the high r-th ranked-order filters ($r > v+1$) enlarge the bright regions and broaden the thin bright lines.

It can be easily proven that monotonic functions $g(x)$, do not alter the performance of the r-th order filters, i.e., the property:

$$g(y_i) = r-th \text{ order statistic of } \{g(x_{i-v}),...,g(x_{i+v})\} \tag{5.4.8}$$

holds for every rank r. Therefore, the threshold decomposition can also be applied to the r-th ranked-order [11]:

$$y_i = \sum_{j=1}^{k-1} y_i^{(j)} \tag{5.4.9}$$

$$y_i^{(j)} = r-th \text{ order statistic of } \{x_{i-v}^{(j)},...,x_{i+v}^{(j)}\}$$
$$= \begin{cases} 1 & \text{if at least } n+1-r \text{ of } x_{i-v}^{(j)},...,x_{i+v}^{(j)} \text{ equal 1} \\ 0 & \text{if at least } r \text{ of } x_{i-v}^{(j)},...,x_{i+v}^{(j)} \text{ equal 0} \end{cases} \tag{5.4.10}$$

$$x_i^{(j)} = g_j(x_i)$$

$$g_j(x) = \begin{cases} 1 & \text{if } x_i \geq i \\ 0 & \text{if } x_i < j \end{cases}$$

Threshold decomposition can be used both for the theoretical analysis and for the practical implementation of the r-th ranked-order filters. Such a VLSI implementation has already been reported in [12,36,42]. It has also been proven that a ranked-order filter can be realized by using one binary processing circuit in k steps, if k is the number of the bits in the input signal samples [36,39]. Such an implementation reduces the time-area complexity to $O(k)$ from $O(2^k)$ required by the classical threshold decomposition realization.

5.5 TRIMMED MEAN FILTERS

It has been observed that the median filter discards impulses (outliers) and preserves edges. However, its performance in the suppression of additive white Gaussian noise, which is frequently encountered in image processing, is inferior to that of the moving average. Therefore, a good compromise between the median and the moving average filter is highly desirable. Such a filter is the α-*trimmed mean*:

$$\overline{x}_\alpha = \frac{1}{n(1-2\alpha)} \sum_{j=\alpha n+1}^{n-\alpha n} x_{(j)} \tag{5.5.1}$$

where $x_{(j)}$, $j=1,...,n$ are the order statistics of $x_{i-v},...,x_i,...,x_{i+v}$ and $\overline{x}_\alpha$ is the output y_i of the α-trimmed mean. The α-trimmed mean is the L-estimator having the maximal asymptotic efficiency for the Huber distribution of the form (2.4.18). For this distribution, the α-trimmed mean has the same influence function and, therefore, the same gross-error sensitivity γ^* and the same asymptotic variance $V(\overline{x}_\alpha,F)$, as the Huber estimator (2.4.14). However, outside this distribution, the two estimators differ. It has been proven [3, p.124] that the α-trimmed mean is the optimal B-robust L-estimator for the Gaussian distribution. Furthermore, it is known that it has breakdown point $\varepsilon^*=\alpha$, i.e., it can reject up to $\alpha\%$ outliers. The α-trimmed mean is an asymptotically normal estimator of the location θ, when the samples $x_1,...,x_n$ are iid distributed according to the distribution $F(x-\theta)$. If $F(x)$ is symmetric about zero, satisfies $F(-c)=0$, $F(c)=1$, $0<c\leq\infty$ and has pdf continuous and positive in $[-c,c]$, then:

$$\sqrt{n}\,(\overline{x}_\alpha-\theta) \to N(0,V(\overline{x}_\alpha,F)) \tag{5.5.2}$$

The asymptotic variance $V(\overline{x}_\alpha,F)$ is given by [2, p. 361]:

$$V(\overline{x}_\alpha,F) = \frac{2}{(1-2\alpha)^2} \left[\int_0^{\xi(1-\alpha)} t^2 f(t)dt + \alpha\xi^2(1-\alpha) \right] \tag{5.5.3}$$

$\xi(\alpha)$ is the unique value for which:

$$F(\xi(\alpha)) = \alpha \tag{5.5.4}$$

The asymptotic relative efficiency of $\overline{x}_\alpha$ to the arithmetic mean $\overline{x}$ is given by:

$$\text{ARE}(\overline{x}_\alpha,\overline{x}) = \frac{V(\overline{x},F)}{V(\overline{x}_\alpha,F)} \tag{5.5.5}$$

Table 5.5.1 shows the asymptotic relative efficiency for the case that F is the t-distribution (4.4.22) with v degrees of freedom. It is seen that a moderate amount of trimming can provide much better protection than $\bar{x}$ against heavy tailed outliers. The same behavior is observed for the Tukey $T(\varepsilon,3)$ (contaminated normal) distribution (4.4.24), as it is seen in Table 5.5.2.

Table 5.5.1: ARE of $\bar{x}_\alpha$ to $\bar{x}$ for t-distributions with v degrees of freedom (adapted from [2]).

α v	.05	.125	.25	.375	.5
3	1.70	1.91	1.97	1.85	1.62
5	1.20	1.24	1.21	1.10	.96

Table 5.5.2: ARE of $\bar{x}_a$ to $\bar{x}$ for Tukey model $T(\varepsilon,3)$ (adapted from [2]).

α ε	.05	.1	.125	.25	.375	.5
.25	1.40	1.62	1.66	1.67	1.53	1.33
.05	1.20	1.21	1.19	1.09	.97	.83

The asymptotic relative efficiency $\mathrm{ARE}(\bar{x}_\alpha,\bar{x})$ possesses the following lower bound for distributions F satisfying the conditions for (5.5.2) [2, p.362]:

$$\mathrm{ARE}(\bar{x}_\alpha,\bar{x}) \geq \begin{cases} (1-2\alpha)^2 \\ \dfrac{1}{1+4\alpha} \qquad \text{for unimodal } F(x) \end{cases} \tag{5.5.6}$$

The absolute efficiency of the α-trimmed mean is given by:

$$e = \frac{V(\bar{x}_\alpha,F)}{J(F)} \tag{5.5.7}$$

where $J(F)$ is the Fisher information. The efficiencies e are particularly high for $\alpha=0.125$ for a wide variety of distributions [2, p.363]. However, this is not the case for extreme distributions. The efficiency of $\bar{x}_\alpha$ is zero for the uniform distribution, where the density is discontinuous at the end points. Thus $J(F)=\infty$.

Generally speaking, $\bar{x}_\alpha$ does not perform well for thin-tailed distributions, e.g., for the uniform distribution. It is known that for the uniform distribution the *midpoint MP*:

$$MP = \frac{1}{2}(x_{(1)} | x_{(n)}) \tag{5.5.8}$$

is the best estimator of location. The statistical properties of the midpoint filter are described in [37]. For such distributions, the α-*trimmed complementary mean* is used [13]:

$$R_\alpha = \frac{1}{2\alpha n}\left[\sum_{j=1}^{\alpha n} x_{(j)} + \sum_{j=n-\alpha n+1}^{n} x_{(j)}\right] \tag{5.5.9}$$

Both $\bar{x}_\alpha$ and R_α are special cases of the so-called *L*-filters which will be discussed in the next section.

A different approach to trimmed filters is to exclude the samples $x_{i+r,j+s}$ in the filter window, which differ considerably either from the local mean $\bar{x}$, or from the central pixel x_{ij}:

$$y_{ij} = \frac{\sum\sum_A a_{rs} x_{i+r,j+s}}{\sum\sum_A a_{rs}} \tag{5.5.10}$$

If a_{rs} are chosen as:

$$a_{rs} = \begin{cases} 1 & \text{if } |x_{i+r,j+s} - \bar{x}_{ij}| \leq q \\ 0 & \text{otherwise} \end{cases} \tag{5.5.11}$$

(5.5.10-11) is the so-called *modified trimmed mean MTM filter*. If a_{rs} are chosen as:

$$a_{rs} = \begin{cases} 1 & \text{if } |x_{i+r,j+s} - x_{ij}| \leq q \\ 0 & \text{otherwise} \end{cases} \tag{5.5.12}$$

the *modified nearest neighbor filter* results [15], [16]. A variation of (5.5.11), called *double window modified trimmed mean (DW MTM) filter*, uses two windows of size $n \times n$ and $m \times m$ with $m > n$. First the median $med(x_{ij})$ is computed in the small window $n \times n$ centered at (i, j). Then an interval $[med(x_{ij})-q, med(x_{ij})+q]$ is determined for an appropriately chosen value q. Finally, the mean of the points lying in the window $m \times m$ and having values in the interval $[med(x_{ij})-q, med(x_{ij})+q]$ is computed and it becomes the filter output [15]. DW MTM filter suppresses additive Gaussian noise effectively because it uses a fairly large window $m \times m$. It can also preserve image details because it rejects pixels having values far away from the median of the small window $n \times n$.

An example of α-trimmed mean filtering of mixed additive Gaussian and impulsive noise is shown in Figure 5.5.1. The impulse noise probability is 10%

and the Gaussian noise standard deviation is 50. The α-trimmed mean filter window has size 3×3. Its coefficients are [0, 1/7, 1/7, 1/7, 1/7, 1/7, 1/7, 1/7, 0], i.e., it trims the largest and the smallest sample in the filter window. As can be seen in Figure 5.5.1, α-trimmed mean filter suppresses both impulsive and Gaussian noise better than the median filter of the same extent, but tends to blur the edges. However, there is no great difference in the output of the median and of the α-trimmed mean filter.

(a)   (b)

(c)   (d)

Figure 5.5.1: (a) Original image; (b) Image corrupted by 10% mixed impulses and white additive Gaussian noise having standard deviation 50; (c) Output of the 3×3 α-trimmed mean filter; (d) Output of the 3×3 median filter.

5.6 L-FILTERS

An important generalization of the median filters are the *L filters* (or *order statistic* filters):

$$y_i = \sum_{j=1}^{n} a_j x_{(j)} \tag{5.6.1}$$

where $x_{(j)}$, $j=1,...,n$ are the ordered samples $x_{i-v},...,x_i,...,x_{i+v}$ and $a_j, j=1,..,n$ are the filter coefficients. The following choices of coefficients give the median, the midpoint, and the α-trimmed mean, respectively:

$$a_j = \begin{cases} 1 & j=v+1 \\ 0 & j\neq v+1 \end{cases} \tag{5.6.2}$$

$$a_j = \begin{cases} 1/2 & j=1,n \\ 0 & \text{otherwise} \end{cases} \tag{5.6.3}$$

$$a_j = \begin{cases} 1/n\,(1-2\alpha) & j=\alpha n+1,...,n-\alpha n \\ 0 & \text{otherwise} \end{cases} \tag{5.6.4}$$

L-filters have very similar structure to that of the FIR linear filters. Their relationship and the similarities in their performance are described in detail in [38]. L-filters can be used as L-estimators of location or scale, by an appropriate choice of the coefficients a_j, $j=1,...,n$, as was described in section 2.7. In the following, we shall concentrate on the performance of the L-filters as minimum mean square error (MMSE) estimators of the location.

The noise model which will be considered is the additive white noise n_i:

$$x_i = s_i + n_i \tag{5.6.5}$$

The problem is to choose the coefficients a_j in such a way that the mean square error:

$$MSE_i = E[(y_i - s_i)^2] = E[(\sum_{j=1}^{n} a_j x_{(j)} - s_i)^2] \tag{5.6.6}$$

becomes minimal. The simplest case arises when the signal s_i is constant, i.e., $s_i = s$. In this case the order statistics $x_{(j)}, n_{(j)}$ are related by the following equation:

$$x_{(j)} = s + n_{(j)} \tag{5.6.7}$$

where $n_{(j)}$, $j=1,...,n$ are formed by ordering the noise samples $n_{i-v},...,n_{i+v}$. Let us use vector notation:

$$\mathbf{a} = [a_1, \ldots, a_n]^T \tag{5.6.8}$$

$$\mathbf{e} = [1, \ldots, 1]^T \tag{5.6.9}$$

$$\mathbf{n} = [n_{(1)}, \ldots, n_{(n)}]^T \tag{5.6.10}$$

A constraint, which should be posed on the estimator, is *location invariance*. This means that the condition:

$$x_i' = x_i + c \tag{5.6.11}$$

implies that:

$$y_i' = y_i + c \tag{5.6.12}$$

The location invariance is satisfied by:

$$\sum_{j=1}^{n} a_j = \mathbf{a}^T \mathbf{e} = 1 \tag{5.6.13}$$

(5.6.13) is an unbiasedness condition. If the noise is symmetrical about zero and the coefficients are constrained to be symmetrical:

$$a_j = a_{n-j+1}, \quad j=1, \ldots, v \tag{5.6.14}$$

By using the vector notation and the unbiasedness condition, the mean square error (5.6.6) becomes:

$$MSE = E[(\sum_{j=1}^{n} a_j n_{(j)})^2] = \sum_{j=1}^{n} \sum_{k=1}^{n} a_j a_k r_{jk} = \mathbf{a}^T \mathbf{R}_{nn} \mathbf{a} \tag{5.6.15}$$

where $\mathbf{R}_{nn}$ is the $n \times n$ correlation matrix of the vector $\mathbf{n}$:

$$\mathbf{R}_{nn} = \left[r_{jk} \right] = \left[E[n_{(j)} n_{(k)}] \right] \tag{5.6.16}$$

The Lagrangian function $F(\lambda, \mathbf{a})$ of the mean square error and the constraint (5.6.13) [18]:

$$F(\lambda, \mathbf{a}) = \mathbf{a}^T \mathbf{R}_{nn} \mathbf{a} + \lambda(\mathbf{a}^T \mathbf{e} - 1) \tag{5.6.17}$$

is minimized by the following choice of the coefficients $\mathbf{a}$:

$$\mathbf{a} = \frac{\mathbf{R}_{nn}^{-1} \mathbf{e}}{\mathbf{e}^T \mathbf{R}_{nn}^{-1} \mathbf{e}} \tag{5.6.18}$$

It can be shown, by using the Cauchy-Schwartz inequality that the corresponding mean square error is always less or equal to that produced by the arithmetic mean [21]. This means that the L-filter will never be worse than the moving average filter in the MSE sense.

In the general case the signal s_i is not constant. Therefore, the MSE (5.6.6) is given by [19]:

$$MSE_2 = E[(s_i - y_i)^2] = \mathbf{a}^T \mathbf{R}_{xx} \mathbf{a} - 2\mathbf{a}^T \mathbf{r} + E[s_i^2] \qquad (5.6.19)$$

where $\mathbf{R}_{xx}$ is the correlation matrix of the ordered statistics $x_{(1)},...,x_{(n)}$:

$$\mathbf{R}_{xx} = \left[E[x_{(j)} x_{(k)}] \right] \qquad (5.6.20)$$

and $\mathbf{r}$ is the vector of the cross-correlations between s_k and the ordered observations $x_{(1)},...,x_{(n)}$:

$$\mathbf{r} = \left[E[s_i x_{(1)}],...,E[s_i x_{(n)}] \right]^T \qquad (5.6.21)$$

The unconstrained solution that minimizes (5.6.19), can be found by direct differentiation with respect to $\mathbf{a}$:

$$\mathbf{a} = \mathbf{R}_{xx}^{-1} \mathbf{r} \qquad (5.6.22)$$

The unbiasedness condition is the following:

$$\mathbf{a}^T \mu = \mu_s \qquad (5.6.23)$$

where μ_s is the mean of s_i and μ is the vector of the means of the ordered statistics $x_{(1)},...,x_{(n)}$:

$$\mu = \left[E[x_{(1)}],...,E[x_{(n)}] \right]^T \qquad (5.6.24)$$

$$\mu_s = E(s_i) \qquad (5.6.25)$$

The Lagrangian function for the MSE (5.6.18) and the unbiasedness constraint (5.6.21):

$$F(\lambda, \mathbf{a}) = \mathbf{a}^T \mathbf{R}_{xx} \mathbf{a} - 2\mathbf{a}^T \mathbf{r} + E[s_i^2] + \lambda(\mu_s - \mathbf{a}^T \mu) \qquad (5.6.26)$$

gives the following equation, if its derivative with respect to $\mathbf{a}$ is set equal to zero:

$$2\mathbf{R}_{xx} \mathbf{a} - 2\mathbf{r} - \lambda \mu = 0 \qquad (5.6.27)$$

By obtaining λ from (5.6.27), the following optimal unbiased solution is found:

$$\mathbf{a} = \mathbf{R}_{xx}^{-1} \mathbf{r} + \frac{\mu_s - \mu^T \mathbf{R}_{xx}^{-1} \mathbf{r}}{\mu^T \mathbf{R}_{xx}^{-1} \mu} \mathbf{R}_{xx}^{-1} \mu \qquad (5.6.28)$$

If the location invariance constraint (5.6.13) is used, the Lagrangian function becomes:

$$F(\lambda, \mathbf{a}) = \mathbf{a}^T \mathbf{R}_{xx} \mathbf{a} - 2\mathbf{a}^T \mathbf{r} + E[s_i^2] + \lambda(\mathbf{a}^T \mathbf{e} - 1) \qquad (5.6.29)$$

The optimal MSE location invariant solution is found by minimizing (5.6.29):

$$\mathbf{a} = \mathbf{R}_{xx}^{-1}\mathbf{r} + \frac{1-\mathbf{e}^T\mathbf{R}_{xx}^{-1}\mathbf{r}}{\mathbf{e}^T\mathbf{R}_{xx}^{-1}\mathbf{e}}\ \mathbf{R}_{xx}^{-1}\mathbf{e} \qquad (5.6.30)$$

The location or time index i has not been used in the solutions (5.6.22), (5.6.28), (5.6.30), for simplicity reason. However, no stationarity assumption for x_i has been made. The same formulae can be used in the nonstationary case, where the filter coefficients i change with the time.

The correlation matrix $\mathbf{R}_{nn}$ used in (5.6.18) can be calculated as follows. If $F_n(x)$ is the distribution of the noise n_i, the joint probability function $f_{rs}(x,y)$ of $n_{(r)}, n_{(s)}$ is given by (2.6.9):

$$f_{rs}(x,y) = \frac{n!}{(r-1)!(s-r-1)!(n-s)!}\ F_n^{r-1}(x)f_n(x)[F_n(y)-F_n(x)]^{s-r-1}$$
$$\cdot f_n(y)[1-F_n(y)]^{n-s} \qquad (5.6.31)$$

The correlations r_{jk} (5.6.16) are computed as follows:

$$r_{jk} = \int\!\!\int_{-\infty}^{\infty} xyf_{jk}(x,y)dxdy \qquad (5.6.32)$$

$$r_{kk} = \int_{-\infty}^{\infty} x^2 f_k(x)dx \qquad (5.6.33)$$

$$f_k(x) = \frac{n!}{(k-1)!(n-k)!}\ F_n^{k-1}(x)[1-F_n(x)]^{n-k}f_n(x) \qquad (5.6.34)$$

Table 5.6.1: Optimal coefficients of the L-filter of length $n=9$ (adapted from [18]).

	a_1	a_2	a_3	a_4	a_5	a_6	a_7	a_8	a_9
Uniform	0.5	0.0	0.0	0.0	0.0	0.0	0.0	0.0	0.5
Gaussian	0.11	0.11	0.11	0.11	0.11	0.11	0.11	0.11	0.11
Laplacian	-.01899	.02904	.06965	.23795	.36469	.23795	.06965	.02904	-.01899

The integrations (5.6.32), (5.6.33) can be done numerically. The number of integrations required is reduced by four due to the following symmetries:

$$r_{jk} = r_{kj} \qquad (5.6.35)$$

$$r_{jk} = r_{n-j+1,n-k+1} \qquad (5.6.36)$$

The symmetry (5.6.36) requires symmetric distribution $F_n(x)$. The correlation matrix $\mathbf{R}_{xx}$ can also be computed in a similar way. The matrix $\mathbf{a}$ has been computed for six distributions and it has been used for the calculation of the optimal coefficients (5.6.18) [18]. As expected, the optimal L-filter for Gaussian noise is the moving average and the optimal L-filter for the uniform noise is the midpoint MP. For the Laplacian distribution larger weights a_j are close to $j=v+1$, as it can be seen in Table 5.6.1 [18].

(a)   (b)

(c)   (d)

Figure 5.6.1: (a) Original image; (b) Image corrupted by 10% mixed impulses and white additive Gaussian noise having standard deviation 50; (c) Output of the 3×3 L-filter; (d) Output of the 3×3 median filter.

An example of the performance of the L-filter in the presence of mixed impulsive and Gaussian noise is shown in Figure 5.6.1. The filter size is 3×3 and the filter coefficients are chosen to be the ones that are suitable for Laplacian noise filtering [18]. The L-filter suppresses impulses and filters Gaussian noise better than the median filter of equal size, as it can be seen by comparing Figures 5.6.1c and 5.6.1d. However, the performance of the L-filter is not significantly better than the one of the median filter.

The probability that an outlier exists in the output of the L-filter having all coefficients nonzero is given [22]:

$$B_n(p) = 1 - (1-p)^n \simeq np \qquad (5.6.37)$$

for data distributions of the Huber type:

$$F = (1-p)F_0 + pH \qquad (5.6.38)$$

$B_n(p)$ is called *breakdown probability*. It should be noted that $B_n(p)$ is always greater than the outlier probability p and that it increases with n. Therefore, trimming is required to improve the robustness of the L-filters:

$$y_i = \sum_{j=\alpha n+1}^{n-\alpha n} a_j x_{(j)} \qquad (5.6.39)$$

The breakdown probability of the trimmed L-filter is given by [22]:

$$B_n(p) = p^{\alpha n+1} \begin{bmatrix} n-1 \\ \alpha n \end{bmatrix} + O(p^{\alpha n+2}) \qquad (5.6.40)$$

which is much less than p. Such a filter can be designed by fixing α and optimizing with respect to the coefficients a_j, $j = \alpha n+1, ..., n - \alpha n$. The optimal design of trimmed L-filters is examined in [30].

One of the main disadvantages of the L-filters and of other filters based on order statistics is that ordering destroys time or neighborhood information. Therefore, their performance deteriorates by increasing the filter length n beyond a certain length [19]. This poor performance is particularly evident in the case of nonstationary signals. Thus a modification of L-filters has been proposed, which takes into account time information. These filters are called C-filters [31] or Ll-filters [19]. Their definition is the following:

$$y_i = \sum_{j=1}^{n} a(R(i-j+v+1), j) x_{i-j+v+1} \qquad (5.6.41)$$

The filter coefficients $a(R(i-j+v+1), j)$, $j = 1, .., n$ depend on the position $i-j+v+1$ of the sample $x_{i-j+v+1}$ as well as on the rank $R(i-j+v+1)$ of this sample in the sample set $\{x_{i-v}, .., x_{i+v}\}$. Thus C-filters can be considered to be time-varying L-filters. Therefore, they can be easily adapted to time-varying environments [31].

5.7 M-FILTERS

M-filters are simply *M*-estimators of the location used in filtering applications. Therefore, their definition is given by (2.3.2), (2.4.2):

$$\sum_{j=i-v}^{i+v} \psi(x_j - y_i) = 0 \qquad (5.7.1)$$

$\psi(x)$ is generally an odd, continuous, and sign-preserving function. y_i is the filter output and x_j, $j = i - v, ..., i + v$ are the input samples in the filter window. A special case of *M*-filters are the *maximum likelihood filters* defined by (5.7.1) and

$$\psi(x) = -\frac{\partial f_\theta(x)/\partial \theta}{f_\theta(x)} \qquad (5.7.2)$$

where $f_\theta(x)$ is the model pdf of the data x_i. The maximum likelihood filter, for Gaussian distributed iid data, is the moving average filter satisfying:

$$\sum_{j=i-v}^{i+v} (x_j - y_i) = 0 \qquad (5.7.3)$$

The maximum likelihood filter of Laplacian distributed iid data is the median satisfying:

$$\sum_{j=i-v}^{i+v} \text{sign}(x_j - y_i) = 0 \qquad (5.7.4)$$

Therefore, both the median and the arithmetic mean are special cases of the *M*-filters. The properties of the *M*-filters depend on the choice of the function $\psi(x)$. It was Huber (1964) that has chosen $\psi(x)$ in such a way that the corresponding *M*-estimator minimizes the maximal asymptotic variance $V(\psi, \sigma)$ over a family of distributions P_ε containing outliers:

$$P_\varepsilon = \{G \mid G = (1-\varepsilon)F + \varepsilon H ; \ H \text{ is a symmetric distribution}\} \quad (5.7.5)$$

where H is the outlier distribution. This estimator, and the corresponding *M*-filter, are defined by

$$\psi(x) = x \min(1, \frac{b}{|x|}) = [x]_{-b}^{b} \qquad (5.7.6)$$

This function is plotted in Figure 2.4.1. The estimator (5.7.6) is the so-called *Huber estimator*. For Gaussian distributed data $(F = \Phi)$ the constant b is given by the implicit equation

$$2\Phi(b) - 1 + 2\phi(b)/b = 1/(1-\varepsilon) \qquad (5.7.7)$$

The Huber estimator, and the corresponding *M*-filter, are *B*-robust and have breakdown point $\varepsilon^* = 1/2$, i.e., they can reject up to 50% outliers. The Huber estimator tends to the median when b tends to zero, because in this case the function (5.7.6) tends to the sign(x) function. It also tends to the arithmetic mean when b tends to infinity. Therefore the M-filter is a compromise between the median and the moving average filters.

A modification of the Huber estimator is the *limiter type M-filter* (LTM) [15]:

$$\psi(x) = \begin{cases} g(b) & x>b \\ g(x) & |x|\leq b \\ -g(b) & x<-b \end{cases} \qquad (5.7.8)$$

where $g(x)$ is a strictly increasing, odd, continuous function and b is a positive constant. When $g(x)=ax$, the *standard type M-filter* (STM) is derived. Without loss of generality, it can be assumed that $a=1/b$. The ψ function of the STM filter is plotted in Figure 5.7.1.

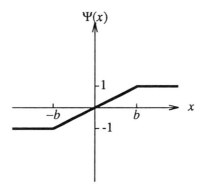

Figure 5.7.1: $\Psi(x)$ function of the standard type M-filter.

It can be proven that the output of the LTM (and, therefore, of the STM) filters is always unique and lies in the range:

$$\text{med}(x_j)-\delta \leq y_i \leq \text{med}(x_j)+\delta \qquad (5.7.9)$$

$$\delta = g^{-1}\left[\frac{vg(b)}{v+1}\right]$$

The proof can be found in [15]. If all the data x_j, $j=i-v,...,i+v$ are far away from the median, i.e., $x_j-\text{med}(x_j)>b$, all the differences x_j-y_i lie in the saturated region of $\psi(x)$ and, therefore, the STM filter coincides with the median filter. On the other hand, if the data x_j, $j=i-v,...,i+v$ are close enough to each other and to the sample median, the differences x_j-y_i lie in the linear region of $\psi(x)$. In this case the STM filter coincides with (5.7.3), i.e., with the moving average filter.

The output of the median filter in the presence of an ideal edge

$$x_i = \begin{cases} 0 & i \leq 0 \\ h & i > 0 \end{cases} \tag{5.7.10}$$

is the following:

$$y_i = \begin{cases} 0 & i < -v \\ \dfrac{wb}{2v+1-w} & -v \leq i < 0 \\ h - \dfrac{(v+1-w)b}{w} & 0 \leq i < v \\ h & i \geq v \end{cases} \tag{5.7.11}$$

$$w = i+v+1$$

when

$$h > (1 + \frac{v}{v+1})b \tag{5.7.12}$$

Thus $2v$ output points differ from their corresponding input values and the edge is blurred. However, the maximal bias $|y_i-x_i|$ is always less than $bv/(v+1)$. If b is sufficiently small, (5.7.12) holds and the STM behaves like a median, i.e., preserves the edge and introduces small smearing. In the presence of a noisy edge, whose height h is large compared to the noise variance, a properly designed STM filter tends to smooth only those samples lying on the side of the edge that includes the sample median.

Impulsive noise can be effectively filtered by the STM filters, because they are robust estimators of location and they limit the influence of very large or very small observations, as can be seen in Figure 5.7.1. Generally speaking, the STM filter behaves like a moving average filter when neither edges nor impulses are present. Its performance though is similar to the median in the presence of edges or impulses. A detailed analysis of the statistical performance of the STM filters can be found in [32].

A critical point in the performance of an M-filter is the choice of the threshold b. It should be at least as large as the noise standard deviation σ, for additive white noise filtering. On the other hand, it should be relatively small to reduce the maximal bias $\dfrac{b\nu}{\nu+1}$ introduced in edge filtering. Therefore, if h/a is the maximal allowable bias, the threshold b should lie in:

$$\sigma \le b \le \frac{h(\nu+1)}{a\nu} \tag{5.7.13}$$

The higher threshold b is, the better is the noise suppression and the worse is the edge preservation. It is explained by the fact that, for large threshold b the STM filter tends to the moving average. Usually the noise standard deviation σ is unknown. In this case a robust estimator of the scale S_n can be used instead of σ. Such an estimator is the median of the absolute deviations $\text{MAD}(x_j)$:

$$S_n = 1.483\,\text{MAD}(x_j) = 1.483\{\,|x_j - \text{med}(x_k)|\,\} \tag{5.7.14}$$

A major disadvantage of M-filters is their implicit definition (5.7.1), which requires iterative techniques for the calculation of the output. However, it has been observed that the iterative Newton's method requires only five iterations with absolute error of less than 0.01 [15].

Another filter, which is closely related to the M-filters, is the W-filter:

$$y_i = \frac{\displaystyle\sum_{j=i-\nu}^{i+\nu} w\,(x_j-y_i)x_j}{\displaystyle\sum_{j=i-\nu}^{i+\nu} w\,(x_j-y_i)} \tag{5.7.15}$$

W-filters are essentially an application of the W-estimator [3]. Its relation to the M-filters can be found from the following relation:

$$\frac{\displaystyle\sum_{j=i-\nu}^{i+\nu} w\,(x_j-y_i)(x_j-y_i)}{\displaystyle\sum_{j=i-\nu}^{i+\nu} w\,(x_j-y_i)} = 0 \tag{5.7.16}$$

which is equivalent to the definition (5.7.1) of an M-filter with function:

$$\psi(x) = xw\,(x) \tag{5.7.17}$$

A reasonable choice of the weighting function $w\,(x)$ is the following [24]:

$$w\,(x) = \begin{cases} 1 & |x|\le c \\ 0 & |x|>c \end{cases} \tag{5.7.18}$$

W-filters have implicit definition. Therefore, they have iterative solutions. However, there exist non-iterative versions of (5.7.15). If the weights $w\,(x_j-\bar{x}_{ij})$ are used instead of $w\,(x_j-y_i)$, the *modified trimmed mean filter* results. If the weights $w\,(x_j-x_i)$ are used instead of $w\,(x_j-y_i)$, the *modified nearest neighbor filter* results. A statistical analysis of the modified nearest neighbor filter:

$$y_i = \frac{\sum\limits_{j=i-v}^{i+v} w(x_j-x_i)x_j}{\sum\limits_{j=i-v}^{i+v} w(x_j-x_i)} \tag{5.7.19}$$

in the presence of a noisy edge can be found in [24].

Of crucial importance for the performance of the W-filter is the choice of the threshold c. It was observed that threshold $c=30$ gives good results for a variety of edge heights h [24]. However, more formal optimization methods can also be used to choose c so that the overall MSE over a wide range of edge heights h is minimized.

5.8 R-FILTERS

R-estimators have already been used in image processing applications [27] resulting in the so-called *R-filters*. An implicit definition of the R-filters is the following [26]. Let x_j, $j=i-v,...,i+v$ be the input samples in the filter window. Let also R_i be the rank of the samples x_j, $2y_i-x_j$, $j=i-v,...,i+v$ in the pooled sample of size $N=2n$. The output y_i of the R-filter is chosen in such a way that:

$$S_N = \frac{1}{n} \sum_{j=1}^{n} a_N(R_j) \simeq 0 \tag{5.8.1}$$

becomes zero when it is computed from the samples x_j, $j=i-v,...,i+v$ and $2y_i-x_j$, $j=i-v,...,i+v$. The coefficients $a_N(i)$ are given by (2.8.4). Definition 5.8.1 is implicit and does not generally lend itself to the numerical calculation of R-filters. However, there exists a special case of R-filters, called *Wilcoxon filter*, which has a simple explicit definition. They correspond to the following choice of coefficients $a_N(R_j)$, according to (2.8.4), (2.8.14):

$$a_N(i) = \frac{2i-N-1}{2N} \tag{5.8.2}$$

In this case definition (5.8.1) takes the following explicit form:

$$y_i = \text{med } \{(x_j+x_k)/2, \quad i-v\leq j\leq i+v, i-v\leq k\leq i+v\} \tag{5.8.3}$$

An asymptotic equivalent definition of Wilcoxon filter is given by [26]:

$$y_i = \text{med } \{(x_j+x_k)/2, \quad i-v\leq j\leq k\leq i+v\} \tag{5.8.4}$$

Definition (5.8.4) is equivalent to:

$$y_i = \text{med } \{(x_{(j)}+x_{(k)})/2, \quad 1\leq j\leq k\leq n\} \tag{5.8.5}$$

where $x_{(j)}, x_{(k)}$, $j=1,...,n$, $k=1,...,n$ are the order statistics of the data samples $x_{i-v},...,x_{i+v}$.

Wilcoxon filters have been proven effective in the filtering of white additive Gaussian noise. However, they do not preserve edges as well as median filters do. The reason for this is that every possible pair is averaged in (5.8.2). If the summation in (5.8.5) is restricted to a maximum distance $j-k \leq D$, the modified Wilcoxon filter results [28]:

$$y_i = \text{med} \{(x_{(j)}+x_{(k)})/2, \quad 1 \leq j \leq k \leq n, \quad k-j<D\} \tag{5.8.6}$$

By letting D range from 1 to n, the modified Wilcoxon filter ranges from the median to the Wilcoxon filter. The modified Wilcoxon filter has better edge preservation than the standard Wilcoxon filter, but it is still worse than the median in this aspect [26-29].

Table 5.8.1: Essential elements in the computation of Wilcoxon filter of length $n=5$.

$x_{(1)}$	$\dfrac{(x_{(1)}+x_{(2)})}{2}$	$\dfrac{(x_{(1)}+x_{(3)})}{2}$	$\dfrac{(x_{(1)}+x_{(4)})}{2}$	$\dfrac{(\mathbf{x}_{(1)}+\mathbf{x}_{(5)})}{2}$
	$x_{(2)}$	$\dfrac{(x_{(2)}+x_{(3)})}{2}$	$\dfrac{(\mathbf{x}_{(2)}+\mathbf{x}_{(4)})}{2}$	$\dfrac{(x_{(2)}+x_{(5)})}{2}$
		$\mathbf{x}_{(3)}$	$\dfrac{(x_{(3)}+x_{(4)})}{2}$	$\dfrac{(x_{(3)}+x_{(5)})}{2}$
			$x_{(4)}$	$\dfrac{(x_{(4)}+x_{(5)})}{2}$
				$x_{(5)}$

One of the main disadvantages of the Wilcoxon filter versus the median is its computational complexity. It requires the ordering of n numbers, $n^2(n^2+1)/2$ additions and the ordering of $n^2(n^2+1)/2$ numbers. The last ordering requires many comparisons, as will be shown in chapter 11. However, there exists an inherent property of the Wilcoxon filter that can be used for the reduction of its computational complexity [29]. Since $x_{(j)}<x_{(k)}$ for $j<k$, the median in (5.8.5) is just the median on or close to the main diagonal of the matrix having elements $(x_{(j)}+x_{(k)})/2$, $1 \leq j \leq k \leq n$. Such a matrix is shown in Table 5.8.1 for $n=5$. Therefore, for $n=5$ the Wilcoxon filter is given by:

$$y_i = \text{med} \{x_{(3)}, (x_{(1)}+x_{(5)})/2, (x_{(2)}+x_{(4)})/2\} \tag{5.8.7}$$

The elements included in (5.8.4) are the so-called *essential* elements. They are 3 out of 15 included in (5.8.5). Thus the computation of the median of 3 instead of 15 numbers is required. This scheme can be extended for larger filter lengths [29], leading to a substantial reduction of the computational load of the Wilcoxon filter.

5.9 DISCUSSION

In this chapter a wide variety of nonlinear image filters has been analyzed both from a statistical and from a deterministic point of view. It is known that the performance of the median filter is excellent close to edges or impulses. However, it is not as good in additive white Gaussian noise filtering in homogeneous image regions. In this case the moving average filter is superior. Several efforts have been made to find a good compromise between the median and the moving average filter. Such successful compromises, according to our opinion, are the α-trimmed mean, the modified trimmed mean, and perhaps the standard type M-filter. However, the M-filters have not been studied thoroughly yet and they are computationally demanding. Another variation of the median filter is the L-filter. Its great advantage over all the previously mentioned filters is the ability to choose the filter coefficients in such a way that it becomes optimal filter, in the mean square error sense, for a variety of noise distributions. Another disadvantage of the median filter is its tendency to destroy image details and to produce streaks and blotches. If fine details are required, space-selective filters, e.g., the median hybrid filters or the multistage median filters, can be used instead. It is clearly seen that a great variety of nonlinear filters exists. Although there are some general rules for their performance in various cases, their success in a specific application depends heavily on the characteristics of the problem. Therefore, the successful application of these filters depends heavily on the experience of the scientist or the application engineer.

REFERENCES

[1]   H.A. David, *Order statistics*, John Wiley, 1981.

[2]   E.L. Lehmann, *Theory of point estimation*, John Wiley, 1983.

[3]   F. Hampel, P.J. Rousseevw, E.M. Ronchetti, W.A. Stahel, *Robust statistics*, John Wiley, 1986.

[4]   E. Ataman, V.K. Aatre, K.M. Wong, "Some statistical properties of median filters", *IEEE Transactions on Acoustics, Speech and Signal Processing*, vol. ASSP-29, no. 5, pp. 1073-1075, Oct. 1981.

[5]   G.R. Arce, M.P. McLoughlin, "Theoretical analysis of the max/median filter", *IEEE Transactions on Acoustics, Speech and Signal Processing*, vol. ASSP-25, no. 1, pp. 60-69, Jan. 1987.

[6]   A. Nieminen, P. Heinonen, Y. Neuvo, "A new class of detail preserving filters for image processing", *IEEE Transactions on Pattern Analysis and Machine Intelligence*, vol. PAMI-9, no. 1, pp. 74-90, Jan. 1987.

[7]   P. Heinonen, Y. Neuvo, "FIR-median hybrid filters", *IEEE Transactions on Acoustics, Speech and Signal Processing*, vol. ASSP-35, no. 6, pp. 832-838, June 1987.

[8]   J. Astola, P. Heinonen, Y. Neuvo, "Linear median hybrid filters", *IEEE Transactions on Circuits and Systems*, under review, 1987.

[9]   E. Heinonen, A. Nieminen, P. Heinonen, Y. Neuvo, "FIR-median hybrid filters for image processing", *Signal Processing III: Theories and Applications*, I.T. Young et al. editors, North Holland, 1986.

[10] G. Heygster, "Rank filters in digital image processing", *Computer, Graphics and Image Processing*, vol. 19, pp. 148-164, 1982.

[11] J.P. Fitch, E.J. Coyle, N.C. Gallagher, "Threshold decomposition of multidimensional ranked-order operations", *IEEE Transactions on Circuits and Systems*, vol. CAS-32, no. 5, pp. 445-450, May 1985.

[12] P.D. Wendt, E.J. Coyle, N.C. Gallagher, "Stack filters", *IEEE Transactions on Acoustics, Speech and Signal Processing*, vol. ASSP-34, no. 4, pp. 898-911, Aug. 1986.

[13] J.B. Bednar, T.L. Watt, "Alpha-trimmed means and their relationship to the median filters", *IEEE Transactions on Acoustics, Speech and Signal Processing*, vol. ASSP-32, no. 1, pp. 145-153, Feb. 1984.

[14] T.A. Nodes, N.C. Gallagher, "Median filters: some modifications and their properties", *Transactions on Acoustics, Speech and Signal Processing*, vol. ASSP-30, no. 5, pp. 739-746, Oct. 1982.

[15] Y.H. Lee, S.A. Kassam, "Generalized median filtering and related nonlinear filtering techniques", *IEEE Transactions on Acoustics, Speech and Signal Processing*, vol. ASSP-33, no. 3, pp. 672-683, June 1985.

[16] Y.H. Lee, S.A. Kassam, "Nonlinear edge preserving filtering techniques for image enhancement", *Proc. 27th Midwest Symp. on Circuits and Systems*, pp. 554-557, 1984.

[17] P.J. Bickel, "On some robust estimates of location", *Ann. Math. Statist.* vol. 36, pp. 847-858, 1965.

[18] A.C. Bovik, T.S. Huang, D.C. Munson, "A generalization of median filtering using linear combinations of order statistics", *IEEE Transactions on Acoustics, Speech and Signal Processing*, vol. ASSP-31, no. 6, pp.

1342-1349, Dec. 1983.

[19] F. Palmieri, C.G. Boncelet, "*Ll*-filters, a new class of order statistic filters", *IEEE Transactions on Acoustics, Speech and Signal Processing*, vol. ASSP-37, no. 5, pp. 691-701, May 1989.

[20] A. Nieminen, Y. Neuvo, "Comments on the theoretical analysis of the max/median filter", *IEEE Transactions on Acoustics, Speech and Signal Processing*, vol. ASSP-35, no. 5, pp. 826-827, May 1988.

[21] E.H. Loyd, "Least-squares estimation of location and scale parameters using order statistics", *Biometrika*, vol. 39, pp. 88-95, 1952.

[22] C.L. Mallows, "Resistant smoothing", O.D. Anderson editor, *Time Series*, pp. 147-155, North Holland, 1980.

[23] C.L. Mallows, "Some theory of nonlinear smoothers", *The Annals of Statistics*, vol. 8, no. 4, pp. 695-715, 1980.

[24] C.A. Pomalaza-Raez, C.D. McGillen, "An adaptive nonlinear edge preserving filter", *IEEE Transactions on Acoustics, Speech and Signal Processing*, vol. ASSP-32, no. 3, pp. 571-576, June 1984.

[25] T.R. Gustafsson, S.M. Pizer, "Non-stationary Metz-filtering", *Proc. 4th Int. Conf. Inf. Proc. in Scintigraphy*, Orsay, France, 1975.

[26] J. Song, S.A. Kassam, "Nonlinear filters based on generalized ranks for edge preserving smoothing", *Proc. IEEE Symposium on Circuits and Systems*, pp. 401-404, 1986.

[27] R.J. Crinon, "The Wilcoxon filter: A robust filtering scheme", *Proc. ICASSP*, 18.5.1-18.5.4, March 1985.

[28] P.P. Gandhi, S.A. Kassam, "Performance of some rank filters for edge preserving smoothing", *Proc. IEEE Symposium on Circuits and Systems*, pp. 264-267, 1987.

[29] J.H. Lee, J.S. Kao, "A fast algorithm for two-dimensional Wilcoxon filtering", *Proc. IEEE Symposium on Circuits and Systems*, pp. 268-271, 1987.

[30] L. Naaman, A.C. Bovik, "Optimal order statistic filters with coefficient censoring", *Proc. ICASSP88*, pp. 820-823, 1988.

[31] P.P. Gandhi, S.A. Kassam, "Nonlinear C-filters for nonstationary signals and deconvolution in impulsive noise", *Proc. 1988 IEEE Symposium on Circuits and Systems*, pp. 1507-1510, Helsinki, Finland, 1988.

[32] S.R Peterson, Y.H. Lee, S.A. Kassam, "Some statistical properties of alpha-trimmed mean and standard type *M*-filters", *IEEE Transactions on Acoustics, Speech and Signal Processing*, vol. ASSP-36, no. 5, pp. 707-713, May 1988.

[33] V. Kim, L. Yaroslawskii, "Rank algorithms for picture processing", *Computer Vision graphics and image processing*, vol. 35, pp. 234-258, 1986.

[34] G.R. Arce, R.E. Foster, "Detail preserving ranked-order based filters for image processing", *IEEE Transactions on Acoustics, Speech and Signal Processing*, vol. ASSP-37, no. 1, pp. 83-98, Jan. 1989.

[35] P. Heinonen, Y. Neuvo, "FIR median hybrid filters with predictive FIR substructures", *IEEE Transactions on Acoustics, Speech and Signal Processing*, vol. ASSP-36, no. 6, pp. 892-899, June 1988.

[36] K. Chen, "Bit-serial realizations of a class of nonlinear filters based on positive Boolean functions", *IEEE Transactions on Circuits and Systems*, vol. CAS-36, no. 6, pp. 785-794, June 1989.

[37] G.R. Arce, S.A. Fontana, "On the midrange estimator", *IEEE Transactions on Acoustics, Speech and Signal Processing*, vol. ASSP-36, no. 6, pp. 920-922, June 1988.

[38] H.G. Longbotham, A.C. Bovik, "Theory of order statistic filters and their relationships to linear FIR filters", *IEEE Transactions on Acoustics, Speech and Signal Processing*, vol. ASSP-37, no. 2, pp. 275-287, Feb. 1989.

[39] K. Chen, "Bit-serial realizations of a class of nonlinear filters using a bit-serial approach", *Proc. 1988 IEEE Symposium on Circuits and Systems*, pp. 1749-1752, Helsinki, Finland, 1988.

[40] A. Restrepo, A.C. Bovik, "Locally monotonic regression", *Proc. IEEE International Conference on Acoustics, Speech and Signal Processing*, pp. 1318-1321, Glasgow, Scotland, 1989.

[41] H.G. Longbotham, A.C. Bovik, A. Restrepo, "Generalized order statistic filters", *Proc. IEEE International Conference on Acoustics, Speech and Signal Processing*, pp. 1610-1613, Glasgow, Scotland, 1989.

[42] J.P. Fitch, "Software and VLSI algorithms for generalized ranked order filtering", *IEEE Transactions on Circuits and Systems*, vol. CAS-34, no. 5, pp. 553-559, May 1987.

CHAPTER 6

MORPHOLOGICAL IMAGE AND SIGNAL PROCESSING

6.1 THE PRINCIPLES OF MATHEMATICAL MORPHOLOGY

The final goal of image processing and analysis is, often, to segment the image into objects in order to analyze the geometric properties (e.g., the size) and the structure of the objects and recognize them. The analysis of the geometric objects must be quantitative, because only such an analysis and description of the geometric objects can provide a coherent mathematical framework for describing the spatial organization. The quantitative description of geometrical structures is the purpose of mathematical morphology [1].

Originally mathematical morphology was developed for the description of binary images and objects. An image object X is represented, in mathematical morphology terminology, as a set inside an n-dimensional Euclidean space $\mathbf{R}^n$. A binary image object X is represented by the set:

$$X = \{z: f(z)=1, \ z=(x,y) \in \mathbf{R}^2\} \tag{6.1.1}$$

The binary object background is the set X^c defined as follows:

$$X^c = \{z: f(z)=0, \ z=(x,y) \in \mathbf{R}^2\} \tag{6.1.2}$$

The function f is called the *characteristic function* of X. If the Euclidean grid $\mathbf{Z}^2$ is used instead of $\mathbf{R}^2$ for digitized binary images, definitions (6.1.1) and (6.1.2) become:

$$X = \{(i,j): f(i,j)=1, \ (i,j) \in \mathbf{Z}^2\} \tag{6.1.3}$$

$$X^c = \{(i,j): f(i,j)=0, \ (i,j) \in \mathbf{Z}^2\} \tag{6.1.4}$$

Later on, mathematical morphology was extended to describe and process multivalued (grayscale) images and objects. Multivalued images are functions $f(x,y)$ of the two spatial coordinates x,y, $(x,y) \in \mathbf{R}^2$:

$$(x,y) \in \mathbf{R}^2 \rightarrow f(x,y) \in \mathbf{R} \tag{6.1.5}$$

Grayscale images can be viewed as subsets of the Cartesian product $\mathbf{R}^2 \times \mathbf{R}$. If the multivalued image is defined over the Euclidean grid $\mathbf{Z}^2$, i.e., it is of the form $f(i,j)$, $(i,j) \in \mathbf{Z}^2$, it can be represented as a subset of $\mathbf{Z}^2 \times \mathbf{R}$. Morphological transformations have applications not only in image analysis but also in image and signal filtering and processing. The so-called morphological filters [11] are a fruitful application of mathematical morphology in image and signal filtering. Also edge detectors based on morphological operators have been proposed [10]. A discussion of such applications can be found in subsequent sections.

An image object is described by a set in mathematical morphology, because its subparts can be described easier in this notation. The group of all the interrelationships among its subparticles gives the structure of the object, according to J. Serra [1]. According to this definition, an image object possesses no information, before an observer studies it. It is the observer, who sees from an image object only what he wants to look at [1]. Therefore, the object structure depends on the observer and it is not objective. The observer interacts with the object and transforms it to another, perhaps more expressive, object and measures certain properties of the new object. The tool of interaction is a simple object called a *structuring element* (e.g., a circle or a square). The mode of interaction is the *morphological transformation* $\psi(X)$ whose operation is described in Figure 6.1.1. Information about the object size, shape, connectivity, convexity, smoothness, and orientation can be obtained by transforming the original object with different structuring elements. Some psychological experiments [66] have revealed that human vision is active and that the mind has a structuring activity on even the simplest perceptive phenomena, i.e., for the mind to perceive an image it has to transform it [1]. Thus, there exists strong psychological motivation for the use of mathematical morphology in image processing and computer vision.

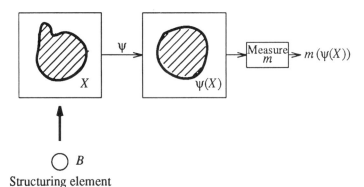

Structuring element

Figure 6.1.1: The methodology of mathematical morphology (adapted from [53]).

The performance of a morphological transformation depends on two factors: the structuring element and the transformation $\psi(X)$. There is a great variety of structuring elements. Their choice depends on the particular task, as will be seen later. The transformation function $\psi(X)$ must satisfy some constraints so that the morphological transformation is quantitative. These constraints form the *principles* of mathematical morphology [1, p.6]:

(1) **Translation invariance.** If X_z is the translate of the set $X \subset \mathbf{R}^2$ by a vector $z \in \mathbf{R}^2$, the translation invariance property can be stated as follows:

$$\psi(X_z) = [\psi(X)]_z \tag{6.1.6}$$

Translation invariance is a very natural requirement for any morphological operation, since the structure of the object does not generally change with its position in space.

(2) **Compatibility with change of scale.** If λ is a scaling factor, a morphological transformation $\psi(X)$ is compatible with change in scale, if a transformation $\psi_\lambda(X)$ can be constructed, such that:

$$\psi_\lambda(X) = \lambda\psi(\lambda^{-1}X) \tag{6.1.7}$$

Compatibility with magnification arises from the fact that, generally, the object structure does not change with scale.

(3) **Local knowledge.** Image objects are usually seen through windows (e.g., through an image frame). If for any bounded window M^* in which we want to know $\psi(X)$, we can find a bounded window M in which the knowledge of X is sufficient to perform the transformation locally (i.e., within M^*):

$$[\psi(X \cap M)] \cap M^* = \psi(X) \cap M^* \tag{6.1.8}$$

the transformation $\psi(X)$ requires only local knowledge of the image object X, and it is called *local*. Otherwise it is called *global*.

(4) **Semicontinuity.** The fourth principle of mathematical morphology describes the continuity properties of the morphological transformations. The definition of continuity is rather difficult and requires the description of some basic notions on limits. Let a_i, i=1,2,.. be a sequence of points of the topological space E. Let us suppose that there exists a point a such that we can find an arbitrary large index n so that $a_n \in N(a)$ for any neighborhood $N(a)$ of a. Such a point is called *adherent point* of this sequence. Let X_i, i=1,2,.. be a sequence of objects of the topologic space E. Let us also suppose that this sequence possesses adherent points. We denote by $\overline{\lim} X_n$ and by $\underline{\lim} X_n$ the union and the intersection of those adherent points, respectively. An example for the clarification of the notion of adherent points is given in Figure 6.1.2. Let X_n, ($n \geq 2$) denote the boundary of the rectangle shown in Figure 6.1.2. When n is even ($n=2i$), a slice of width w/i and height $h_i = h(1-1/i)/2$ is removed from

the top of the rectangle. When n is odd ($n=2i+1$), a slice of the same width and height is removed from the bottom of the rectangle. The union of all adherent points of X_n is the rectangle and the vertical line segment passing through its center. The intersection of the adherent points is the rectangle and its center. Having given those definitions, we can describe the fourth principle of mathematical morphology.

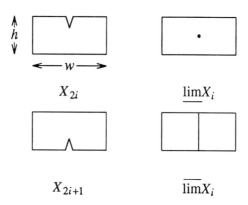

X_{2i} $\underline{\lim}X_i$

X_{2i+1} $\overline{\lim}X_i$

Figure 6.1.2: Example of adherent points (adapted from [1]).

Let X_n, $n=1,2,...$ be a sequence of closed objects tending toward a limit image object X. Let $\psi(X_n)$ be the sequence of the morphologically transformed objects. The morphological transformation is *semicontinuous*, if the sequence of transformed objects tends to $\psi(X)$. There are two types of semicontinuity: *upper* and *lower semicontinuity*. A morphological transformation $\psi(X)$ is upper semi-continuous if:

$$\overline{\lim} \, \psi(X_n) \subset \psi(X) \tag{6.1.9}$$

The upper semi-continuity property of a transformation is essentially a guarantee for the robustness of the transformation against small variations in the object structure. An example of a discontinuous transformation is shown in Figure 6.1.3. The transformation shown is the following:

$$\psi(X): \quad X \rightarrow \text{maximal inscribable disk in } X$$

Let us suppose that X is a rectangle and X_i is a decreasing sequence of objects, which tend to X. In this case, the limit $\psi(X_i)$ is different from $\psi(X)$. A similar discontinuity exists for increasing sequences X_i, as can also be seen in Figure 6.1.3.

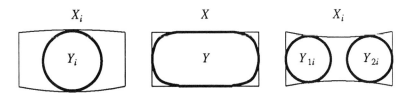

Figure 6.1.3: An example of discontinuous morphological transformation (adapted from [1]).

A much weaker notion is lower semicontinuity:

$$\underline{\lim}\, \psi(X_n) \supset \psi(X) \qquad (6.1.10)$$

where $\underline{\lim}\, X_n$ is the intersection of all adherent points A. The principles (1)-(4) form a general framework for mathematical morphology. However, there exist several morphological operations (e.g., skeletonization) that do not conform with all above mentioned principles. Serra has dropped semicontinuity as a basic property of morphological transformations in [5]. Generally speaking, those principles can be followed as long as they do not pose problems in the development of morphological techniques.

All morphological transformations are non-reversible operations, except for some subclasses of the object X. In fact, the philosophy of mathematical morphology is not to restore the image but to *manage the loss of information* of an image through successive transformations [5]. The following properties are related to the ability of the morphological transformations to perform such a controlled loss of information.

(1) **Increasing.** $\psi(X)$ is increasing, when it preserves inclusion:

$$X \subset Y \Rrightarrow \psi(X) \subset \psi(Y) \qquad (6.1.11)$$

(2) **Anti-extensivity.** $\psi(X)$ is anti-extensive, if it shrinks X:

$$\psi(X) \subset X \qquad (6.1.12)$$

(3) **Idempotence.** $\psi(X)$ is idempotent, when $\psi(X)$ is unchanged by the reapplication of the transformation:

$$\psi[\psi(X)] = \psi(X) \qquad (6.1.13)$$

(4) **Homotopy.** Each object X is associated with an *homotopy tree*. Its root corresponds to the background X and its leaves to the components of X, as is seen in Figure 6.1.4. Homotopy is more severe than connectivity: a disk and a ring are both connected but not homotopic. A transformation is called

homotopic if it preserves the homotopy tree of X.

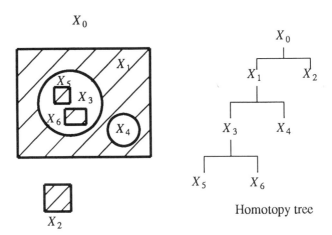

Figure 6.1.4: Homotopy tree.

These four properties are very important for the analysis of morphological transformations. However, other properties exist too, e.g., symmetrical treatment of the background, connectivity, and monotonicity.

Every morphological transformation ψ has its *dual* ψ^*. The transformation $\psi(X)$ transforms also the image object background X^c to $\psi(X^c)$. This operation is the dual $\psi^*(X)$ with respect to complementation:

$$\psi^*(X) = [\psi(X^c)]^c \tag{6.1.14}$$

Several dual morphological transformations will be discussed in the subsequent sections. Duality applies also to the properties of the morphological transformations. The dual properties of anti-extensivity, idempotence, and homotopy are extensivity, idempotence, and homotopy [1, p.588]. It is important to note that the dual of the increasing property is the same property. If $\psi(X)$ is increasing, then $\psi^*(X)$ is increasing also.

Before we proceed to the definition of morphological transformations it is important to review some notions of set theory. Let E be an arbitrary set and $p(E)$ be its *power set*, i.e., the set of all subsets of E:

$$p(E) = \{X : X \subset E\} \tag{6.1.15}$$

Throughout this chapter, E will denote either the Euclidean space $\mathbf{R}^n$, or its digital counterpart $\mathbf{Z}^n$. The set $p(E)$ is a *Boolean algebra* and has the following properties [2]:

(1) $p(E)$ is a *complete lattice*, i.e., it is provided with a partial ordering relation called *inclusion* and denoted by $\subset$. Every family of sets $X_i \in p(E)$ has a least upper bound, namely the union $\cup X_i$, and a greatest lower bound, namely the intersection $\cap X_i$, which both belong to $p(E)$.

(2) The lattice $p(E)$ is *distributive:*

$$X \cup (Y \cap Z) = (X \cup Y) \cap (X \cup Z), \qquad \forall X,Y,Z \in p(E) \qquad (6.1.16)$$

(3) The lattice $p(E)$ is *complemented*, i.e., for each set X exists its *complement X^c*:

$$X \cup X^c = E \qquad X \cap X^c = \varnothing \qquad\qquad (6.1.17)$$

The set difference in $p(E)$ is defined as follows:

$$X - Y = X \cap Y^c \qquad\qquad (6.1.18)$$

i.e., $X-Y$ is the part of X that does not belong to Y. For two given sets $B,X \in p(E)$, we may have:

(a) B is included in X: $B \subset X$ meaning $B \cap X = B$

(b) B *hits* X: $B \uparrow X$ meaning $B \cap X \neq \varnothing$

(c) B *misses* X: $B \subset X^c$ meaning $B \cap X = \varnothing$.

These definitions are illustrated in Figure 6.1.5. Although they are rather abstract, they will be useful in the subsequent sections.

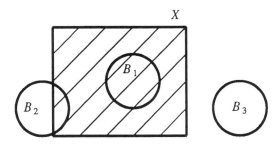

Figure 6.1.5: Set B_1, B_2, B_3 is included, hits, and misses object X respectively.

Another notion which is useful in mathematical morphology is the notion of the topologically open and closed set, which is defined as follows [1, p. 66]. A topological space is a pair consisting of a set E and a collection of subsets of

E called open sets, satisfying the three following properties:

(1)   Every union of open sets is open.

(2)   Every *finite* intersection of open sets is open.

(3)   The set E and the empty set $\varnothing$ are open.

   A subset X of E is closed if X^c is open.

   If $E=\mathbf{R}$, then the open sets are the open intervals of the form $Y=(a,b)$. If Y_i, $i=1,...,n$ are open sets in $\mathbf{R}$, the set $Y_1 \times Y_2 \times \cdots \times Y_n$ is an *open interval* in $\mathbf{R}^n$. Every union of open intervals in $\mathbf{R}^n$ is an open set. There are sets that are both open and closed. Such sets are the subsets of $\mathbf{Z}^n$.
Finally, the De Morgan's laws will be stated:

$$(X \cup Y)^c = X^c \cap Y^c \tag{6.1.19}$$

$$(X \cap Y)^c = X^c \cup Y^c \tag{6.1.20}$$

Having defined all basic mathematical notions, we proceed to the definitions of the basic morphological transformations.

6.2 EROSION AND DILATION IN THE EUCLIDEAN SPACE

   The analysis of the elementary morphological transformations is restricted in the Euclidean space, i.e., $E=\mathbf{R}^n$ or $\mathbf{Z}^n$, because this space is important for image processing applications. In this section we shall work only with binary images. The binary objects X are of the form (6.1.1) or (6.1.3) and E is either $\mathbf{R}^2$ or $\mathbf{Z}^2$. In the second case, the binary image object X can be viewed as a set of integer pairs (i,j). Each pair gives the coordinates of an image pixel, with respect to two basis unit vectors. The length of the unit vectors is equal to the sampling period along each direction. If the angle between the basis vector is $60°$, the Euclidean grid is the hexagonal one. If the angle is $90°$, the Euclidean grid is the rectangular one, which is used in the vast majority of image processing applications.

   The simplest morphological transformation is *erosion* and *dilation* [1]. These two operations are based on *Minkowski set addition and subtraction* [1,7]. Let us suppose that a, b are two vectors, which are members of the sets $A,B \in E$ respectively. If z is the result of the vector addition of a and b, the Minkowski addition of A,B is given by:

$$A \oplus B = \{z \in E: \quad z=a+b, \ a \in A, \ b \in B\} \tag{6.2.1}$$

If A_b is the translate of the set A by the vector b:

$$A_b = \{z \in E: \quad z=a+b, \ a \in A\} \tag{6.2.2}$$

the Minkowski addition is also given by:

$$A \oplus B = \bigcup_{b \in B} A_b \qquad (6.2.3)$$

The equivalence of (6.2.1) and (6.2.3) is easily proven:

$$A \oplus B = \bigcup_{b \in B} A_b = \bigcup_{b \in B} \{z \in E: \ z = a + b, \ a \in A\} \qquad (6.2.4)$$

$$= \{z \in E: z = a + b, \ a \in A, b \in B\}$$

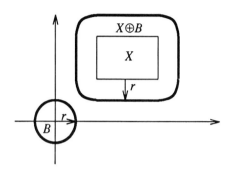

Figure 6.2.1: Minkowski addition of a rectangle and a disk.

The Minkowski set subtraction of B from A, denoted by $A \ominus B$, is defined by:

$$A \ominus B = \bigcap_{b \in B} A_b \qquad (6.2.5)$$

Minkowski subtraction is the dual operation of Minkowski addition:

$$A \oplus B = (A^c \ominus B)^c$$
$$A \ominus B = (A^c \oplus B)^c \qquad (6.2.6)$$

Proof: Let $z \in (A^c \ominus B)^c$. Then $z \notin A^c \ominus B$. This happens if and only if there exists a $b \in B$ such that $z \notin A_b^c$. This also happens if and only if there exists a $b \in B$ such that $z \in A_b$, i.e. $z \in A \oplus B$.

The notions of Minkowski addition and subtraction are clarified in Figures 6.2.1 and 6.2.2.

Let us suppose that there exists an image object X and a structuring element B. The set B^s is called symmetric of B with respect to the origin if:

$$B^s = \{-b: \quad b \in B\}$$

Therefore, B^s is obtained by rotating B 180 degrees in the plane as is seen in

Figure 6.2.3. If B is symmetric about the origin, there is no difference between B and B^s. Some symmetric structuring elements are shown in Figure 6.2.4.

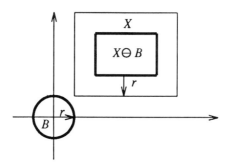

Figure 6.2.2: Minkowski subtraction of a disk from a rectangle .

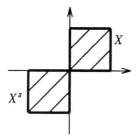

Figure 6.2.3: Definition of the symmetric set X^s of the set X about the origin.

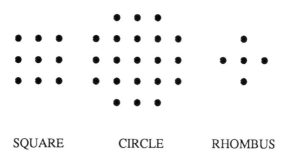

SQUARE   CIRCLE   RHOMBUS

Figure 6.2.4: Symmetric structuring elements.

The *dilation* of X by B is the Minkowski set addition of X with B^s:

$$X \oplus B^s = \bigcup_{b \in B} X_{-b} \tag{6.2.7}$$

Another definition of dilation is the following:

$$X \oplus B^s = \{z \in E: B_z \cap X \neq \emptyset\} = \{z: B_z \uparrow X\} \tag{6.2.8}$$

Proof: The equivalence of (6.2.7) and (6.2.8) is proven as follows:

$$\{z \in E: B_z \cap X \neq \emptyset\} = \{z \in E: \text{exists } a \in X \text{ such that } a \in B_z\} =$$

$$\{z \in E: \text{exists } a \in X \text{ such that } a-z \in B\} =$$

$$\{z \in E: \text{exists } a \in X, \ b \in B \text{ such that } z=a-b\} =$$

$$\{z \in E: \text{exists } a \in X, \ b' \in B^s \text{ such that } z=a+b'\} = X \oplus B^s \tag{6.2.9}$$

Definition (6.2.8) suggest that the dilation $X \oplus B^s$ includes all the translates B_z of B which have common points with X. Therefore, dilation is an expanding operator, as it is seen in Figure 6.2.5.

The *erosion* of X by B is the Minkowski set subtraction of B^s from X:

$$X \ominus B^s = \bigcap_{b \in B} X_{-b} \tag{6.2.10}$$

A fully equivalent definition of erosion is the following:

$$X \ominus B^s = \{z \in E: B_z \subset X\} \tag{6.2.11}$$

Proof:

$$\{z \in E: B_z \subset X\} = \bigcap_{b \in B} \{z: b+z \in X\} = \bigcap_{b \in B} X_{-b} = X \ominus B^s \tag{6.2.12}$$

Definition (6.2.11) shows that erosion $X \ominus B^s$ includes all the translates B_z which are included in X. Therefore, it is a shrinking operator. An example of erosion is shown in Figure 6.2.5.

The definitions of dilation and erosion used are the same as those given by Serra [1,5]. Other authors, e.g., Haralick et al. [4], Giardina et al. [8,9], give slightly different definitions and they make no distinction between dilation and Minkowski set addition. The differences between dilation, Minkowski addition and erosion, Minkowski subtraction, according to definitions (6.2.7), (6.2.10), are illustrated in Figure 6.2.6.

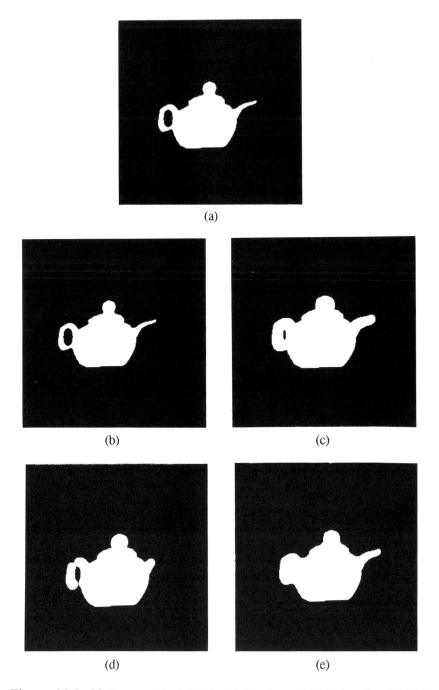

Figure 6.2.5: (a) Image object POT; (b) Erosion of POT by the CIRCLE

structuring element; (c) Dilation of POT by the CIRCLE structuring element; (d) Opening of POT by the structuring element $CIRCLE \oplus CIRCLE$; (e) Closing of POT by the structuring element $CIRCLE \oplus CIRCLE$.

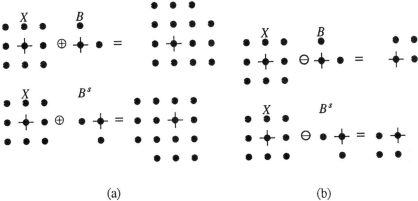

(a)   (b)

Figure 6.2.6: Example of the difference between: (a) Minkowski addition and dilation, (b) Minkowski subtraction and erosion.

Minkowski set addition and subtraction, erosion and dilation possess several interesting properties. They are:

(1) Commutative

$$A \oplus B = B \oplus A \tag{6.2.13}$$

(2) Associative

$$A \oplus (B \oplus C) = (A \oplus B) \oplus C \tag{6.2.14}$$

The proofs of (6.2.13-14) are very simple and they are omitted.

(3) Translation invariance of Minkowski addition and dilation

$$A_z \oplus B = (A \oplus B)_z \tag{6.2.15}$$

Proof: Every element $y \in A_z \oplus B$ has the form $y=(a+z)+b$, $a \in A$, $b \in B$. Therefore, $y=(a+b)+z$ and it is also an element of $(A \oplus B)_z$.

Corollaries

$$A \oplus B_1 \oplus B_2 \oplus \cdots \oplus (B_n)_z \oplus \cdots \oplus B_N = (A \oplus B_1 \oplus \cdots \oplus B_N)_z \tag{6.2.16}$$

$$A_z \oplus B_{-z} = A \oplus B \tag{6.2.17}$$

(4) Increasing property

$$A \subseteq B \Rightarrow A \oplus D \subseteq B \oplus D \tag{6.2.18}$$

Proof: Let $x \in A \oplus D$. Thus $x = a + d$, for $a \in A$, $d \in D$. Since $A \subset B$, it implies $a \in B$. Therefore, $x \in B \oplus D$ and (6.2.18) is valid.
According to (6.2.18) dilation is an increasing transformation.

(5) Distributive properties of Minkowski addition and dilation

$$(A \cup B) \oplus C = (A \oplus C) \cup (B \oplus C) \tag{6.2.19}$$

$$A \oplus (B \cup C) = (A \oplus B) \cup (A \oplus C) \tag{6.2.20}$$

Proof:

$$(A \cup B) \oplus C = \bigcup_{x \in A \cup B} C_x = (\bigcup_{x \in A} C_x)(\bigcup_{x \in B} C_x) = (A \oplus C) \cup (B \oplus C) \tag{6.2.21}$$

(6.2.20) follows from (6.2.19) from the commutativity of Minkowski addition.
(6.2.19) suggests that dilation possesses the distributivity with set union. (6.2.20) is more important since it suggests that, if a structuring element B can be decomposed in a union of structuring elements, the image can be dilated independently by the two structuring elements and that the results can be combined by set union.
However, Minkowski addition is not distributive with set intersection:

$$(A \cap B) \oplus C \subseteq (A \oplus C) \cap (B \oplus C) \tag{6.2.22}$$

$$A \oplus (B \cap C) \subseteq (A \oplus B) \cap (A \oplus C) \tag{6.2.23}$$

Proof: If $x \in (A \cap B) \oplus C$, it is given by $x = y + c$, $y \in (A \cap B)$, $c \in C$. The fact $y \in (A \cap B)$ implies that $y \in A$ and $y \in B$. Therefore, x belongs to $A \oplus C$ and to $B \oplus C$. Hence it belongs to $(A \oplus C) \cap (B \oplus C)$. Thus (6.2.22) is valid. (6.2.23) comes from (6.2.22) by using the commutativity of Minkowski addition.

(6) Translation invariance of Minkowski subtraction and erosion

$$A_z \ominus B = (A \ominus B)_z \tag{6.2.24}$$

$$A \ominus B_z = (A \ominus B)_z \tag{6.2.25}$$

Proof:

$$A_z \ominus B = ((A_z)^c \oplus B)^c = ((A^c)_z \oplus B)^c = [(A^c \oplus B)^c]_z = (A \ominus B)_z \qquad (6.2.26)$$

$$A \ominus B_z = (A^c \oplus B_z)^c = (A^c \oplus B)^c_z = (A \ominus B)_z \qquad (6.2.27)$$

The duality property (6.2.6) and the translation invariance of Minkowski addition have been used in (6.2.26-27). (6.2.24-25) suggest that erosion is translation invariant.

(7) Increasing property of Minkowski subtraction and erosion

$$A \subseteq B \Rightarrow A \ominus D \subseteq B \ominus D \qquad (6.2.28)$$

Proof:

$$A \ominus D = \bigcap_{d \in D} A_d \subseteq \bigcap_{d \in D} B_d = B \ominus D \qquad (6.2.29)$$

It should be mentioned that both erosion and dilation are increasing transformations, although they are dual operations.

(8) Distributivity properties of Minkowski subtraction and erosion

$$A \ominus (B \cup C) = (A \ominus B) \cap (A \ominus C) \qquad (6.2.30)$$

$$(B \cap C) \ominus A = (B \ominus A) \cap (C \ominus A) \qquad (6.2.31)$$

Proof: By applying DeMorgan's law and duality, it is found that:

$$A \ominus (B \cup C) = [A^c \oplus (B \cup C)]^c = [(A^c \oplus B) \cup (A^c \oplus C)]^c$$

$$= (A^c \oplus B)^c \cap (A^c \oplus C)^c = (A \ominus B) \cap (A \ominus C) \qquad (6.2.32)$$

By applying DeMorgan's law and duality, we find:

$$(B \cap C) \ominus A = [(B \cap C)^c \oplus A]^c = [(B^c \cup C^c) \oplus A]^c = [(B^c \oplus A) \cup (C^c \oplus A)]^c$$

$$= (B^c \oplus A)^c \cap (C^c \oplus A)^c = (B \ominus A) \cap (C \ominus A) \qquad (6.2.33)$$

Property (6.2.31) asserts a right distributivity of Minkowski subtraction over intersection. However, left distributivity does not hold:

$$A \ominus (B \cap C) \supseteq (A \ominus B) \cup (A \ominus C) \qquad (6.2.34)$$

Proof: By using DeMorgan's law and property (6.2.23), we find

$$A \ominus (B \cap C) = [A^c \oplus (B \cap C)]^c \supseteq [(A^c \oplus B) \cap (A^c \oplus C)]^c$$

$$= (A \ominus B) \cup (A \ominus C) \qquad (6.2.35)$$

(9) Distributivity between Minkowsky addition and subtraction

$$A \ominus (B \oplus C) = (A \ominus B) \ominus C \tag{6.2.36}$$

Proof:

$$A \ominus (B \oplus C) = [A^c \oplus (B \oplus C)]^c = [(A^c \oplus B) \oplus C]^c$$

$$= (A^c \oplus B)^c \ominus C = (A \ominus B) \ominus C \tag{6.2.37}$$

The property (6.2.36) is very important for the practical implementation of the erosion. If a structuring element B^s can be decomposed as the dilation of K structuring elements

$$B^s = B_1^s \oplus B_2^s \oplus \cdots \oplus B_K^s \tag{6.2.38}$$

the erosion $X \ominus B^s$ can be implemented in K steps as follows:

$$X \ominus B^s = (...((X \ominus B_1^s) \ominus B_2^s) \cdots \ominus B_K^s) \tag{6.2.39}$$

The same decomposition can be used for the implementation of the dilation $X \oplus B^s$, by virtue of (6.2.14):

$$X \oplus B^s = (\cdots ((X \oplus B_1^s) \oplus B_2^s) \oplus \cdots \oplus B_K^s) \tag{6.2.40}$$

Both decompositions (6.2.39-40) will be used in subsequent sections. A structuring element decomposition of the form (6.2.38) is shown in Figure 6.2.7. The theory for optimal structuring element decomposition is presented in [59].

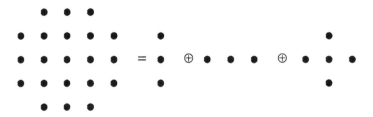

Figure 6.2.7: Decomposition of the structuring set CIRCLE in the Minkowski addition of smaller structuring elements.

(6.2.36) assures the left distributivity of erosion with dilation. The opposite is not true:

$$A \oplus (B \ominus C) \subseteq (A \oplus B) \ominus C \tag{6.2.41}$$

Proof: By using the translation invariance (6.2.15) and the inequality (6.2.23) it is found that

$$A \oplus (B \ominus C) = A \oplus (\bigcap_{z \in C} B_z) \subseteq \bigcap_{z \in C} (A \oplus B_z) = \bigcap_{z \in C} (A \oplus B)_z = (A \oplus B) \ominus C \quad (6.2.42)$$

Erosions and dilations are special cases of the *hit-or-miss transformation*. Let X be an image object and A, B two structuring elements. The hit-or-miss transformation $X \circledast (A,B)$ is defined as [1, pp. 39,270],[5]:

$$X \circledast (A,B) = (X \ominus A^s) - (X \oplus B^s) = (X \ominus A^s) \cap (X^c \ominus B^s) \quad (6.2.43)$$

The general hit-or-miss transformation is the logical ancestor of all other transformations involved in mathematical morphology. For example, if $B = \varnothing$, $X \circledast (A,B)$ equals to the erosion $X \ominus A^s$. However, it is neither increasing, nor extensive, nor anti-extensive and, therefore, it has limited applications.

6.3 CLOSINGS AND OPENINGS

Erosion is generally a non-reversible operation. Therefore, an erosion followed by a dilation does not generally recover the original object X. Instead it defines a new morphological transformation called *opening* X_B:

$$X_B = (X \ominus B^s) \oplus B \quad (6.3.1)$$

Its dual operation is the *closing* X^B:

$$X^B = (X \oplus B^s) \ominus B \quad (6.3.2)$$

The duality between opening and closing is expressed by the following relations:

$$(X^B)^c = (X^c)_B \quad (6.3.3)$$

$$(X_B)^c = (X^c)^B \quad (6.3.4)$$

Proof:

$$(X^B)^c = [(X \oplus B^s) \ominus B]^c = (X \oplus B^s)^c \oplus B = (X^c \ominus B^s) \oplus B = (X^c)_B$$

$$(X_B)^c = [(X \ominus B^s) \oplus B]^c = (X \ominus B^s)^c \ominus B = (X^c \oplus B^s) \ominus B = (X^c)^B$$

The opening X_B consists of the union of those translates of B that lie inside X:

$$X_B = \cup \{B_z : B_z \subset X\} \quad (6.3.5)$$

Proof: The opening X_B is given by $X_B = \bigcup_{z \in (X \ominus B^s)} B_z$. We have to prove that $B_z \subset X$, when $z \in (X \ominus B^s)$. Each element b' of B_z has the form $b' = b + z$. For each member $z \in (X \ominus B^s)$ an element X can be found such that $z = x - b$. Therefore, each element of B_z satisfies $b' = b + z = x \in X$. Thus $B_z \subset X$ and

$$X_B = \bigcup_{z \in (X \ominus B')} B_z = \cup\{B_z : B_z \subset X\}$$

(6.3.5) is an important property because it shows that the opening smooths the contours of X, cuts the narrow isthmuses, suppresses the small islands and the sharp capes of X. This behavior can be seen in Figure 6.2.5.

By duality the closing X^B consists of the intersection of the complements of the translates of B that lie outside X:

$$X^B = \cap\{(B_z)^c : B_z \subset X^c\} \tag{6.3.6}$$

Proof: By using the duality property:

$$X^B = [(X^B)^c]^c = [(X^c)_B]^c = [\cup\{B_z : B_z \subset X^c\}]^c = \cap\{(B_z)^c : B_z \subset X^c\}$$

By duality, closing blocks up narrow channels, small holes and thin gulfs of X, as is seen in Figure 6.2.5.

The opening is an antiextensive, idempotent, and increasing transformation:

$$\text{(1)} \quad X_B \subset X \quad \text{(antiextensive)} \tag{6.3.7}$$

$$\text{(2)} \quad X_1 \subset X_2 \Rightarrow (X_1)_B \subset (X_2)_B \quad \text{(increasing)} \tag{6.3.8}$$

$$\text{(3)} \quad (X_B)_B = X_B \quad \text{(idempotent)} \tag{6.3.9}$$

Proof:

(1) $X_B = \cup\{B_z : B_z \subset X\} \subset X$

(2) $(X_1)_B = \cup\{B_z : B_z \subset X_1\} \subset \cup\{B_z : B_z \subset X_2\} = (X_2)_B$

(3) $(X_B)_B = [[(X \ominus B^s) \oplus B] \ominus B^s] \oplus B = (X \ominus B^s)^{B'} \oplus B \supset (X \ominus B^s) \oplus B = X_B$

The containment comes from the fact $X^B \supset X$ (see (6.3.10)). On the other hand, by using (6.3.7) $(X_B)_B \subset X_B$. Therefore, (6.3.9) is valid.

Similar properties are valid for closing due to duality:

$$\text{(1)} \quad X^B \supset X \quad \text{(extensive)} \tag{6.3.10}$$

$$\text{(2)} \quad X_1 \subset X_2 \Rightarrow (X_1)^B \subset (X_2)^B \quad \text{(increasing)} \tag{6.3.11}$$

$$\text{(3)} \quad (X^B)^B = X^B \quad \text{(idempotent)} \tag{6.3.12}$$

Proof:

(1) $(X^B)^c = (X^c)_B \subset X^c$.   Therefore, $X^B \supset X$.

(2) $[(X_1)^B]^c = (X_1^c)_B \supset (X_2^c)_B = [(X_2)^B]^c$.   Therefore, $(X_1)^B \subset (X_2)^B$.

(3) $(X^B)^B = \{[(X^B)^B]^c\}^c = \{[(X^B)^c]_B\}^c = \{[(X^c)_B]_B\}^c = \{(X^c)_B\}^c = X^B$.

The extensive, increasing, and idempotent properties of opening and closing are very simple but powerful and important for theoretical and practical reasons.

In many cases it is important to know which objects X are invariant to opening with B. Such objects are said to be *open with respect to B*. The objects that are invariant to closing with B are called *closed with respect to B*. The following property is valid due to duality:

$$X_B = X \Rightarrow (X^c)^B = X^c \tag{6.3.13}$$

$$X^B = X \Rightarrow (X^c)_B = X^c \tag{6.3.14}$$

i.e., if X is open with respect to B, X^c is closed with respect to B.

Proof: By duality

$$(X^c)^B = (X_B)^c = X^c$$

$$(X^c)_B = (X^B)^c = X^c$$

The following two propositions define the structure of the open and closed sets X with respect to a structuring element B.

Proposition 1: X is B-open, if and only if there exists a set A such that $X = A \oplus B$.

Proof: (Necessity) If X is B-open, then $X = (X \ominus B^s) \oplus B$ and A is given by $A = X \ominus B^s$.
(Sufficiency) Let us suppose that $X = A \oplus B$. Its opening is given by

$$X_B = (X \ominus B^s) \oplus B = [(A \oplus B) \ominus B^s] \oplus B = A^{B'} \oplus B \supset A \oplus B = X$$

On the other hand, by antiextensivity $X_B \subset X$. Therefore, $X_B = X$.

Proposition 2: X is B-closed, if and only if there exists a set A such that $X = A \ominus B$.
Proof: (Necessity) If X is B-closed, then $X = (X \oplus B^s) \ominus B$ and A is given by $A = X \oplus B^s$.
(Sufficiency) Let us suppose that $X = A \ominus B$. Then $X^c = A^c \oplus B$ and it is B-open according to proposition 1. Therefore, X is B-closed according to (6.3.14).

The idempotence of opening can be generalized as follows. If A is B-open, then:

$$(X_B)_A = X_A \tag{6.3.15}$$

$$(X_A)_B = X_A \tag{6.3.16}$$

Similarly the idempotence of closing can be generalized. If A^c is B-closed then:

$$(X^B)^A = X^A \tag{6.3.17}$$

$$(X^A)^B = X^A \tag{6.3.18}$$

Furthermore, if A is B-open the following inclusion properties hold:

$$X_A \subset X_B \subset X \subset X^B \subset X^A \tag{6.3.19}$$

The proof of the properties (6.3.15-19) can be found in [9, p.111].

Until now only the morphological transformations of binary images have been considered. The analysis of gray-level morphological transformations is considered in the next section.

6.4 GRAYSCALE MORPHOLOGY

The notions and morphological transformations of a binary image can also be extended to grayscale images. Such images can be viewed as three-dimensional surfaces that usually have very interesting and nice landscapes. These surfaces can be modified by sliding structuring elements, leading to a variety of morphological transformations. Historically, the literature on the morphology of (graytone) functions starts from the second half of the seventies. J. Serra (1975) extended the hit-or-miss transformation and the size distributions to functions. Lantuejoul (1977) described the functions by their watersheds. Sternberg (1979) introduced the notion of umbra and several related transformations. Meyer (1977) developed contrast descriptors based on the top-hat transformation. Goetcherian contributed to the notion of lower skeleton (1979) and to the connection of mathematical morphology with fuzzy logic (1980). Rosenfeld (1977) proposed a generalization of connectivity functions. Recently Preston (1983) introduced the Ξ Filters, Lantuejoul and Serra (1982) introduced the morphological filters, and Maragos (1985) and Stevenson et al. (1987) have studied their connection to order statistic filters and their applications in image coding.

The bridge to use mathematical morphology for grayscale functions is the creation of a link between functions and sets. Such links are important because an image is a two-dimensional function $f(x,y)$, $(x,y) \in \mathbf{R}^2$ or $\mathbf{Z}^2$. Therefore, an effort will be made to describe an image by sets, and then to apply the known morphological transformations on those sets.

6.5 LINKS BETWEEN FUNCTIONS AND SETS

One of the most important links between sets and functions is the notion of *umbra*, introduced by Sternberg [17,18] and shown in Figure 6.5.1. If a function $f(x)$ has domain $D \subset \mathbf{R}^n$ or $D \subset \mathbf{Z}^n$ and takes values in $\mathbf{R}$:

$$x \in D \quad \rightarrow \quad f(x) \in \mathbf{R} \tag{6.5.1}$$

its umbra $U(f)$ is a subset of the Cartesian product $D \times \mathbf{R}$ consisting of those points of $D \times \mathbf{R}$, which occupy the space below the graph of $f(x)$ and down to $-\infty$:

$$U(f) = \{(x,y) \in D \times \mathbf{R} \colon f(x) \geq y\} \tag{6.5.2}$$

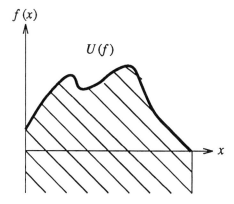

Figure 6.5.1: Definition of the umbra of a function.

In usual images the region of support D is a subset of $\mathbf{R}^2$. Therefore, their umbra is a subset of $\mathbf{R}^2 \times \mathbf{R} = \mathbf{R}^3$. A unique umbra $U(f)$ corresponds to any real-valued upper semicontinuous (u.s.c.) function $f(x)$. The function $f(x)$ can be reconstructed from its umbra as follows:

$$f(x) = \max\{y \in \mathbf{R} \colon (x,y) \in U(f)\} \tag{6.5.3}$$

(6.5.3) would be more mathematically concise if *supremum* (*sup*) were used instead of max, because y is a continuous variable and $y \in \mathbf{R}$. However, we use the max operator throughout, because in all practical cases y is discretized. The careful reader should substitute *sup* to *max* in appropriate circumstances. A concept related to umbra is the *top surface* $T(X)$ of a set $X \subset D \times \mathbf{R}$:

$$T[X](x) = \max\{y \in \mathbf{R} \colon (x,y) \in X\} \tag{6.5.4}$$

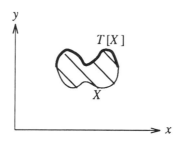

Figure 6.5.2: Definition of the top surface of a set.

which is illustrated in Figure 6.5.2. By combining (6.5.4) and (6.5.3), it is found that:

$$T[U(f)](x) = f(x) \tag{6.5.5}$$

Therefore, the top surface and the umbra inverse each other. In the following some properties of the umbra are given.

Proposition 1: Let $X \subseteq D \times \mathbf{R}$. Then $X \subseteq U[T(X)]$.

Proposition 2: If A is an umbra, then $A = U[T(A)]$.

Let us define the following ordering of functions:

$$f \leq g \longleftrightarrow f(x) \leq g(x) \quad \forall x \in D$$

Then the following property is valid:

Proposition 3: If $f \leq g$, then $U(f) \subseteq U(g)$.
Two new pointwise operations can be defined between two functions f, g:

$$(f \wedge g)(x) = \min(f(x), g(x)), \quad \forall x \in D \tag{6.5.6}$$

$$(f \vee g)(x) = \max(f(x), g(x)), \quad \forall x \in D \tag{6.5.7}$$

Their umbrae possess the following form:

Proposition 4:

$$U(f \wedge g) = U(f) \cap U(g) \tag{6.5.8}$$

$$U(f \vee g) = U(f) \cup U(g) \tag{6.5.9}$$

as can be seen in Figure 6.5.3.

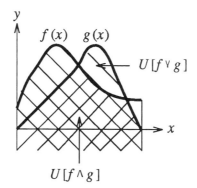

Figure 6.5.3: Umbrae of the maximum and the minimum of two functions.

Proposition 5: Suppose that A,B are umbrae. Then $A \oplus B, A \ominus B$ are umbrae.

Finally, a proposition similar to proposition 3 is valid for the top of surface:

Proposition 6: If $A \subseteq B$, then $T[A](x) \leq T[B](x)$.

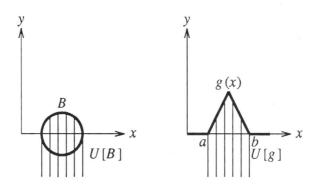

Figure 6.5.4: Examples of the umbra of a structuring element and of a structuring function.

The notion of umbra can easily be extended to the umbra of a set and of a structuring element, as is seen in Figure 6.5.4. In the next section, the importance of umbra in the definition of morphological transformations will be shown. Before doing this, another important link between functions and sets will be presented.

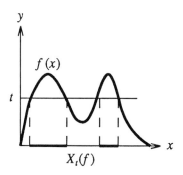

Figure 6.5.5: Definition of the cross-section of a function.

The *cross-section* of a function $f(x)$ at level t is given by:

$$X_t(f) = \{x \in D : f(x) \geq t\}, \quad -\infty < t < \infty \qquad (6.5.10)$$

and is presented in Figure 6.5.5. It greatly resembles the *threshold decomposition* of ranked-order filters discussed in chapter 5. Obviously, $X_t(f) \subseteq D$. Cross-sections $X_t(f)$ are sets. Therefore, a function f can be decomposed to the sets $X_t(f)$, $-\infty < t < \infty$. Given the function cross-sections, the function itself can be reconstructed as follows:

$$f(x) = \max\{t \in \mathbf{R} : x \in X_t\} \qquad (6.5.11)$$

The function cross-sections possess similar properties with umbra.

In a previous section some general properties of the morphological transformations have been given, namely the translation-invariant, increasing, extensive, and idempotent properties. Their generalization to the grayscale morphological operations is simple. Having defined all important links between sets and functions, we are ready for the transition to grayscale morphological transformations.

6.6 GRAYSCALE MORPHOLOGICAL TRANSFORMATIONS

Let us denote by $g(x)$ a simple function called *structuring function*. The simplest form of such a function is of the form:

$$g(x) = 0, \qquad x \in G \tag{6.6.1}$$

where G is its domain, which is a subset of $\mathbf{R}^n$. The umbra of a function having $G=[a,b]\subset\mathbf{R}$ is shown in Figure 6.5.4. Functions of the form (6.6.1) are equivalent to structuring sets. The symmetric function $g^s(x)$ with respect to the origin is given by

$$g^s(x) = g(-x) \tag{6.6.2}$$

If D is the domain of f and G is the domain of the structuring function g, the Minkowski addition of a function $f(x)$ by $g(x)$, denoted by $f \oplus g$ is defined by:

$$[f \oplus g](x) = \max_{\substack{z \in D \\ z-x \in G}} \{f(z)+g(x-z)\} \tag{6.6.3}$$

The Minkowski subtraction of $f(x)$ by $g(x)$ is defined by:

$$[f \ominus g](x) = \min_{\substack{z \in D \\ z-x \in G}} \{f(z)-g(x-z)\} \tag{6.6.4}$$

The grayscale dilation $f \oplus g^s$ and erosion $f \ominus g^s$ are given by:

$$[f \oplus g^s](x) = \max_{\substack{z \in D \\ z-x \in G}} \{f(z)+g(z-x)\} \tag{6.6.5}$$

$$[f \ominus g^s](x) = \min_{\substack{z \in D \\ z-x \in G}} \{f(z)-g(z-x)\} \tag{6.6.6}$$

Morphological operations (6.6.5-6) greatly resemble linear convolution. Definitions (6.6.5-6) are of great practical importance since they introduce a method of numerical computation of erosion and dilation. Definitions (6.6.5-6) can be greatly simplified when the structuring function $g(x)$ is of the form (6.6.1):

$$[f \oplus G^s](x) = [f \oplus g^s](x) = \max_{\substack{z \in D \\ z-x \in G}} \{f(z)\} \tag{6.6.7}$$

$$[f \ominus G^s](x) = [f \oplus g^s](x) = \min_{\substack{z \in D \\ z-x \in G}} \{f(z)\} \tag{6.6.8}$$

The grayscale erosion $f \ominus G^s$ and dilation $f \oplus G^s$ are called *erosion* and *dilation of function by set* [1].

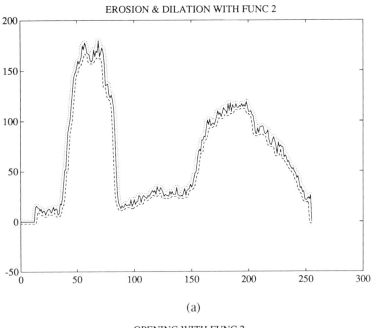

(a)

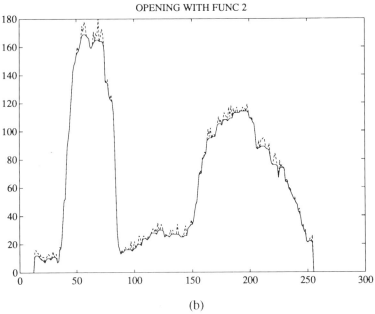

(b)

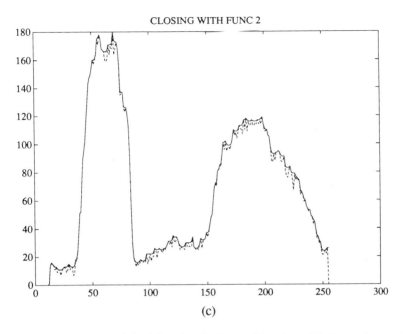

(c)

Figure 6.6.1: (a) The original function is denoted by the solid line, whereas its dilation and erosion by a 5-point structuring function are denoted by dotted and dashed lines respectively; (b) Opening of the original function (solid line); (c) Closing of the original function (solid line).

If the domain $D = \mathbf{Z}$ and the domain G is a subset of $\mathbf{Z}$:

$$G = \{-v,...,0,...,v\} = G^s$$

the erosion and dilation of function by set are given by:

$$[f \oplus G^s](i) = \max\{f(i-v),...,f(i),...,f(i+v)\} \qquad (6.6.9)$$

$$[f \ominus G^s](i) = \min\{f(i-v),...,f(i),...,f(i+v)\} \qquad (6.6.10)$$

Therefore, they are completely equivalent to the n-th ranked order filter and to the 1st ranked-order filter, respectively ($n = 2v+1$), which are analyzed in chapter 5. An example of such an erosion and dilation is shown in Figure 6.6.1. The structuring function used has a 5-point domain $G = \{-2, -1, 0, 1, 2\}$.

Grayscale erosion and dilation possess similar properties to the binary erosion and dilation. They are presented here without proof, since their proofs are very simple and can be based on the corresponding properties of the binary operators [41],[1, p.442]:

(1)	Commutativity	$f \oplus g = g \oplus f$	(6.6.11)
(2)	Associativity	$f \oplus (g \oplus k) = (f \oplus g) \oplus k$	(6.6.12)
(3)	Distributivity	$(f \vee g) \oplus h = (f \oplus h) \vee (g \oplus h)$	(6.6.13)
		$(f \wedge g) \ominus h = (f \ominus h) \wedge (g \ominus h)$	(6.6.14)
(4)	Parallel composition	$f \ominus (g \vee h) = (f \ominus g) \wedge (f \ominus h)$	(6.6.15)
(5)	Serial composition	$f \oplus (g \oplus h) = (f \oplus g) \oplus h$	(6.6.16)
		$f \ominus (g \oplus h) = (f \ominus g) \ominus h$	(6.6.17)
(6)	Umbra homomorphism	$U[f \oplus g] = U[f] \oplus U[g]$	(6.6.18)
		$U[f \ominus g] = U[f] \ominus U[-g]^c$	(6.6.19)

Umbra homomorphism is a very important property from a theoretical point of view because it links grayscale morphology with umbrae and binary morphology. Serial composition is another important property of grayscale erosion and dilation because it can be used for its fast implementation when a structuring function can be decomposed to simpler ones of the form:

$$g = g_1 \oplus g_2 \oplus \cdots \oplus g_k \qquad (6.6.20)$$

In this case, erosion and dilation become:

$$f \oplus g = (...((f \oplus g_1) \oplus g_2) \oplus \cdots \oplus g_k) \qquad (6.6.21)$$

$$f \ominus g = (...((f \ominus g_1) \ominus g_2) \ominus \cdots \ominus g_k) \qquad (6.6.22)$$

Grayscale erosion and dilation are dual operations. Their duality is expressed by the following relation:

$$f \ominus g^s = -[(-f) \oplus g^s] \qquad (6.6.23)$$

Proof:

$$-[(-f) \oplus g^s](x) = - \max_{\substack{z \in D \\ z-x \in G}} \{-f(z) + g(z-x)\}$$

$$= \min_{\substack{z \in D \\ z-x \in G}} \{f(z) - g(z-x)\} = [f \ominus g^s](x)$$

Another set of dual operations is grayscale opening:

$$f_g(x) = [(f \ominus g^s) \oplus g](x) = [f(x) \ominus g(-x)] \oplus g(x) \qquad (6.6.24)$$

and grayscale closing:

$$f^g(x) = [(f \oplus g^s) \ominus g](x) = [f(x) \oplus g(-x)] \ominus g(x) \qquad (6.6.25)$$

Lantuejoul and Serra [11] introduced the name *M-filters* for grayscale opening and closing. We shall use instead the term *morphological filters* to avoid any confusion with the *M*-estimators and *M*-filters described in chapters 2 and 5.

If $g(x)$ is a structuring function of the form (6.6.1), (6.6.24-25) are equivalent to opening and closing of function by set:

$$f_G(x) = [(f \ominus G^s) \oplus G](x) \qquad\qquad (6.6.26)$$

$$f^G(x) = [(f \oplus G^s) \ominus G](x) \qquad\qquad (6.6.27)$$

An example of such openings and closings for $G = \{-2, -1, 0, 1, 2\} \in \mathbf{Z}$ is shown in Figure 6.6.1.

Opening is antiextensive, whereas closing is extensive. Thus, if $g(0) > 0$ [41]:

$$f \ominus g \leq f_g \leq f \leq f^g \leq f \oplus g \qquad\qquad (6.6.28)$$

Furthermore, opening and closing are:

(1) translation invariant $(f_g)(x-t) = [f(x-t)]_g$ (6.6.29)

$(f^g)(x-t) = [f(x-t)]^g$ (6.6.30)

(2) increasing $f_1 \leq f_2 \Rightarrow (f_1)_g \leq (f_2)_g$ (6.6.31)

$f_1 \leq f_2 \Rightarrow (f_1)^g \leq (f_2)^g$ (6.6.32)

(3) idempotent $(f_g)_g = f_g$ (6.6.33)

$(f^g)^g = f^g$ (6.6.34)

The proof of translation invariance is very simple. The proofs of increasing property and idempotence are omitted because they are similar to the proofs of (6.3.7-12).

Finally, the duality theorem for opening and closing can be stated as follows:

$$-f_g = (-f)^g \qquad\qquad (6.6.35)$$

$$f^g = -(-f)_g \qquad\qquad (6.6.36)$$

Proof:

$$-f_g = -[(f \ominus g^s) \oplus g] = [-(f \ominus g^s)] \ominus g = [(-f) \oplus g^s] \ominus g = (-f)^g$$

A function f is called *open with respect to g* if and only if $f = f_g$. Likewise, f is called closed with respect to g iff $f = f^g$. Thus f is open, if it is a root of the opening transformation. It is closed, if it is a root of the closing transformation. The following theorem gives the roots of opening and closing [41].

Theorem 6.6.1: A function f is open (respectively, closed) with respect to a function g if and only if $f = h \oplus g$ (respectively $f = h \ominus g$) where h is an arbitrary function.

The proof is omitted because it is similar to the proof of Proposition 1 (section 6.3). The threshold decomposition described in chapter 4 can also be applied to the grayscale morphological operations and used for their fast implementation [58].

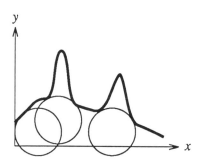

Figure 6.6.2: Rolling ball transformation.

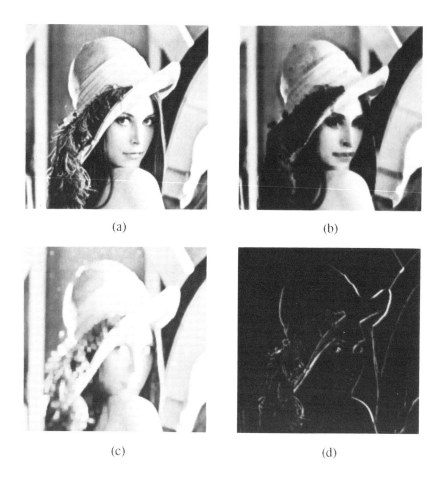

(a)   (b)

(c)   (d)

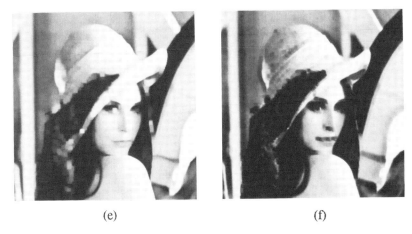

(e)   (f)

Figure 6.6.3: (a) Original image; (b) Erosion by a 3×3 structuring set; (c) Dilation by a 3×3 structuring set; (d) Difference between the original image and the eroded image; (e) Opening by a 3×3 structuring set; (f) Closing by a 3×3 structuring set.

Opening and closing are the basis of the definitions of morphological filters. Therefore it would be interesting to give a graphical representation of their operation. An 1-d grayscale function is shown in Figure 6.6.2. The opening is essentially a rolling ball transformation. The rolling ball exactly traces the smoothly varying contours, where positive impulses exist the rolling ball deletes them, and where negative impulses exist, the rolling ball leaves gaps. The effect of the closing is exactly opposite to that of opening: it deletes negative impulses and broadens the positive ones. An example of erosion, dilation, opening and closing of a gray-valued image with a 3×3 structuring set is shown in Figure 6.6.3. It is clearly seen that erosion broadens the image black region (see the girl's eyes) because of the min operator used in (6.6.6), whereas dilation broadens the image bright regions because of the max operator used in (6.6.5).

6.7 MORPHOLOGICAL FILTERS

Morphological filters are nonlinear filters based on morphological transformations of functions by sets. Such filters are the grayscale erosion and dilation, defined by (6.6.7-8) and the opening and closing of a function by set defined (6.6.26-27). Furthermore, another set of morphological filters studied in the literature [19,41,42] are the *close-opening (CO)* and the *open-closing (OC) filters:*

$$y = [(f^G)_G](x) \tag{6.7.1}$$

$$y = [(f_G)^G](x) \tag{6.7.2}$$

where f_G, f^G are given by (6.6.26-27). If $f(x)$ is a sampled function and $f_i = f(i)$, $i \in \mathbf{Z}$ and G is the structuring set (6.6.8) the output of one-dimensional erosion and dilation filters is given by (6.6.9-10):

$$y_i = [f \oplus G^s](i) = \max\{f_{i-v}, ..., f_i, ..., f_{i+v}\} \tag{6.7.3}$$

$$y_i = [f \ominus G^s](i) = \min\{f_{i-v}, ..., f_i, ..., f_{i+v}\} \tag{6.7.4}$$

These two filters will not be studied further here because they correspond to ranked-order filters, whose study is included in chapter 5. Opening and closing are two-step operations. Therefore, the output of the one-dimensional opening filter is given by

$$y_i = f_G(i) = \max\{f_{i-v}^{(1)}, ..., f_i^{(1)}, ..., f_{i+v}^{(1)}\} \tag{6.7.5}$$

$$f_i^{(1)} = \min\{f_{i-v}, ..., f_i, ..., f_{i+v}\} \tag{6.7.6}$$

Similarly, the one-dimensional closing filter is defined as:

$$y_i = \min\{f_{i-v}^{(1)}, ..., f_i^{(1)}, ..., f_{i+v}^{(1)}\} \tag{6.7.7}$$

$$f_i^{(1)} = \max\{f_{i-v}, ..., f_i, ..., f_{i+v}\} \tag{6.7.8}$$

The one-dimensional close-opening filter is a four-step operation:

$$y_i = [(f^G)_G](i) = \max\{f_{i-v}^{(3)}, ..., f_i^{(3)}, ..., f_{i+v}^{(3)}\} \tag{6.7.9}$$

$$f_i^{(3)} = \min\{f_{i-v}^{(2)}, ..., f_i^{(2)}, ..., f_{i+v}^{(2)}\} \tag{6.7.10}$$

$$f_i^{(2)} = \min\{f_{i-v}^{(1)}, ..., f_i^{(1)}, ..., f_{i+v}^{(1)}\} \tag{6.7.11}$$

$$f_i^{(1)} = \max\{f_{i-v}, ..., f_i, ..., f_{i+v}\} \tag{6.7.12}$$

The definition of an open-closing filter is the same as (6.7.9-12) if the max and min operators interchange positions.

Morphological filters possess certain nice syntactic and statistical properties, which are related to the corresponding properties of the median filter. Let A be the window set of the median filter having size $n = 2v+1$ and let $m = v+1$ be the size of the set G. Let $\mathrm{med}(f;A)$ denote the median filtered sequence. This sequence is bounded from below and above by opening and closing, respectively [42]:

Proposition 6.7.1:

$$f_G \leq \mathrm{med}(f;A) \leq f^G \tag{6.7.13}$$

$$X_G \leq \mathrm{med}(X;A) \leq X^G \tag{6.7.14}$$

where $\mathrm{med}(X;A)$ denotes the binary median filtering of a binary sequence X

(i.e., of a binary set).

The property 6.7.1 is very important because it can be used to find a relation between the roots of one-dimensional OC and CO filters and those of the median filter [42]:

Proposition 6.7.2: A set X or a function f, of finite extent, is a median root with respect to A, iff it is a root of both opening and the closing by G.

Let us define by $\text{med}^n(f;A)$ the output of a median filter iterated n times and by $\text{med}^\infty(f;A)$ the median root. It can be proven that the median root is bounded by close-opening and open-closing [42]:

Proposition 6.7.3:

$$(f_G)^G \le \text{med}^\infty(f;A) \le (f^G)_G \qquad (6.7.15)$$

$$(X_G)^G \le \text{med}^\infty(X;A) \le (X^G)_G \qquad (6.7.16)$$

The outputs of the CO and OC filters are median roots themselves [42]:

Proposition 6.7.4: The open-closing and the close-opening by G of any signal f of finite extent are median roots with respect to A. That is:

$$(f_G)^G = \text{med}[(f_G)^G;A] \qquad (6.7.17)$$

$$(f^G)_G = \text{med}[(f^G)_G;A] \qquad (6.7.18)$$

Proposition 6.7.4 is a very important practical result since it can give a median root in just one pass.

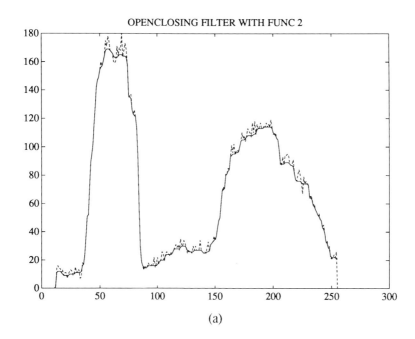

OPENCLOSING FILTER WITH FUNC 2

(a)

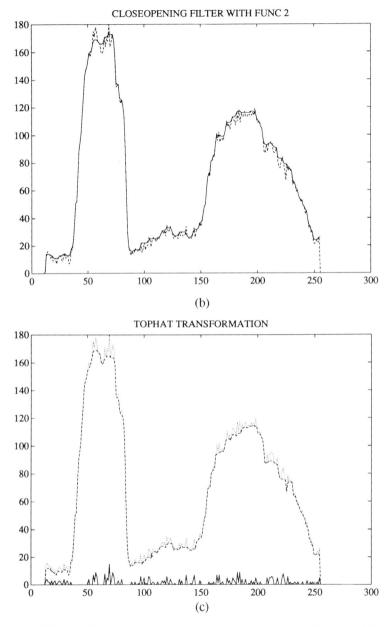

Figure 6.7.1: (a) Open-closing of the function shown in Figure 6.6.1a; (b) Close-opening of the function shown in Figure 6.6.1a; (c) Top-hat transformation of the function shown in Figure 6.6.1a.

An example of one-dimensional 3-point OC and CO filtering is shown in Figure 6.7.1. The median root has also been obtained by using a median filter having window size 5 in four iterations. The bounds of proposition 6.7.3 are satisfied. However, the difference between the median root and the open-closing and close-opening is very small. Thus the median root is not displayed.

A statistical analysis of the morphological filters is presented in [19]. The output distributions of the close-opening and open-closing filters are given by:

$$F_{CO}(y) = mF(y)^m - (m-1)F(y)^{m+1} + F(y)^{2m}(1-F(y))$$

$$+ \frac{(m+1)m}{2} F(y)^{2m}(1-F(y))^2 \tag{6.7.19}$$

$$F_{OC}(y) = 1 - [m(1-F(y))^m - (m-1)(1-F(y))^{m+1} + (1-F(y))^{2m}F(y)$$

$$+ \frac{m(m+1)}{2} (1-F(y))^{2m}F(y)^2] \tag{6.7.20}$$

for $m = v+1$

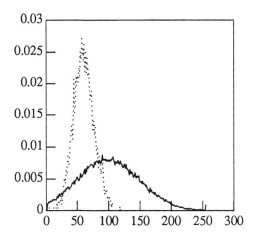

Figure 6.7.2: Gaussian probability input distribution having mean at 100 (solid curve) and the probability distribution of the output of a 3×3 openclosing filter (dotted curve). The shift of the output mean to smaller values is evident.

The proofs are rather lengthy and they are omitted here. $F(y)$ is the probability distribution of the input signal. The output pdf of a 3×3 two-dimensional OC filter having Gaussian input density is shown in Figure 6.7.2. It is seen that the output of the OC filter is strongly biased toward small output values. This is explained by the min operator in (6.7.12). The output density function of the

CO filter is not plotted because it is the symmetric of the pdf of the OC filter output. The output of the CO filter is biased toward large output values. The morphological filters can be easily extended to two dimensions with respect to a two dimensional structuring set G. However, in this case, propositions similar to 6.7.1 and 6.7.2 are not valid in general for any structuring element.

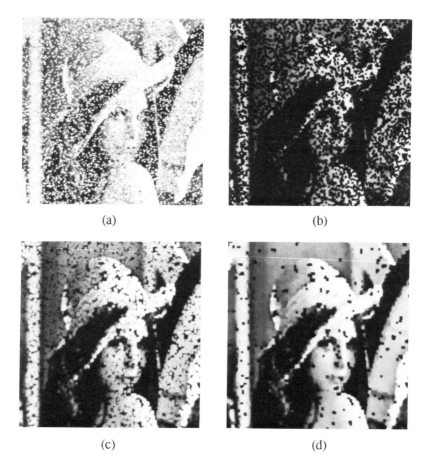

(a)   (b)

(c)   (d)

Figure 6.7.3: (a) Original image corrupted by 10% mixed impulsive noise; (b) Output of the erosion by a 3×3 structuring set; (c) Output of the opening by a 3×3 structuring set; (d) Output of the OC filter by a 3×3 structuring set.

An example of two-dimensional OC filtering (6.7.1), by using a 3×3 square structuring set, is shown in Figure 6.7.3. Image 6.7.3a is corrupted by 10%

salt-pepper noise. The opening operation of the OC filter removes positive impulses but enhances the negative ones. This mode of operation is easily explained by the opening definition (6.7.5-6). The min operator in (6.7.6) suppresses any positive impulse. Most of the negative impulses are removed by the closing operation of the OC filter. However, the negative impulses have already been enhanced by opening. Therefore, some negative impulses remain in the OC filtered image. In contrast, some positive impulses can remain in an CO filtered image. The OC filter can be compared to the 3×3 median filter by comparing Figures 6.7.3d and 4.7.1c. Clearly, the median filter has superior performance in the salt-pepper noise filtering than the OC filter.

Digital filters based in morphological operations have also been used in impulsive noise filtering and background normalization of ECG signals [72]. Morphological filters can also be used for edge detection. The following morphological filter:

$$y = f(x) - [f \ominus B](x) \tag{6.7.21}$$

has been proposed [10] as a robust edge detector. It has also been found that this morphological edge detector is more robust than other linear edge detection schemes. An example of the performance of this edge detector is shown in Figure 6.6.3d. An extension of this edge detector is the following:

$$y = f(x) - [f \ominus nB](x) \tag{6.7.22}$$

where the structuring set nB is defined as:

$$nB = B \oplus B \oplus \cdots \oplus B \quad (n \text{ times}) \tag{6.7.23}$$

and the symbol - in (6.7.22) denotes algebraic difference. Parameter n controls the thickness of the edges, i.e., the larger n, the thicker the edges. The orientation of B controls the orientation of the edges. Other morphological edge detectors are presented in [63-65].

The opening f_{nB} of a function is a low-pass nonlinear filter, because it destroys the high frequencies of an image, as has already been shown by the illustrative rolling ball interpretation shown in Figure 6.6.2. Therefore, the following filter, called *top-hat transformation* [20], is a high-pass filter:

$$y \doteq f(x) - f_{nB}(x) \tag{6.7.24}$$

The opening f_{nB} erases all peaks in which the structuring element nB cannot enter. Therefore, only those peaks appear in $f - f_{nB}$, and the background is eliminated as it is seen in Figure 6.7.1c. Combinations of top-hat transformations by anisotropic structuring elements have been used for cleaning images from astronomy [22]. The top-hat transformation has also been used to find galaxies in astronomical images [11].

6.8 MORPHOLOGICAL SKELETONS

Mathematical morphology is very rich in providing means for the representation and analysis of binary and grayscale images. The morphological representation of images is very well suited for the description of the geometrical properties of the image objects. Therefore, it can be used for pattern recognition, robotic vision, image coding, etc.

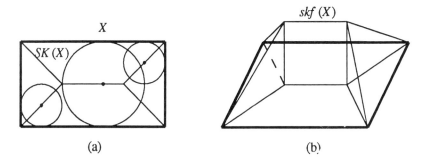

(a)   (b)

Figure 6.8.1: (a) Skeleton of a rectangle; (b) Skeleton function of the rectangle.

One of the most important morphological representation schemes is the *skeleton* of an image, shown in Figure 6.8.1. The idea of transforming an image to a skeleton is due to Blum [25,26], who called it *medial axis*. Blum's initial procedure to obtain the medial axis was to set up *grass fires* along the object boundary, simultaneously at $t=0$, and to let the fire wavefronts to propagate toward the image object center by Huygen's principle at uniform speed. The medial axis consists of those points, where the fire wavefronts intersect and extinguish, together with their arrival times, which constitute the *medial axis function*. Blum tried to relate this grass fire propagation mechanism to a hypothesized brain mechanism of shape recognition in animal vision. Subsequently, a mathematical theory for the skeletons of continuous images [27,28] and for discrete images [29, 30, 31] has been developed. In reviewing the previous research, Blum [26] showed that the medial axis is the locus of the centers of the *maximal discs* inscribable inside an image object. Lantuejoul [32,33] connected skeletons with mathematical morphology and proved that skeletons can be obtained by morphological transformations. Such a skeleton will be called *morphological*, to be distinguished from skeletons obtained by other mathematical descriptions, because there are some differences in the various skeleton definitions. These differences are discussed in [1, p. 382].

In this section the definition of the skeleton of binary images will be given first and it will be generalized subsequently for graytone images. Let X be an image object in $\mathbf{R}^2$. The skeleton $SK(X)$ is the set of the centers of the maximal inscribable disks inside X. Such a disk is maximal if it is not properly contained in any other disk totally included in X. A maximal disk touches the object boundary in at least two different points. This definition of the skeleton is exemplified in Figure 6.8.1. Each point of the skeleton is associated with the radius r of the corresponding maximal disk. Therefore, a function $skf(z)$ can be defined, whose domain is $SK(X)$ and which takes values in $\mathbf{R}$:

$$z \in SK(X) \; \rightarrow \; skf(z) \in \mathbf{R} \qquad (6.8.1)$$

This function is called *skeleton function* or *quench function*. Let us denote by $S_r(X)$ the subset of $SK(X)$ whose points correspond to centers of maximal inscribable disks having the same radius $r > 0$. Lantuejoul [33] has proven that the morphological skeleton is given by:

$$SK(X) = \bigcup_{r>0} S_r(X) = \bigcup_{r>0} [(X \ominus rB) - (X \ominus rB)_{drB}] \qquad (6.8.2)$$

where rB denotes an open disk of radius r and drB denotes a closed disk of infinitesimally small radius dr. This process is illustrated in Figure 6.8.2.

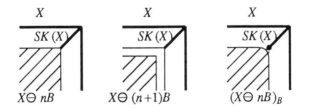

Figure 6.8.2: Illustration of the skeletonization process.

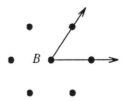

Figure 6.8.3: Symmetric hexagonal structuring element of radius 1 on the hexagonal grid.

The transition from the skeleton definition (6.8.2) in $\mathbf{R}^n$ to the skeleton in $\mathbf{Z}^n$ is quite troublesome. The major problem is the approximation of the disks rB and drB in the grid $\mathbf{Z}^n$. In a hexagonal grid, the circle can be approximated by the symmetric hexagon shown in Figure 6.8.3. This seven pixel structuring set can be said to have a *discrete radius* $n=1$. The element nB:

$$nB = B \oplus B \oplus \cdots \oplus B \quad (n \text{ times}) \tag{6.8.3}$$

has discrete radius n. By using the hexagon structuring element, Serra [1, p. 389] provided the following algorithm for the morphological skeleton on a hexagonal grid:

$$S_n(X) = (X \ominus nB^s) - (X \ominus nB^s)_B \quad n=0,1,2,...,N \tag{6.8.4}$$

$$SK(X) = \bigcup_{n=0}^{N} S_n(X) \tag{6.8.5}$$

The discrete binary object X can be reconstructed exactly from its discrete skeleton as follows:

$$X = \bigcup_{n=0}^{N} [S_n(X) \oplus nB] \tag{6.8.6}$$

However, hexagonal grids are of limited use. Therefore, Maragos [24] proposed the use of definitions (6.8.4-6) for the rectangular grid. In this grid the closest approximation to the circle of discrete radius one is the SQUARE structuring element. A better approximation of the circle is the CIRCLE structuring element shown in Figure 6.2.4. Its discrete radius is 2. A skeleton obtained by using the CIRCLE is shown in Figure 6.8.4.

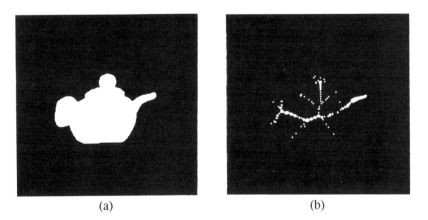

(a)   (b)

Figure 6.8.4: (a) Object POT; (b) Skeleton of POT obtained by using the CIRCLE.

Furthermore, Maragos [24] proposed a serial decomposition of (6.8.4-6.8.5) for the fast calculation of the skeleton:

$$S_0(X) = X - X_B \tag{6.8.7}$$

$$S_1(X) = (X \ominus B^s) - (X \ominus B^s)_B \tag{6.8.8}$$

$$\vdots$$

$$S_N(X) = [(X \ominus (N-1)B^s) \ominus B^s] - [X \ominus (N-1)B^s]_B \tag{6.8.9}$$

Definition (6.8.4) is fully equivalent to (6.8.7-9) if the distributivity property (6.2.36) is considered. A similar serial decomposition of the reconstruction procedure (6.8.6) is given by:

$$X = [...[[S_N(X) \oplus B \cup S_{N-1}(X)] \oplus B] \cup S_{N-2}(X)...] \oplus B \cup S_0(X) \tag{6.8.10}$$

The fast skeletonization and reconstruction procedures are shown in Figure 6.8.5.

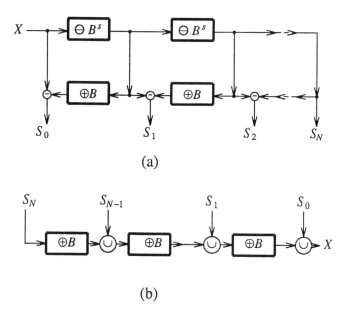

(a)

(b)

Figure 6.8.5: (a) Fast skeletonization process; (b) Fast object reconstruction from skeleton subsets (adapted from [24]).

Partial reconstructions of X can be obtained if only the $N-k$ skeleton subsets $S_n(X)$, $n = k, ..., N$ are used:

$$X' = \bigcup_{n=k}^{N} S_n(X) \oplus nB \qquad\qquad (6.8.11)$$

The result of the partial reconstruction (6.8.11) is the opening of X by kB, X_{kB} [24]:

$$X_{kB} = (X \ominus kB^s) \oplus kB = \left[\left[\bigcup_{n=0}^{N} S_n(X) \oplus nB \right] \ominus kB^s \right] \oplus kB$$

$$= \left[\bigcup_{n=k}^{N} S_n(X) \oplus (n-k)B \right] \oplus kB = X'$$

Also eroded versions of X of the form $X \ominus kB^s$ can be obtained from the partial reconstruction of X:

$$X \ominus kB^s = \bigcup_{n=k}^{N} [S_n(X) \oplus (n-k)B] \qquad\qquad (6.8.12)$$

if the disks of size n are substituted with disks of size $n-k$. A dilated version of X can also be obtained from its skeleton:

$$X \oplus kB^s = \bigcup_{n=0}^{N} [S_n(X) \oplus (n+k)B] \qquad\qquad (6.8.13)$$

if the disks of size n are replaced with disks of size $n+k$. If some skeleton subsets $S_n(x)$, $0 \le n \le k$ are omitted and the result of the reconstruction is dilated, the original object X cannot be reconstructed:

$$X' = \bigcup_{n=k}^{N} [S_n(X) \oplus (n+k)B] \ne X$$

This result is important in image coding applications because it states that the loss of some fine details described by $S_n(X)$, $0 \le n \le k$ cannot be compensated by a final dilation.

Morphological skeletons of binary images possess certain interesting properties [1,24]:

Proposition 6.8.1: Skeletons are translation invariant.

Proposition 6.8.2: Skeletons in the Euclidean space $\mathbf{R}^2$ are invariant in the change of scale.

Proposition 6.8.3: The mapping $X \rightarrow \overline{SK(X)}$ is lower semicontinuous, i.e., very point x of $\overline{SK(X)}$ is a limit of a sequence of points x_i with $x_i \in SK(X_i)$. $\overline{SK(X)}$ is the adherence of $SK(X)$ [1, p.378].

The definition of adherent points is given in section 6.1. Note that $\overline{SK(X)}$ is not upper-semicontinuous transformation, which is a requirement to be a morphological transformation.

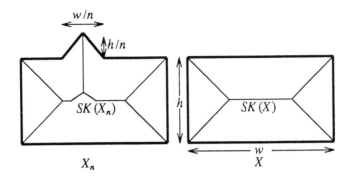

Figure 6.8.6: Illustration of the continuity properties of skeleton (adapted from [53]).

The poor continuity properties of skeleton are depicted in Figure 6.8.6. A rectangle has a small bump of height h/n and width w/n. This bump causes a disturbance to the skeleton of the rectangle. When n tends to infinity the bump disappears. However, the transition in the skeleton is not continuous. Furthermore, skeleton requires global knowledge for its computation. Thus *it is not* a morphological transformation and it is sensitive to noise, especially to small holes in an image object.

Proposition 6.8.4: Skeleton is antiextensive (if $\{0\} \in B$), translation-invariant, and idempotent.

Proposition 6.8.5:

$$\text{If } |B| = 1 \Rightarrow SK(X) = \varnothing$$
$$\text{If } X \ominus B^s = \varnothing \Rightarrow SK(X) = X$$

This property is important because it shows that any zero thickness set is a skeleton itself.

Proposition 6.8.6: If $SK(X)$ is the skeleton of X using B as a structuring element, then B_z gives $[SK(X)]_{-z}$ as the skeleton of X.

Proposition 6.8.7: The skeleton subsets are disjoint.

Proposition 6.8.8: If the skeleton has been obtained by using a convex and bounded structuring element B, $(nB)_z$ is the maximal inscribable structuring element in X, iff $z \in S_n(X)$.

Proposition 6.8.9: If B is bounded and convex, X is equal to X_{kB} iff $S_n(X) = \varnothing$, $n = 0,1,..,k-1$.

Proposition 6.8.8 is important because it justifies the fact that the skeleton given by (6.8.4-5) is the locus of the centers of the maximal inscribable disks in X. Proofs of all previously mentioned properties can be found in [24,53].

Skeletons defined by (6.8.4-6) over a rectangular grid are translation and scale invariant but not rotation invariant, even in the case when the CIRCLE structuring element is employed. Therefore, a generalized definition of skeletons has been proposed [43] to accommodate rotation invariance:

$$SK(X) = \bigcup_{n=0}^{N} S_n(X) \qquad (6.8.14)$$

$$S_n(X) = (X \ominus B(n)^s) - (X \ominus B(n)^s)_{B_n(1)}$$

$$N = \max\{n : X \ominus B(n) \neq \varnothing\}$$

where $B(n)$ is a structuring element of size n and $B_n(1)$ is a structuring element related to $B(n)$ and having size 1. Several skeleton subclasses can be derived from (6.8.14) by specifying the types of structuring elements $B(n)$, $B_n(1)$ used. For example, if $B(n)$ is equal to nB of (6.8.3) and $B_n(1)=B$, the skeleton definition (6.8.4) results. If a multitude of structuring elements is used, the structuring element of size n can be defined as follows:

$$B(n) = \begin{cases} B_1 \oplus B_2 \oplus \cdots \oplus B_m & n \leq m \\ pB(m) \oplus B(q) & n = pm + q > m \,, \, q < m \end{cases} \qquad (6.8.15)$$

$$B_n(1) = \{B_i : i = (n-1) \bmod m + 1\}$$

The resulting skeleton (6.8.14) is the so-called *periodic uniform-step distance (PUSD) skeleton*. PUSD skeleton is more isotropic than the original skeleton (6.8.4), if the structuring elements are chosen properly [43]. If the structuring elements in (6.8.15) are chosen in such a way that they approximate a disk, the resulting skeleton (6.8.14) is approximately rotation invariant and it is called *pseudo-Euclidean skeleton*. A systematic method to generate quasi-circular structuring elements of all sizes has been found [52]. They can always be decomposed into small structuring elements of size 2 or 3 pixels, as is seen in Figure 6.8.7. Therefore, the computation of the pseudo-Euclidean skeleton can be done in a way similar to the one described by (6.8.7-9). An example of the rotation invariance properties of the pseudo-Euclidean skeleton is shown in Figure 6.8.8. Figures 6.8.8a and 6.8.8b show the skeleton (6.8.4) and the pseudo-Euclidean skeleton, respectively, of an airplane figure. The skeletons of the airplane figure rotated by 210 degrees is shown in Figures 6.8.8c and 6.8.8d, respectively. The pseudo-Euclidean skeleton shows better rotation invariance than the classical skeleton definition (6.8.4).

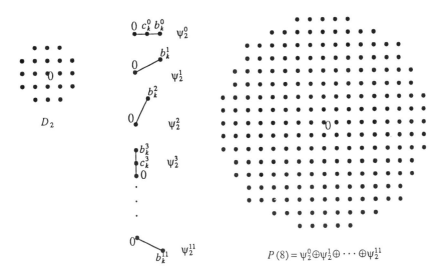

Figure 6.8.7: Decomposition of a disk of size 8 to structuring sets having size 2 or 3 pixels.

Sometimes skeletons carry more information for an object than what is needed for certain applications e.g., for object recognition. In those cases it is possible to do only partial object reconstruction by discarding some skeleton subsets that describe either details or object regions which are also described by

other skeleton subsets having larger skeleton values. Such a partial object reconstruction is shown in Figure 6.8.9 for the airplane figure [44].

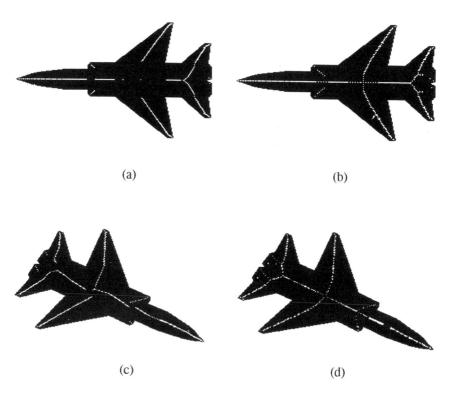

(a)   (b)

(c)   (d)

Figure 6.8.8: (a) Uniform skeleton of the object AIRPLANE; (b) Pseudo-
Euclidean skeleton of the object AIRPLANE; (c) Uniform skeleton of the object
AIRPLANE rotated by 210 degrees; (d) Pseudo-Euclidean skeleton of the object
AIRPLANE rotated by 210 degrees.

A generalization of the concept of the skeleton to graytone images has
already been suggested by various researchers [34,35,36]. As will be seen, the
generalization will not be straightforward since there exist certain asymmetries
between the binary and grayscale images. Let $g(x)$ be a structuring function of
finite support and ng the n-fold Minkowksi addition [53]:

$$ng = g \oplus g \oplus \cdots \oplus g \tag{6.8.16}$$

The grayscale *skeleton subfunctions* $S_n(f)$ of a function f are given by:

$$S_n(f) = (f \ominus ng^s) - (f \ominus ng^s)_g \tag{6.8.17}$$

where - denotes algebraic difference. It should be noted that the definition (6.8.17) is analogous with definition (6.8.4). If g has the simple form (6.6.1), this definition is equivalent to the "min-max" approach to grayscale skeletonization [34]. Given the skeleton subfunctions $S_n(f)$, there are two ways to define the image skeleton:

$$SK_{sum}(f)(x) = \sum_{n=0}^{N} [S_n(f)](x) \quad x \in \mathbf{Z}^n \tag{6.8.18}$$

$$SK_{max}(f)(x) = \max_{n=0}^{N} \{[S_n(f)](x)\} \quad x \in \mathbf{Z}^n \tag{6.8.19}$$

Figure 6.8.9: Partial reconstruction of the AIRPLANE by using some skeleton subfunctions.

An example of a 1-D skeleton obtained by using a 5-point structuring set $G = \{-2, -1, 0, 1, 2\}$ is shown in Figure 6.8.10. However, the full knowledge of each skeleton subfunction $S_n(f)$, $n = 1, ..., N$ is required for the reconstruction of the original function:

$$f = [...[[S_N(f) \oplus g] + S_{N-1}(f)] \oplus g + \cdots] \oplus g + S_0(f) \tag{6.8.20}$$

$$\neq \sum_{n=0}^{N} [S_n(f) \oplus ng]$$

The last inequality stems from the fact that function dilation does not generally commute with function addition:

$$(f + g) \oplus h \le (f \oplus h) + (g \oplus h) \tag{6.8.21}$$

Therefore, reconstruction (6.8.20) is not analogous with the binary image reconstruction (6.8.6). Modifications of the skeleton definition so that a reconstruction scheme similar to that of (6.8.20) exists can be found in [53].

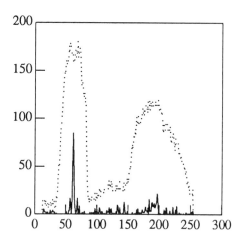

Figure 6.8.10: Skeleton $SK_{sum}(f)$ (solid curve) of an 1-D function (dotted curve).

Skeletons have already been used in several applications, e.g., in biological shape description, pattern recognition, industrial inspection, quantitative metallography, and image coding. Block-Huffman coding, run-length coding, or Elias coding schemes can be used for the coding of the skeleton subsets and Huffman code can be used for the coding of the skeleton function $skf(z)$ [24]. Compression rates up to 11:1 can be obtained by such a skeleton coding scheme. A combination of threshold decomposition and binary skeletonization has also been proposed for grayscale image coding [60].

6.9 MORPHOLOGICAL SHAPE DECOMPOSITION

Shape decomposition is a very common technique in object description [50]. It decomposes a "complex" object X into a union of "simple" subsets $X_1,...,X_n$. Such a decomposition must have the following desirable properties:

1.   It should conform with our intuitive notion of simple components of a complex object.

2.   It should have a well-defined mathematical characterization.

3.   Its characterization must be object independent.

4.   The complexity of the representation must be comparable with the complexity of the original description of X.

5.   It must be invariant under translation, scaling, and rotation.

6.   It should allow arbitrary amounts of detail to be computed and also allow abstraction from detail.

7.   It should be fast and unique.

8.   It should be stable under noise.

Usually the simple components are convex polygons [50]. However, in this case the object decomposition does not always correspond to our notion of "simple" components. Another approach to the problem could be the following. The "body" X_1 of the object X is found. This can be easily described in terms of either the maximal inscribable disk or square or rectangle or triangle in the object. The "body" X_1 is subsequently subtracted from the object X. The maximal inscribable disk or square or rectangle or triangle X_2 in the object $X - X_1$ is found. This process is repeated until all the details of the object X are described. This process can be easily described in the notation of mathematical morphology. Let B be the structuring element that represents the geometric primitive to be used in the description of X. The maximal inscribable element B in X has the form $X_1 = X_{r_1 B}$ and it has size r_1. It can be found by eroding the object X by B many times, until it vanishes. The number of erosions is the size r_1 of the maximal inscribable element B in X. This processes can be repeated for $X - X_1$. The whole procedure is called *morphological shape decomposition* and it can be described by the following recursive formula [45,46]:

$$X_i = (X - X'_{i-1})_{r_i B} \tag{6.9.1}$$

$$X'_i = \bigcup_{j=1}^{i} X_j$$

$$X'_0 = \varnothing$$

Stopping condition : $(X - X'_K) \ominus B^s = \varnothing$

Morphological shape decomposition decomposes an object X in a series of simple objects $X_1,...,X_n$. The object X can be reconstructed from its decomposition as follows:

$$X'_i = \bigcup_{j=1}^{K} X_j \tag{6.9.2}$$

The simple objects are of the form:

$$X_i = L_i \oplus r_iB \tag{6.9.3}$$

where L_i is either a point or a line. The objects described by (6.9.2) are the so-called *Blum ribbons* [51]. Such a Blum ribbon is shown in Figure 6.9.1. The line L is called the *spine* of the Blum ribbon.

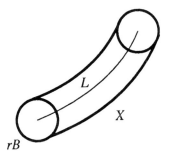

Figure 6.9.1: Example of a Blum ribbon.

Morphological shape decomposition is unique, translation and scale invariant, antiextensive, and idempotent. If the disk is used as a structuring element it is also rotation invariant. The morphological decomposition of the binary image POT, shown in Figure 6.9.2a, by using the CIRCLE and the SQUARE structuring elements is shown in Figures 6.9.2b and 6.9.2c, respectively. Morphological shape decomposition creates a tree structure for the description of an object X. The leaves of this tree are the simple components of the the object. This morphological shape decomposition tree has been successfully used for binary object recognition [48].

Usually objects can not be represented well by using one geometric primitive. The use of a multitude of structuring elements $S = \{B_1,..,B_N\}$ e.g., square, disk, triangle, gives much more descriptive power in the object representation scheme. In this case, the maximal inscribable sets r_jB_j, $j=1,..,N$ in the object X are found. Let $m(X)$ be a measure of the object X, e.g. the area of X. The maximal inscribable element r_iB_i that minimizes $m(X-r_iB_i)$ describes the object body in the best way and it can be used as a representation of the object body. The same procedure can be repeated for the object $X-r_iB_i$. The whole procedure is described by the following recursive relation [47,55]:

$$X_i = (X - X'_{i-1})_{r_iB_i} \tag{6.9.4}$$

$$r_iB_i = \{ B_i \in S : m(X-X'_{i-1}-r_iB_i) \qquad \text{is minimal} \}$$

$$X'_i = \bigcup_{j=1}^{i} X_j$$

$$X'_0 = \varnothing$$

The morphological representation of the POT by using two structuring elements, namely the CIRCLE and the SQUARE, is shown in Figure 6.9.2d. Another morphological decomposition scheme has been proposed recently for binary shape recognition [73]. It decomposes a binary object into triangles.

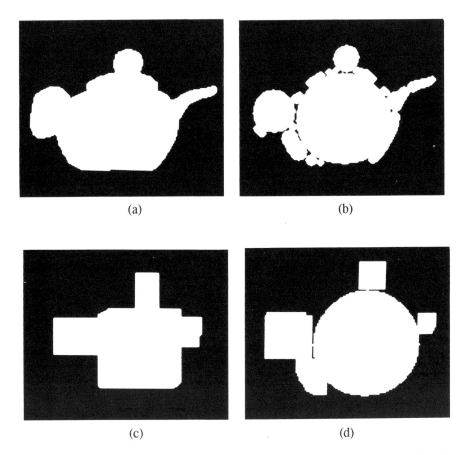

(a)   (b)

(c)   (d)

Figure 6.9.2: (a) POT; (b) Morphological decomposition of POT in a union of non-overlapping disks; (c) Morphological decomposition of POT in a union of non-overlapping squares; (d) Morphological decomposition of POT in a union of non-overlapping squares and disks.

Finally, mathematical morphology has been used for shape representation of

two-dimensional and three-dimensional objects as an extension of the Construc-
tive Solid Geometry (CSG) which is extensively used in graphics [47,55]. A
similar shape representation scheme is also proposed in [74].

The notion of decomposition can be easily extended to the morphological
analysis of multivalued signal and grayscale images. In this case, a signal $f(x)$
can be decomposed in simple functions of the form:

$$f_i(x) = [l_i \oplus rg](x) \tag{6.9.5}$$

where $g(x)$ is a structuring function and $l_i(x)$ is a function having constant value
$l > 0$ in its region of support L_i. A structuring function and a simple function are
shown in Figures 6.9.3 and 6.9.4, respectively.

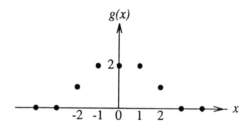

Figure 6.9.3: Structuring function.

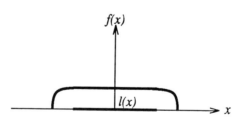

Figure 6.9.4: Simple function.

A structuring function $rg(x)$ is called maximal in the signal $f(x)$ if :

$$f(x) \ominus (r+dr)g^s(x) < 0, \qquad \forall x \in D \tag{6.9.6}$$

Having defined these two basic notions, the morphological analysis of $f(x)$ is
given by the following recursive formula:

$$f_i(x) = (f - f'_{i-1})_{r_i g}(x) \qquad (6.9.7)$$

$$f'_i = \sum_{j=1}^{i} f_j(x)$$

$$f'_0(x) = 0$$

$$[(f - f'_K) \ominus g](x) < 0 \qquad \forall x \in D$$

The function $f(x)$ can be easily reconstructed from its simple components:

$$f(x) = \sum_{j=1}^{K} f_j(x) \qquad (6.9.8)$$

An example of the morphological signal representation of the signal of Figure 6.9.5a is shown in Figure 6.9.5b. The structuring function used is shown in Figure 6.9.3. A partial reconstruction of this function by using only its first six simple components is shown in Figure 6.9.5b.

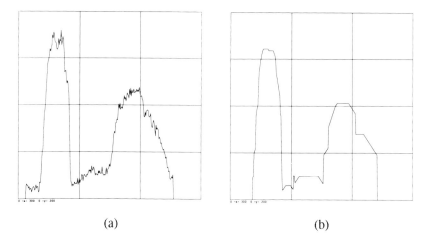

(a)   (b)

Figure 6.9.5: (a) Original function; (b) Partial reconstruction of the original function by using its first six simple components.

6.10 THINNINGS AND THICKENINGS

Thinnings and thickenings are operations that are closely related to the notion of skeletonization. Their relation to the digital skeletons has been already studied in [37], [38]. Later on their relation to the morphological operations has been investigated [1, p.390],[5]. In the following a description of thinning and thickening in terms of mathematical morphology will be given. Let X be an image object and T a compact structuring element consisting of two subsets T_1, T_2, i.e., $T=\{T_1,T_2\}$. *Thinning* $X \bigcirc T$ of X by T is the set subtraction:

$$XOT = X - X \circledcirc T = X \cap [(X \ominus T_1^s) \cap (X^c \ominus T_2^s)]^c \qquad (6.10.1)$$

where $X \circledcirc T$ is the hit-or-miss transformation (6.2.43):

$$X \circledcirc T = (X \ominus T_1^s) - (X \oplus T_2^s) = (X \ominus T_1^s) \cap (X^c \ominus T_2^s) \qquad (6.10.2)$$

Its dual operation is *thickening*:

$$X \circledbullet T = X \cup (X \circledcirc T) \qquad (6.10.3)$$

Thinning and thickening are dual to each other by complementation:

$$(X \circledbullet T)^c = X^c O T^* \qquad (6.10.4)$$

where

$$T^* = \{T_2, T_1\}.$$

Definitions (6.10.1-2) have a meaning if the subsets T_1, T_2 do not intersect each other, i.e., if $T_1 \cap T_2 = \varnothing$. If they intersect each other, both thinning and thickening are equivalent to X:

$$T_1 \cap T_2 \neq \varnothing \Rightarrow XOT = X \circledbullet T = X \qquad (6.10.5)$$

If T_2 contains the origin, the thinning XOT coincides to X. If T_1 contains the origin, the thickening $X \circledbullet T$ coincides with X:

$$\{0\} \in T_2 \Rightarrow XOT = X \quad \text{for every } T_1 \qquad (6.10.6)$$

$$\{0\} \in T_1 \Rightarrow X \circledbullet T = X \quad \text{for every } T_2 \qquad (6.10.7)$$

The two parts of the structuring element T are chosen to be disjoint and not to contain the origin (in T_1 for thickening and in T_2 for thinning) in order to avoid trivial results. By definition, the thickenings are extensive and the thinning antiextensive:

$$T' \subset T \Rightarrow XOT' \subset XOT \subset X \subset X \circledbullet T \subset X \circledbullet T' \qquad (6.10.8)$$

where the set inclusion $T' \subset T$ means also subset inclusion: $T_1' \subset T_1$ and $T_2' \subset T_2$.

Thinnings and thickenings are translation and scale invariant since their definitions (6.10.1), (6.10.2) employ erosion which is translation and scale invariant. Furthermore the operations in their definition require only local knowledge of object X since they are local operations. The semicontinuity of erosion implies that, for a closed set X and compact structuring elements T, the thinning XOT is an upper semicontinuous transformation [1]. The thickening $X \circledbullet T$ is a lower semicontinuous transformation if X is an open set. Therefore, thinning and thickening are morphological transformations.

An extension of thinning is the *sequential thinning*. Let $T_i=\{T_{1i},T_{2i}\}$ be a sequence of structuring elements. The sequential thinning is defined as [5]:

$$...(...((X \text{O} T_1) \text{O} T_2) \cdots \text{O} T_i) \cdots = X \text{O} \{T_i\} \tag{6.10.9}$$

It can be proven that if T_i is infinite, $X \text{O} \{T_i\}$ is idempotent. If T_{i+1} is derived from T_i by a rotation of 60 degrees, the corresponding sequence is called *standard* and the sequential thinning (6.10.9) is denoted by $X \text{O} \{T\}$.

6.11 GRANULOMETRIES AND THE PECSTRUM

The fundamental approach of morphological image analysis is to transform an image object X by a morphological transformation ψ and then to measure the transformed image:

$$X \rightarrow \psi(X) \rightarrow m(\psi(X)) \tag{6.11.1}$$

This approach is clearly illustrated in Figure 6.1.1. One of the most basic measures of the transformed image $\psi(X)$ is the area of $\psi(X)$. Usually a family of transformations $\psi_r(X)$ is used, where r is a parameter. Such a family consists of the openings X_{rB} of object X with a structuring element rB of size r. The mapping

$$r \in \mathbf{R} \rightarrow X_{rB} \tag{6.11.2}$$

is called *granulometry*. Matheron [3] has introduced a general theory of granulometries, with axiomatic basis, whose special case is the opening (6.11.2). However, the present discussion will be limited to (6.11.2) because it is very interesting from a practical point of view. A consequence of (6.11.2) is that there exists a real-valued function $u(r)$ that gives the measure of X_{rB}:

$$r \in \mathbf{R} \rightarrow u(r) = m(X_{rB}) \in \mathbf{R} \tag{6.11.3}$$

This function is called *size distribution* of X_{rB}. From a physical point of view, the size distribution of X gives the area of X which can be covered by a disk rB of radius r, when this disk moves *inside* the object X. The greater the radius is, the smaller the area. Therefore, the size distribution $u(r)$ is a strictly monotone function. Furthermore, $u(r)$ depends greatly on the shape of the structuring element B. However, the interest lies in the fact that different objects X give different size distributions, when opened by the same structuring element B. Therefore, granulometries can be used for pattern recognition, as will be seen later on. From another point of view, granulometric processes can be considered as a type of filtering that removes the sections of the image object X, which are not sufficiently large to include the structuring element rB. Such an interpretation of granulometries is shown in Figure 6.11.1. Therefore, a granulometry can be thought as a one-parameter family of low-pass filters.

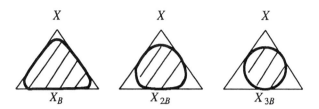

Figure 6.11.1: Opening of a set X by disks of increasing size.

Granulometries $\psi_r(X)$ possess several useful algebraic properties. Some of them will be repeated here without proof, when $\psi_r(X)=X_{rB}$ [9, pp. 120, 163]:

(1) $\psi_r(X) \subset X$ for any $r > 0$.

(2) $r \geq s$ implies that $\psi_r(X) \subset \psi_s(X)$.

(3) $A \subset X$ implies that $\psi_r(A) \subset \psi_r(X)$.

       This property simply states that the granulometry X_{rB} is increasing transformation.

(4) $\psi_r[\psi_s(X)] = \psi_s[\psi_r(X)] = \psi_{\max(r,s)}(X)$ (6.11.4)

       This operation states that opening by an iteration of sB and rB is equivalent to opening by tB, where t is the maximum of r,s. The properties (2)-(4) are the axioms that must be satisfied by every granulometry.

(5)   Granulometry X_{rB} is translation and scale invariant:

$$\psi_r(X_z) = [\psi_r(X)]_z \qquad\qquad\qquad (6.11.5)$$

$$\psi_r(X) = \lambda\psi_r\left(\frac{1}{\lambda}X\right) \qquad\qquad\qquad (6.11.6)$$

Granulometries satisfying (6.11.5-6) are called *Euclidean granulometries*.

The size distribution $u(r)$ is a strictly monotonically decreasing function. Therefore, its normalized negative derivative $f(r)$ can be used instead:

$$f(r) = \frac{-\dfrac{d}{dr}[m(\psi_r(X))]}{m(X)} \qquad\qquad\qquad (6.11.7)$$

The division by the area $m(X)$ of X is a simple scaling, because $m(X)$ does not depend on r. Function $f(r)$ is called *pattern spectrum* or *pecstrum* [39,40,53,54,56,81]. An example of the pecstrum is shown in Figure 6.11.2. The structuring element B used in this example was a square of unit dimensions. The object shown in Figure 6.11.2a has simple shape. Its pecstrum is a delta function as can be seen in Figure 6.11.2c. The object shown in Figure 6.11.2d has a more complex shape. Its pecstrum has another delta function at a smaller value of r, as can be seen in Figure 6.11.2f. This component of the pecstrum is due to the "jagginess" of the object contour.

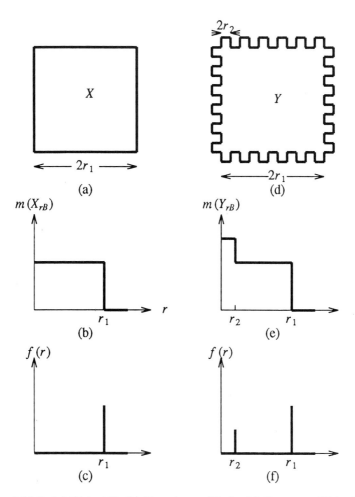

Figure 6.11.2: (a) Object X; (b) Function $m(X_{rB})$; (c) Pecstrum $f(r)$ of X; (d) Object Y; (e) Function $m(Y_{rB})$; (f) Pecstrum $f(r)$ of Y.

A digital version of the pecstrum is the following:

$$f(n) = \frac{m(\psi_n(X)) - m(\psi_{n+1}(X))}{m(X)} = \frac{m(X_{nB}) - m(X_{(n+1)B})}{m(X)} \qquad (6.11.8)$$

There exist two conceptual similarities between frequency spectrum and pecstrum:

(1)   The spectrum is obtained by transforming the signal through a multiplication by a complex exponential $e^{i\omega t}$ of a certain frequency ω, whereas the pecstrum is obtained by transforming the signal by taking its opening with respect to a certain structuring element.

(2)   The magnitude of the frequency spectrum of a signal at a frequency ω is the *spectral content* of the signal in this frequency, whereas the pecstrum gives the pattern content of the signal.

The area under the pecstrum is one:

$$\int_0^\infty f(r)dr = 1 \tag{6.11.9}$$

This property can be easily proved by using the definition (6.11.7). It can also be proven by using Proposition 6.8.9 and (6.11.8) that if an object X has the k minimal skeleton subsets empty, its first k size pecstrum samples are zero [56]. Rotation and size invariant modifications of the pecstrum are described in [61].

Pecstra can be used for pattern recognition [40]. If $f_R(n)$ is a reference pecstrum and $f(n)$ is the pecstrum of a new object X, the distance d:

$$d = \left[\sum_{n=0}^{N-1} c_n (f(n) - f_R(n))^2 \right]^{1/2} \tag{6.11.10}$$

can be used to decide if the new object coincides with a reference pattern.

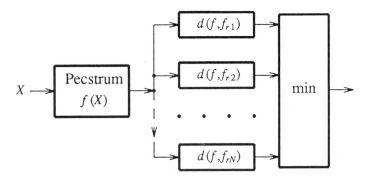

Figure 6.11.3: Pattern recognition scheme using comparisons of the pecstrum of an object with reference pecstra.

Such a pattern recognition scheme is shown in Figure 6.11.3. The weights c_n can be chosen to be equal to unity, if no specific parts of pecstrum differences are emphasized. The weights c_n can also be used to emphasize certain parts of the object pecstrum. If the large differences from the reference pecstrum are to be emphasized, the weights can be chosen as follows:

$$c_n = exp\,(a\,[f\,(n){-}f_R(n)]^2)\,, \qquad a > 0 \qquad\qquad (6.11.11)$$

or

$$c_n = exp\,(a\,|f\,(n){-}f_R(n)|)\,, \qquad a > 0 \qquad\qquad (6.11.12)$$

The choice of the appropriate value of the parameter a is usually application dependent.

6.12 DISCUSSION

Mathematical morphology comes from a completely different mathematical background than the rest of the digital signal and image processing techniques. Thus it has found many applications in the cases where the description of geometrical and topological information about images or objects is important. Such cases usually are found in image analysis applications. However, it has been proven that mathematical morphology techniques can also be applied in digital image processing. The use of mathematical morphology in image processing has been greatly enhanced by extending the notions of morphology from binary images to multivalued signals and images (grayscale morphology). Another reason for the recent popularity of mathematical morphology is that it employs simple mathematical operations which can be programmed easily and computed very fast on any computer. Furthermore, the morphological operations can be easily implemented in parallel or pipelined architectures, thus tremendously increasing the speed of their execution. Such architectures are described in a subsequent chapter.

Mathematical morphology is one of the most active research areas in non-linear image processing. This chapter covers most of the basic theory of mathematical morphology as well as several recent research results. The interested reader can find more theoretical results in a recent book edited by Serra [6] and in a special issue on mathematical morphology in *Signal Processing* [75-80]. More information on the skeleton properties can be found in [6], [77]. Extensions of mathematical morphology to general image algebras can be found in [68]. Its use in L_1 norm image matching and correlation is described in [62]. Its application in the texture analysis of seismic images, in the computation of non-planar point neighborhoods on cellular automata and in the 3-d scene segmentation can be found in [69, 70, 71], respectively. The processing of graphs by mathematical morphology tools is addressed in [78]. Finally, speckle noise removal and image modelling by mathematical morphology are described in [76, 80], respectively.

REFERENCES

[1]   J. Serra, *Image analysis and mathematical morphology*, Academic Press, 1983.

[2]   James and James, *Mathematical dictionary*, Van Nostrand, 1976.

[3]   G. Matheron, *Random sets and integral geometry*, J. Wiley, 1975.

[4]   R.M. Haralick, S.R. Sternberg, X. Zhuang, "Image analysis using mathematical morphology", *IEEE Transactions on Pattern Analysis and Machine Intelligence*, vol. PAMI-9, no. 4, pp. 532-550, July 1987.

[5]   J. Serra, "Introduction to mathematical morphology", *Computer Vision, Graphics and Image Processing*, vol. 35, pp. 283-305, 1986.

[6]   J. Serra (editor), *Image analysis and mathematical morphology: theoretical advances*, vol. 2, Academic Press, 1988.

[7]   H. Minkowski, "Volumen und oberflaechen", *Math. Annalen*, vol. 57, pp. 447-495, 1903.

[8]   C.R. Giardina, E.R. Dougherty, *Morphological methods in image and signal processing*, Prentice Hall, 1988.

[9]   E.R. Dougherty, C.R. Giardina, *Image processing - continuous to discrete*, vol. 1, Prentice-Hall, 1987.

[10] V. Goetcherian, "From binary to gray level tone image processing using fuzzy logic concepts", *Pattern Recognition*, vol. 12, pp. 7-15, 1980.

[11] C. Lantuejoul, J. Serra, "M-filters", *Proc. IEEE Int. Conf. on ASSP*, pp. 2063-2066, 1982.

[12] F. Meyer, "Contrast feature extraction", in *Quantitative Analysis of Microstructures in Material Sciences, Biology and Medicine*, Special issue of Practical Metallography, J.L. Charmont editor, Reidner-Verlag, 1977.

[13] A. Rosenfeld, "Connectivity in digital pictures", *J. Assoc. Comput. Mach.*, vol. 17, no. 1, pp. 146-160, 1970.

[14] S.R. Sternberg, "Parallel architecture for image processing", *Proc. 3rd COMPSAC*, Chicago, 1979.

[15] J. Serra, *Morphologie pour les fonctions "a peu pres en tout ou rien"*, Tech. report Centre de Morphologie Mathematique, Fontainebleau, France, 1975.

[16] C. Lantuejoul, *Sur le modele de Johnson-Mehl generalise*, Tech. report Centre de Morphologie Mathematique, Fontainebleau, France, 1977.

[17] S.R. Sternberg, "Biological image processing", *Computer*, pp. 22-34, Jan. 1983.

[18] S.R. Sternberg, "Grayscale morphology", *Computer Vision, Graphics and Image Processing*, vol. 35, pp. 333-355, 1986.

[19] R.L. Stevenson, G.R. Arce, "Morphological filters: statistics and further syntactic properties", *IEEE Transactions on Circuits and Systems*, vol. CAS-34, pp.1292-1305, Nov. 1987.

[20] F. Meyer, "Iterative image transformations for an automatic screening of cervical smears", *J. Histoch. Cytochem.*, vol. 27, pp. 128-135, 1979.

[21] F. Meyer, "Automatic screening of cytological specimens", *Computer Vision, Graphics and Image Processing*, vol. 35, pp. 356-369, 1986.

[22] S.D. Pass, "Segmentation by shape discrimination using spatial filtering techniques", in *Digital Image Processing*, J.C. Simon and R.M. Haralick editors, D. Reidel Publ. Co., 1981.

[23] H.P. Kramer, J.B. Bruckner, "Iterations of nonlinear transformation for enhancement of digital images", *Pattern Recognition*, vol. 7, 1975, pp. 53-58.

[24] P.Maragos, R.W. Schafer, "Morphological skeleton representation and coding of binary images", *IEEE Transactions on Acoustics, Speech and Signal Processing*, vol. ASSP-34, no.5, pp. 1228-1244, Oct. 1986.

[25] H. Blum, "A transformation for extracting new descriptors of shape", *Models for the Perception of Speech and Visual Forms*, W. Wathen-Dunn editor, pp. 362-380, MIT Press, 1967.

[26] H. Blum, "Biological shape and visual sciences (part I)", *J. Theoret. Biology*, vol. 38, pp. 205-287, 1973.

[27] J.C. Kotelly, "A mathematical model of Blum's theory of pattern recognition", Rep. 63-164, Airforce Cambridge Res. Labs, Mass., 1963.

[28] L. Calabi, "A study of the skeleton of plane figures", Res. SR20-60429, Parke Mathematical Labs., Mass., 1965.

[29] A. Rosenfeld, J.L. Pfalz, "Sequential operations in digital picture processing", *J. Assoc. Comput. Mach.*, vol. 13, pp. 471-494, Oct. 1966.

[30] J.C. Mott-Smith, "Medial axis transformations", in *Picture Processing and Psychopictorics*, B.S. Lipkin and A. Rosenfeld editors, Academic,

1970.

[31] U. Montanari, "A method for obtaining skeletons using a quasi-Euclidean
      distance", *J. Assoc. Comput. Mach.*, vol. 15, pp. 600-624, Oct. 1968.

[32] C. Lantuejoul, *La squelettisation et son application aux mesures topolo-
      giques des mosaiques polycristallines*, These de Docteur-Ingenieur, Ecole
      de Mines, Paris, 1978.

[33] C. Lantuejoul, "Skeletonization in quantitative metallography", in *Issues
      of Digital Image Processing*, R.M. Haralick and J.C. Simon editors, Sitjh-
      off and Noordhoff, 1980.

[34] S. Peleg, A. Rosenfeld, "A min-max medial axis transformation", *IEEE
      Trans. on Pat. Anal. Machine Intelligence*, vol. PAMI-4, no. 4, pp. 419-
      421, July 1982.

[35] N. Ahuja, L.S. Daris, D.L. Milgram, A. Rosenfeld, "Piecewise approxi-
      mation of pictures using maximal neighborhoods", *IEEE Trans. Comput.*,
      vol. C-27, pp. 375-379, 1978.

[36] S. Lobregt, P.W. Verbeek, F.C.A. Groen, "Three dimensional skeletoniza-
      tion: principles and algorithms", *IEEE Transactions on Pattern Analysis
      and Machine Intelligence*, vol. PAMI-2, no.1, pp.75-77, Jan. 1980.

[37] A. Rosenfeld, A.C. Kak, *Digital picture processing*, Academic Press,
      1976.

[38] S. Levialdi, "Parallel pattern processing", *IEEE Trans. System, Man and
      Cybern.*, SMC-1, pp. 292-296, 1971.

[39] J.F. Bronskill, A.N. Venetsanopoulos, "The pecstrum", *3rd ASSP
      Workshop on Spectral Estimation and Modelling*, Boston, 1986.

[40] J.F. Bronskill, A.N. Venetsanopoulos, "Multidimensional shape recogni-
      tion using mathematical morphology", *Proc. Int. Workshop on Time-
      Varying Image Processing and Moving Object Recognition*, Florence,
      Italy, pp. 3-18, 1986.

[41] P.Maragos, R.W. Schafer, "Morphological filters, part I: their set
      theoretic analysis and relations to linear shift invariant filters", *IEEE
      Transactions on Acoustics, Speech and Signal Processing*, vol. ASSP-35,
      no.8, pp. 1153-1169, Aug. 1987.

[42] P.Maragos, R.W. Schafer, "Morphological filters, part II: their relations
      to median, order statistic and stack filters", *IEEE Transactions on Acous-
      tics, Speech and Signal Processing*, vol. ASSP-35, no.8, pp. 1170-1184,
      Aug. 1987.

[43] Z. Zhou, A.N. Venetsanopoulos, "Analysis and implementation of mor-
      phological skeleton transforms", *IEEE Transactions on Acoustics, Speech
      and Signal Processing*, in press.

[44] Z. Zhou, A.N. Venetsanopoulos, "Morphological skeleton representation and shape recognition", *Proc. IEEE Int. Conf. on Acoustics, Speech and Signal Processing*, pp. 948-951, New York, 1988.

[45] I. Pitas, A.N. Venetsanopoulos, "Morphological shape decomposition", *IEEE Transactions on Pattern Analysis and Machine Intelligence*, in press, Jan. 1990.

[46] I. Pitas, A.N. Venetsanopoulos, "Morphological shape decomposition", *Proc. IEEE Int. Conf. on Computer Vision*, London, 1987.

[47] I. Pitas, A.N. Venetsanopoulos, "Morphological shape representation", *Computer Vision, Graphics and Image Processing*, under review.

[48] I. Pitas, N. Sidiropoulos, "Pattern recognition of binary image objects by using morphological shape decomposition", *Computer Vision, Graphics and Image Processing*, under review.

[49] I. Pitas, "Morphological signal analysis", *IEEE Transactions on Acoustics, Speech and Signal Processing*, under review.

[50] T. Pavlidis, "A review of algorithms for shape analysis", *Computer Graphics and Image Processing*, vol. 7, no.2, pp. 243-258, April 1978.

[51] A. Rosenfeld, "Axial representations of shape", *Computer Vision, Graphics and Image Processing*, vol. 33, pp. 156-173, 1986.

[52] Z. Zhou, A.N. Venetsanopoulos, "Pseudo-euclidean morphological skeleton transform for machine vision", *Proc. 1989 IEEE Int. Conf. on Acoustics, Speech and Signal Processing*, Glasgow, England, 1989.

[53] P. Maragos, *A unified theory of translation invariant systems with applications to morphological analysis and coding of images*, Ph.D. Thesis, Georgia Institute of Technology, 1985.

[54] J.F. Bronskill, A.N. Venetsanopoulos, "Multidimensional shape description and recognition using mathematical morphology", *Journal of Intelligent and Robotic Systems*, vol. 1, pp. 117-143, 1988.

[55] I. Pitas, A.N. Venetsanopoulos, "Morphological shape representation and recognition", *Proc. 3rd Workshop on Time-varying Image Processing and Object Recognition*, Florence, Italy, 1989.

[56] P. Maragos, "Pattern spectrum of images and morphological shape-size complexity", *Proc. 1987 IEEE Int. Conf. on Acoustics, Speech and Signal Processing*, pp. 241-244, Dallas, USA, 1987.

[57] K. Preston Jr., "Ξ filters", *IEEE Transactions on Acoustics, Speech and Signal Processing*, vol. ASSP-31, pp.861-876, Aug. 1983.

[58] F.Y.C. Shih, O.R. Mitchell, "Threshold decomposition of grayscale morphology into binary morphology", *IEEE Trans. on Pat. Anal. Machine*

Intelligence, vol. PAMI-11, no. 1, pp. 31-42, Jan. 1989.

[59] X. Zhuang, R.M. Haralick, "Morphological structuring element decomposition", *Computer Vision, Graphics and Image Processing*, vol.35, pp. 370-382, 1986.

[60] S.A. Rajala, H.A. Peterson, E.L. Delp, "Binary morphological coding of grayscale images", *Proc. IEEE Int. Symp. on Circuits and Systems*, pp. 2807-2811, Helsinki, Finland, 1988.

[61] M. Binaghi, V. Cappellini, C. Raspolini, "Description and recognition of multidimensional signals using rotation and scale invariant morphological transformations", *Proc. European Signal Processing Conference*, 1988.

[62] P. Maragos, "Optimal morphological approaches to image matching and object detection", *Proc. IEEE Int. Conf. on Computer Vision*, pp. 695-699, Tampa, Florida, 1988.

[63] J.S.J Lee, R.M. Haralick, L.G. Shapiro, "Morphological edge detection", *International Journal of Robotics and Automation*, vol. RA-3, no. 2, pp. 142-156, 1987.

[64] R.J. Feehs, G.R. Arce, "Multidimensional morphological edge detection", *SPIE Visual Computing and Image Processing II*, pp.285-292, 1987.

[65] J.A. Noble, "Morphological feature detection", *Proc. IEEE Int. Conf. on Computer Vision*, pp. 112-115, Tampa, Florida, 1988.

[66] W. Kohler, *Gestalt psychology*, Liveright Pub. Co., 1970.

[67] R.M. Haralick, X. Zhuang, C. Lin, J. Lee, "The digital morphological sampling theorem", *Proc. IEEE Int. Symp. on Circuits and Systems*, pp. 2789-2791, Helsinki, Finland, 1988.

[68] G.X. Ritter, J.L. Davidson, J.N. Wilson, "Beyond mathematical morphology", *SPIE Visual Computing and Image Processing II*, vol. 845, pp.260-268, 1987.

[69] K. Kotropoulos, I. Pitas, "Texture analysis and segmentation of seismic images", *Proc. 1989 IEEE Int. Conf. on Acoustics, Speech and Signal Processing*, pp. 1437-1440, Glasgow, England, 1989.

[70] M.M. Skolnik, S. Kim, R. O'Bara, "Morphological algorithms for computing nonplanar point neighborhoods on cellular automata", *Proc. IEEE Int. Conf. on Computer Vision*, pp. 106-111, Tampa, Florida, 1988.

[71] C.D. Brown, R.W. Marvel, G.R. Arce, C.S. Ih, D.A. Fertell, "Morphological 3-d segmentation using laser structured light", *Proc. IEEE Int. Symp. on Circuits and Systems*, pp. 2803-2805, Helsinki, Finland, 1988.

[72] C.-H. H. Chu, E.J. Delp, "Impulsive noise suppression and background normalization using morphological operators", *IEEE Transactions on Biomedical Engineering*, vol. BME-36, no. 2, pp. 262-273, Feb. 1989.

[73] Y. Zhao, R.M. Haralick, "Binary shape recognition based on automatic morphological shape decomposition", *Proc. IEEE International Conference on Acoustics, Speech and Signal Processing*, pp. 1691-1694, Glasgow, Scotland, 1989.

[74] P.K. Ghosh, "A mathematical model for shape description using Minkowski operators", *Computer Vision Graphics and Image Processing*, vol. 43, no. 3, pp. 239-269, Dec. 1988.

[75] F. Meyer, J. Serra, "Contrasts and activity lattice", *Signal Processing*, vol. 16, no. 4, pp. 303-317, April 1989.

[76] F. Safa, G. Flouzat, "Speckle removal on radar imagery based on mathematical morphology", *Signal Processing*, vol. 16, no. 4, pp. 319-333, April 1989.

[77] F. Meyer, "Skeletons and perceptual graphs", *Signal Processing*, vol. 16, no. 4, pp. 335-363, April 1989.

[78] L. Vincent, "Graphs and mathematical morphology", *Signal Processing*, vol. 16, no. 4, pp. 365-388, April 1989.

[79] M. Schmitt, "Mathematical morphology and artificial intelligence: an automatic programming system", *Signal Processing*, vol. 16, no. 4, pp. 389-401, April 1989.

[80] D. Jeulin, "Morphological modelling of images by sequential random functions" *Signal Processing*, vol. 16, no. 4, pp. 403-431, April 1989.

[81] P. Maragos, "Pattern spectrum and multiscale shape representation", *IEEE Transactions on Pattern analysis and Machine Intelligence*, vol. PAMI-11, no. 7, pp. 701-716, July 1989.

CHAPTER 7

HOMOMORPHIC FILTERS

7.1 INTRODUCTION

In many applications signals are combined in a rather complicated way. Convolved signals are encountered in seismic signal processing, digital speech processing, digital echo removal and digital image restoration. Signals combined in a nonlinear way are encountered in digital signal processing for communication systems and in digital image filtering. Classical linear processing techniques are not so useful in those cases because the superposition property does not hold any more. Therefore, a special class of filters has been developed for the processing of convolved and nonlinearly related signals. They are called *homomorphic filters*. Their basic characteristic is that they use nonlinearities (mainly the logarithm) to transform convolved or nonlinearly related signals to additive signals and then to process them by linear filters. The output of the linear filter is transformed afterwards by the inverse nonlinearity. Homomorphic filtering has found many applications in digital image processing. It is recognized as one of the oldest nonlinear filtering techniques applied in this area. The main reason for its application is the need to filter multiplicative and signal-dependent noise, whose form was described in chapter 3. Linear filters fail to remove such types of noise effectively. Furthermore, the nonlinearity (logarithm) in the human vision system suggests the use of classical homomorphic filters. Homomorphic filtering can also be used in image enhancement. As we saw in chapter 3, object reflectance and source illumination contribute to the image formation in a multiplicative way. Ideally, the source illumination is constant over the entire image. However, in many practical cases, e.g., in outdoor scenes, source illumination is not constant over the entire scene. Therefore, it can be modeled as noise in the low spatial frequencies. If this noise is removed, the object reflectance is enhanced. Homomorphic filtering has found various practical applications, e.g., in satellite image processing and in the identification of fuzzy fingerprints. Homomorphic filtering has also found several applications in other areas, e.g., in speech processing and in geophysical signal processing. In the following, the theory and several applications of homomorphic filters will be given.

7.2 ALGEBRAIC THEORY OF HOMOMORPHIC SIGNAL PRO-CESSING

One very important property of linear filters is the superposition property:

$$L [ax_1(i)+bx_2(i)] = aL [x_1(i)]+bL [x_2(i)] \qquad (7.2.1)$$

which allows them to process signals that have been added. However, sometimes signals are combined in other ways, e.g., in a multiplicative way:

$$y(i) = x(i)s(i) \qquad (7.2.2)$$

In this case the superposition property does not hold any more and the linear filters are of limited use. A useful idea is to map them to additive signals and then to process them in a linear way.

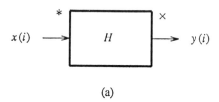

(a)

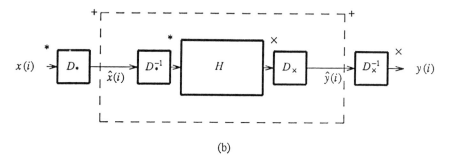

(b)

Figure 7.2.1: (a) Definition of a homomorphic system; (b) Canonical decomposition of a homomorphic filter.

Let $*$ and $\times$ be two operations in the *input signal space* and in the *output signal space*, respectively. Suppose that a homomorphic system H transforms the input signal space to the output signal space, as is shown in Figure 7.2.1.

Two isomorphisms $D_*, D_\times$ can be found, which transform the input and the output signal spaces to the standard additive signal space. Let also $D_*^{-1}, D_\times^{-1}$ be their inverse transformations. In this case the homomorphic system H can take the form of Figure 7.2.1b because the transformations are canceled by their inverses. The cascade of $D_*^{-1}, H, D_\times$ is a linear system that operates on the additive signal space.

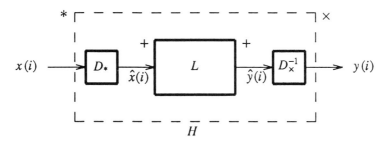

Figure 7.2.2: Canonical decomposition of a homomorphic system. Linear filter L determines the characteristics of the homomorphic system.

If this linear system is denoted by L, the homomorphic system takes its *canonical decomposition form*, shown in Figure 7.2.2. Let two signal vectors $x_1(i)$, $x_2(i)$ be combined by the operation *:

$$x(i) = x_1(i) * x_2(i) \tag{7.2.3}$$

If $x(i)$ passes through the homomorphic system H, it is transformed as follows:

$$\hat{x}(i) = D_*[x(i)] = D_*[x_1(i)] + D_*[x_2(i)] = \hat{x}_1(i) + \hat{x}_2(i) \tag{7.2.4}$$

The system L is linear. Therefore:

$$\hat{y}(i) = L[\hat{x}(i)] = L[\hat{x}_1(i) + \hat{x}_2(i)] = \hat{y}_1(i) + \hat{y}_2(i) \tag{7.2.5}$$

The output $y(i)$ of the homomorphic system H is given by:

$$y(i) = D_\times^{-1}[\hat{y}_1(i) + \hat{y}_2(i)] = y_1(i) \times y_2(i) \tag{7.2.6}$$

Thus the homomorphic system satisfies the following *generalized principle of superposition*:

$$H[x_1(i) * x_2(i)] = H[x_1(i)] \times H[x_2(i)] \tag{7.2.7}$$

In many image processing applications the operations *, $\times$ are just the usual pointwise vector multiplications:

$$x_1(i) * x_2(i) = x_1(i) x_2(i) = x_1(i) \times x_2(i) \tag{7.2.8}$$

In this case the functions D_*, D_x^{-1} are given by:

$$D_*(x) = \ln(x) \qquad\qquad (7.2.9)$$

$$D_x^{-1}(x) = e^x \qquad\qquad (7.2.10)$$

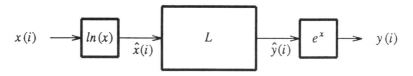

Figure 7.2.3: Homomorphic system using the logarithmic and exponential non-linearities.

This is the familiar homomorphic system shown in Figure 7.2.3. Another homomorphic system that is also useful in image processing applications is the following:

$$D_*(x) = D_x(x) = x^\gamma \qquad\qquad (7.2.11)$$

It is clear that, after the choice of the characteristic functions D_*, D_x, the filtering properties of the homomorphic filter lie in the appropriate choice of the linear filter L. In the next section the choice of such a filter L suitable for image processing applications will be discussed.

7.3 HOMOMORPHIC FILTERING IN IMAGE ENCHANCEMENT

It has been described in chapter 3 that object reflectance $r(x,y)$ and source illumination $E(x,y)$ contribute to the observed image in a multiplicative way:

$$i(x,y) = r(x,y)E(x,y) \qquad\qquad (7.3.1)$$

If the illumination is constant $E(x,y)=E$ over the entire scene, the observed image is linearly related to the reflectivity $r(x,y)$. However, illumination is usually uncontrollable, especially in outdoor scenes. Therefore, it can be described as a slowly varying random process. Its spectrum usually occupies the low spatial frequency regions. Linear techniques are not suitable for the removal of illumination effects because it is a multiplicative process. However, a logarithmic function can be used to transform multiplicative noise to additive one:

$$\ln[i(x,y)] = \ln[r(x,y)] + \ln[E(x,y)] \qquad\qquad (7.3.2)$$

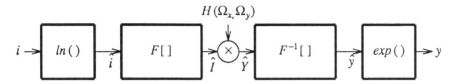

Figure 7.3.1: Homomorphic system for image enhancement.

A linear filter can be used to remove the illumination component, as is seen in Figure 7.3.1. The filter characteristics $H(\Omega_x,\Omega_y)$ can be easily described in the spatial frequency domain. Let F denote the two-dimensional Fourier transform. The illumination spectrum $E(\Omega_x,\Omega_y)$:

$$E(\Omega_x,\Omega_y) = F[\ln[E(x,y)]] \tag{7.3.3}$$

occupies the region of the low spatial frequencies, whereas the reflectance spectrum:

$$R(\Omega_x,\Omega_y) = F[\ln[r(x,y)]] \tag{7.3.4}$$

occupies the regions of higher frequencies. Furthermore, both illumination and reflectance spectra are assumed to be spatially isotropic.

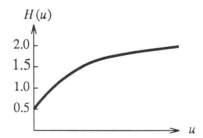

Figure 7.3.2: The radial cross-section of the frequency response of the linear part of the homomorphic filter for image processing.

Therefore, the filter characteristics must be high-pass. The following form is very suitable:

$$H(\Omega_x,\Omega_y) = \gamma \tag{7.3.5}$$

The coefficient γ must be greater than unity and smaller than unity in the low and high spatial frequency regions, respectively. The values $\gamma = \dfrac{1}{2}$ and 2 have been suggested [1,14]. However, a smooth transition from 0.5 to 2 is desirable. Therefore, a filter has been used whose radial characteristics are shown in Figure 7.3.2. The filter operates on the frequency domain:

$$\hat{Y}(\Omega_x,\Omega_y) = H(\Omega_x,\Omega_y)\hat{I}(\Omega_x,\Omega_y) \qquad (7.3.6)$$

where $\hat{I}(\Omega_x,\Omega_y)$ denotes the Fourier transform of $\ln[i(x,y)]$. The homomorphic filter output is given by:

$$y(x,y) = exp[F^{-1}(\hat{Y}(\Omega_x,\Omega_y))] \qquad (7.3.7)$$

The output of the homomorphic filter can be roughly approximated as follows:

$$y(x,y) \simeq E(x,y)^{\frac{1}{2}} r(x,y)^2 \qquad (7.3.8)$$

Therefore the reflectance is enhanced, whereas the illumination is attenuated.

(a)   (b)

Figure 7.3.3: (a) Original image; (b) Image processed by a homomorphic filter. The image contrast has increased.

An example of homomorphic filtering is shown in Figure 7.3.3. The output of the homomorphic filter has higher contrast than the original image. Thus, it is more pleasant to the human eye. The above interpretation of the performance of the homomorphic filter has been challenged in [3]. However, we think that the high-pass characteristics of the filter are responsible for its pleasant output, even in the absence of the illumination term.

Homomorphic filtering can be used for the removal of multiplicative noise of any kind [1, 17,18]:

$$i(x,y) = f(x,y)n(x,y) \qquad (7.3.9)$$

The same structure of Figure 7.2.3 can be used. The linear filter L can be a classical low-pass filter or a Wiener filter [4]. Both can be implemented either in the frequency domain or in the space domain. The frequency response of the Wiener filter can be found as follows. If $S_{ff}(\Omega_x,\Omega_y)$ and $S_{nn}(\Omega_x,\Omega_y)$ are the power spectral densities of the original image $ln(\mathbf{f})$ and of the noise $ln(\mathbf{n})$, respectively, the frequency response of the Wiener filter is given by:

$$H(\Omega_x,\Omega_y) = \frac{S_{ff}(\Omega_x,\Omega_y)}{S_{ff}(\Omega_x,\Omega_y)+S_{nn}(\Omega_x,\Omega_y)} \qquad (7.3.10)$$

The Wiener filter can be implemented either in the frequency domain in the form (7.3.6) or as an Infinite Impulse Response (IIR) two-dimensional digital filter. Let $i(m,n)$, $\hat{y}(m,n)$ denote $\ln[i(m,n)]$ and the Wiener filter output respectively. In this case, the Wiener filter is described by the following difference equation:

$$\hat{y}(m,n) = \sum a_{kl}\hat{y}(m-k,n-l) + \sum b_{kl}\hat{i}(m-k,n-l) \qquad (7.3.11)$$

Such homomorphic filters employing Wiener filters for illumination and noise removal have been designed in [4]. The Wiener filter has been implemented as a second order IIR filter having transfer function:

$$H(z_1,z_2) = H_0(z_1,z_2)+H_0(z_1^{-1},z_2)+H_0(z_1,z_2^{-1})+H_0(z_1^{-1},z_2^{-1}) \qquad (7.3.12)$$

$$H_0(z_1,z_2) = \frac{b_{00}+b_{01}(z_1^{-1}+z_2^{-1})+b_{11}(z_1^{-1}z_2^{-1})+b_{02}(z_1^{-2}+z_2^{-2})}{1+a_{01}(z_1^{-1}+z_2^{-1})+a_{11}(z_1^{-1}z_2^{-1})+a_{02}(z_1^{-2}+z_2^{-2})} \qquad (7.3.13)$$

Table 7.3.1: Coefficients of the Wiener filter (adapted from [4]).

Shadow characteristics	a_{01}	a_{11}	a_{02}	b_{00}	b_{01}	b_{11}	b_{02}
Soft	-0.865	0.783	-0.024	0.2376	-0.1957	0.1831	-0.0147
Medium	-0.8503	0.7743	-0.0326	0.2681	-0.2216	0.2072	-0.016
Harsh	-0.8817	0.7959	-0.0118	0.5109	-0.4718	0.4458	-0.0065

Typical parameters of the filter coefficients can be found in Table 7.3.1. Three sets of parameters are given for soft, medium, and strong shadow characteristics. The stronger the high-pass filter characteristics are, the harsher the shadow effects it produces. The homomorphic Wiener filter has also been applied in speckle noise filtering [19]. Experimental studies have shown that its performance was better than that of the linear filtering. An example of multiplicative noise filtering by using homomorphic filtering is shown in Figure 7.3.4. The linear filter L used in this case is a 3×3 moving average filter implemented in the space domain.

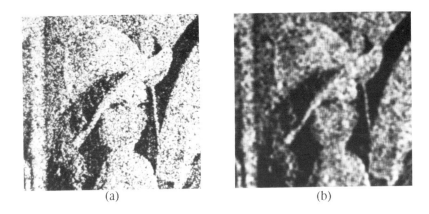

(a)   (b)

Figure 7.3.4: (a) Original image corrupted by multiplicative noise; (b) Output of the logarithmic homomorphic filter employing a moving average filter in its linear part.

Another application of homomorphic filtering is for cloud cover removal from satellite images [6-8]. The cloud distortion process can be described in Figure 7.3.5. The received image at the satellite scanner is given by:

$$s(x,y) = aLr(x,y)t(x,y)+L(1-t(x,y)) \tag{7.3.14}$$

The received image is denoted by $s(x,y)$. The ground reflectance, which is also the desired image, is denoted by $r(x,y)$. The attenuation due to clouds is denoted by $t(x,y)$. Finally, L,a denote the sun illumination and the sunlight attenuation, respectively. By subtracting $s(x,y)$ from L and taking the logarithm, (7.3.14) becomes:

$$ln[L-s(x,y)] = ln[t(x,y)]+ln[L-aLr(x,y)] \tag{7.3.15}$$

The signal is represented by $ln[L-aLr(x,y)]$, whereas noise is represented by $ln[t(x,y)]$. The noise has low spatial frequency content in comparison to the ground reflectance. Therefore, a low-pass filter can be used to obtain the noise term $ln[t(x,y)]$. The cloud noise is obtained by exponentiation. If cloud noise is known, the desired ground reflectance can be found:

$$r(x,y) = \frac{s(x,y)-L\,[1-t(x,y)]}{aLt(x,y)} \qquad (7.3.16)$$

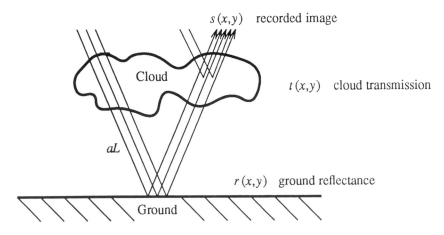

$s(x,y)$ recorded image

$t(x,y)$ cloud transmission

$r(x,y)$ ground reflectance

Figure 7.3.5: Image formation at the satellite scanners.

Such a low-pass filter can be the filter having a Kaiser window function response [9]:

$$h(m,n) = \begin{cases} I_0(\alpha(1-(m^2+n^2)/\tau^2)^{1/2})/I_0(\alpha) & (m^2+n^2)^{1/2} \le \tau \\ 0 & elsewhere \end{cases} \qquad (7.3.17)$$

where $I_0(x)$ is the modified zeroth-order Bessel function. Another possibility is to design a Wiener filter H that can remove the cloud transmission noise $ln\,[t(x,y)]$. Let $S_{ss}(\Omega_x,\Omega_y)$, $S_{tt}(\Omega_x,\Omega_y)$ be the power spectra of the processes $ln\,[L-s(x,y)]$ and $ln\,[t(x,y)]$, respectively. Let also m_r, m_t be the means of the signal and the cloud noise processes. The Wiener filter has the following frequency characteristics [7]:

$$H(\Omega_x,\Omega_y) = \frac{S_{rs}(\Omega_x,\Omega_y)}{S_{ss}(\Omega_x,\Omega_y)} = \frac{S_{ss}(\Omega_x,\Omega_y)-S_{tt}(\Omega_x,\Omega_y)-m_r m_t \delta(\Omega_x,\Omega_y)}{S_{ss}(\Omega_x,\Omega_y)} \qquad (7.3.18)$$

An example of the application of the homomorphic filters for the removal of cloud cover in satellite images is shown in Figure 7.3.6. The homomorphic filter used employs a linear bandstop filter. This filter suppresses the frequencies having most cloud noise energy. The result of cloud noise removal is shown in Figure 7.3.6b. Most cloud noise has been removed. However, bandpass filtering has also smoothed the lake contours because it has removed part of their high frequency content. Finally, homomorphic filtering has been applied in the enhancement of noisy fingerprints [5].

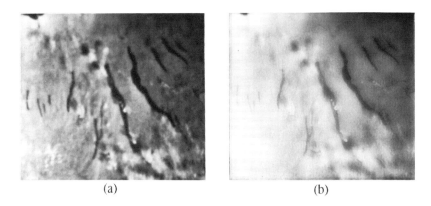

(a)   (b)

Figure 7.3.6: (a) Cloudy satellite image (b) Cloud free satellite image after homomorphic filtering.

7.4 HOMOMORPHIC FILTERING OF SIGNAL-DEPENDENT NOISE

Signal-dependent noise is often encountered in image formation, as was described in chapter 3. Film-grain noise and photoelectronic noise are special kinds of signal-dependent noise. Another kind of noise is *speckle noise* which is encountered in images formed by laser beams. Its probability density function is given by [9]:

$$p(i) = \frac{M^M}{\Gamma(M)m_i}(\frac{i}{m_i})^{M-1}exp(\frac{-Mi}{m_i}) \qquad (7.4.1)$$

M is a parameter which is approximately equal to the number of speckles in the integrating aperture, $\Gamma(M)$ is the gamma function, and m_i is the mean intensity. Thus speckle noise has gamma distribution with variance:

$$\sigma^2 = \frac{m_i^2}{M} \qquad (7.4.2)$$

Speckle variance is proportional to the square of the signal mean m_i. Therefore, speckle noise is can be modeled as multiplicative noise.

A useful representation of the signal-dependent noise is given by the following relation:

$$x = t(s) + r(s)n \tag{7.4.3}$$

as it has already been described in chapter 3. s denotes the original image, $t(.), r(.)$ are pointwise nonlinearities, n is a white noise process, usually uncorrelated with the signal s, and x is the observed image. Multiplicative noise is a special case of (7.4.3):

$$x = s + csn \tag{7.4.4}$$

Photoelectron noise is also a special case of (7.4.3):

$$x = c_1 s^l + c_2 s^m n \tag{7.4.5}$$

Linear filtering is not very useful for signal-dependent noise filtering, due to the nonlinear coupling of signal and noise. Therefore, decoupling of the noise from signal is highly desirable [10, 11]. This can be obtained by passing the observed image x through a nonlinearity $g(x)$, which is chosen in such a way that [12]:

$$g(x) = g[t(s) + r(s)n] = u(s) + N(n) \tag{7.4.6}$$

where $u(s)$ is a nonlinear function of s and $N(n)$ is a signal-independent noise term. By using Taylor series expansion of $g(x)$ around the point $t(s)$, it is found that:

$$g(x) = g[t(s)] + g^{(1)}[t(s)]r(s)n + \frac{1}{2}g^{(2)}[t(s)]r^2(s)n^2 + .. \tag{7.4.7}$$

$$+ \frac{1}{k!}g^{(k)}[t(s)]r^k(s)n^k + ...$$

where $g^{(k)}(x)$ is the k-th derivative of $g(x)$. Thus, $u(s), N(n)$ are given by:

$$u(s) = g[t(s)] \tag{7.4.8}$$

$$N(n) = \sum_{k=1}^{\infty} \frac{g^{(k)}[t(s)]}{k!} r^k(s)n^k \tag{7.4.9}$$

The function g is chosen as follows, in order to decouple the noise term from signal s:

$$g^{(k)}[t(s)]\frac{r^k(s)}{k!} = a_k , \qquad k=1,2,... \tag{7.4.10}$$

where a_k are constants. The set of equations (7.4.10) to be satisfied is infinite. It can be proven [12] that (7.4.10) is satisfied only when:

$$r(s) = bt(s) \tag{7.4.11}$$

where b is a constant. (7.4.11) corresponds to multiplicative noise:

$$x = t(s)(1+bn) \tag{7.4.12}$$

Thus only the multiplicative noise can be transformed completely to additive noise, if $g(x)$ is chosen according to (7.4.10). It is well known that, in this case, the nonlinearity $g(x)$ takes the form:

$$g(x) = ln(x) \tag{7.4.13}$$

This choice of $g(x)$ satisfies (7.4.10). In all other cases, signal dependent noise is only approximately transformed to additive noise. As it is seen in (7.4.9), $N(n)$ consists of various terms whose contribution decreases with their order. The most important term is $g^{(1)}[t(s)]r(s)n$. Thus, $g(x)$ is chosen in such a way so that this term becomes signal-independent:

$$g^{(1)}[t(s)] = \frac{1}{r(s)} \tag{7.4.14}$$

A similar choice of $g(x)$ is given in [10], although different methodology has been used. The noise term $N(n)$ becomes:

$$N(n) = n + \sum_{k=2}^{\infty} \frac{g^{(k)}[t(s)]}{k!} r^k(s)n^k \tag{7.4.15}$$

The higher order terms of (7.4.15) can be neglected and the noise becomes additive:

$$g(x) \simeq g[t(s)] + n \tag{7.4.16}$$

The additive noise term n can be removed by any convenient linear technique, i.e. by a moving average filter, a low-pass linear filter, or a Wiener filter. After additive noise filtering, the distorted image $g[t(s)]$ can be restored by using another nonlinear function $f(.)$ such that:

$$f\{g[t(s)]\} = s \tag{7.4.17}$$

Thus the filter for the removal of signal dependent noise has a homomorphic structure shown in Figure 7.4.1.

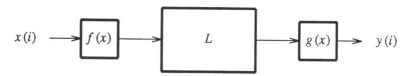

Figure 7.4.1: Structure of a homomorphic filter for signal-dependent noise filtering.

For the case of the signal-dependent noise of (7.4.5) the nonlinear function $g(x)$, $f(x)$ must be chosen as follows [12]:

$$g(x) = \frac{l}{c_2(l-m)c_1^{-m/l}} x^{(l-m)/l} \qquad (7.4.18)$$

$$f(x) = \left[\frac{l-m}{l}\frac{c_2}{c_1}x\right]^{1/(l-m)} \qquad (7.4.19)$$

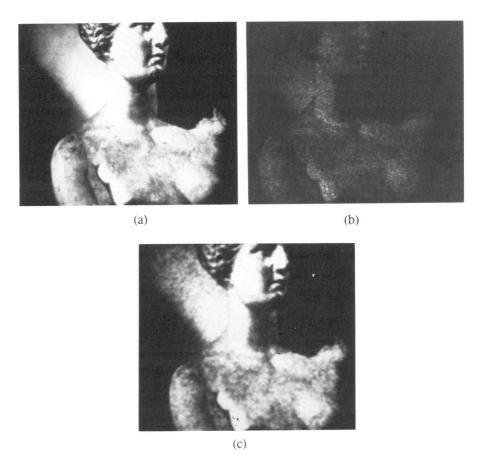

(a)   (b)

(c)

Figure 7.4.2: (a) Venus of Milos; (b) Image corrupted by signal-dependent noise; (c) Output of the homomorphic filter.

An example of signal-dependent noise filtering is shown in Figure 7.4.2. The image in Figure 7.4.2a corrupted by signal-dependent noise:

$$x = s^{0.7} + s^{0.35}n \qquad (7.4.20)$$

is shown in Figure 7.4.2b. (7.4.20) describes sufficiently well the characteristics of a commercial orthicon or vidicon TV camera. The dynamic range of the image is compressed, due to the first term in (7.4.20), and the noise is more evident on the bright image regions. The nonlinear functions $g(x)$, $f(x)$ are given by (7.4.18-19):

$$g(x) = 2x^{0.5} \qquad (7.4.21)$$

$$f(x) = (0.5x)^{2.857} \qquad (7.4.22)$$

The homomorphic filter employs a moving average filter of 3×3 extent. The output of the homomorphic filter is shown in Figure 7.4.2c. The signal-dependent noise has been removed.

7.5 NONLINEAR MEAN FILTERS

Consider the numbers x_i, $i=1,..,n$. Their nonlinear mean is the following function:

$$y = f(x_1,..,x_n) = g^{-1}\left[\frac{\sum\limits_{i=1}^{n} a_i g(x_i)}{\sum\limits_{i=1}^{n} a_i}\right] \qquad (7.5.1)$$

where $g(x)$ is a single-valued analytic nonlinear function and a_i are weights. If the weights are constants, the nonlinear filters reduce to homomorphic filters. The properties of the nonlinear means depend on the function $g(x)$ and on the weights a_i, $i=1,..,n$. The following choices of $g(x)$ produce filters that are of interest in image processing:

$$g(x) = \begin{cases} x & \text{arithmetic mean } \bar{x} \\ 1/x & \text{harmonic mean } y_H \\ logx & \text{geometric mean } y_G \\ x^p, \ p \in Q - \{-1,0,1\} & L_p \text{ mean } y_L, \end{cases} \qquad (7.5.2)$$

If the weights a_i are not constants, other classes of nonlinear means can be obtained by appropriate choice of parameters a_i. Such a filter is the contraharmonic filter:

$$y_{CH,} = \frac{\sum\limits_{i=1}^{n} x_i^{p+1}}{\sum\limits_{i=1}^{n} x_i^{p}} \qquad (7.5.3)$$

Contraharmonic filters are generated by choosing:

$$g(x) = x, \qquad a_i = x_i^p, \quad i=1,..,n \qquad (7.5.4)$$

i.e., it can be interpreted as an arithmetic mean having data-dependent coefficients. The nonlinear means have the following well-known property:

$$min(x_i) \le y_{CH_-} \le y_{L_-} \le y_H \le y_G \le \bar{x} \le y_{L_+} \le y_{CH_+} \le max(x_i) \qquad (7.5.5)$$

The nonlinear mean filters have the following definition:

$$y_i = f(x_{i-v},..,x_i,x_{i+v}) \qquad i \in Z, \quad n=2v+1 \qquad (7.5.6)$$

Similar definitions can be given for $n=2v$ and for two-dimensional nonlinear mean filters.

The statistical properties of the nonlinear mean filters depend on the output mean m_y and variance σ_y^2. Let m and σ_x^2 be the mean and the variance of the independent identically distributed input variables $x_i, i=1,..,n$. Let also $f_i', f_i'', i=1,..,n$ be the first and second order partial derivatives of f evaluated at the point $(x_1,\ldots,x_n) = (m,..,m)$. By taking only a first order Taylor series approximation of f, the output mean and variance are given by:

$$m_y \simeq f(m,..,m) + \frac{\sigma_x^2}{2} \sum_{i=1}^{n} f_i'' \qquad (7.5.7)$$

$$\sigma_y^2 \simeq \sigma_x^2 \sum_{i=1}^{n} f_i'^2 \qquad (7.5.8)$$

for independent, identically distributed variables $x_i, i=1,..,n$. The performance of the arithmetic mean, geometric mean, contraharmonic mean, L_p mean and median, expressed as the ratio σ_y^2/σ_x^2, is summarized in Table 7.5.1. The nonlinear mean filters are clearly superior to either the moving average or the median filters for the uniform and the Gaussian noise distributions.

Table 7.5.1: Performance of nonlinear mean filters expressed as σ_y^2/σ_x^2

pdf	$\bar{x}$	Geometric mean	L_p mean	CH_p mean	Median
Uniform	$\frac{1}{n}$	$\frac{1}{n} - \frac{(n-1)^2\sigma_x^2}{4n^2 m^2}$	$\frac{1}{n} - \frac{(p-1)^2(n-1)^2\sigma_x^2}{4n^2 m^2}$	$\frac{1}{n} - \frac{p^2(n-1)^2\sigma_x^2}{n^2 m^2}$	$\frac{3}{n+2}$
Gaussian	$\frac{1}{n}$	$\frac{1}{n} - \frac{(n-1)^2\sigma_x^2}{4n^2 m^2}$	$\frac{1}{n} - \frac{(p-1)^2(n-1)^2\sigma_x^2}{4n^2 m^2}$	$\frac{1}{n} - \frac{p^2(n-1)^2\sigma_x^2}{n^2 m^2}$	$\frac{\pi}{2(n+\pi/2-1)}$

The $L_p, L_{-p}, CH_p, CH_{-p}$ filters can also be used for the removal of impulsive noise [11,13]. Let the background image have a mean m and the spikes a mean $M = bm$ ($b \gg 1$). Let also have the following impulsive noise model for positive (white) spikes:

$$z_{ij} = \begin{cases} bm & \textit{with probability } q \\ m & \textit{with probability } 1-q \end{cases} \qquad (7.5.9)$$

The impulsive noise is considered to be suppressed if the filter output satisfies:

$$E(|y|) \le am , \qquad a = 1 + \varepsilon , \quad \varepsilon \ge 0 \qquad (7.5.10)$$

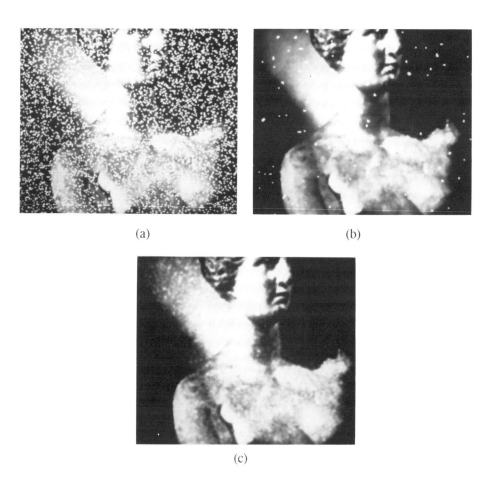

(a)   (b)

(c)

Figure 7.5.1: (a) Image corrupted by positive impulse noise having probability of occurrence 30%; (b) Output of the 3×3 median filter; (c) Output of a harmonic 3×3 filter.

If an L_{-p} mean filter has sufficiently high p that the following relation is satisfied:

$$q \leq \frac{a^{-p}-1}{b^{-p}-1} \qquad (7.5.11)$$

it can be proven that it removes the positive spikes and that it satisfies (7.5.10) [11]. Similarly, an CH_{-p} filter, whose coefficient p satisfies:

$$q \leq \frac{a-1}{a-1+b^{-p}(b-a)} \qquad (7.5.12)$$

removes the positive spikes having probability of occurrence p [12]. Both L_{-p} and CH_{-p} filters tend to $min(x_i)$ for sufficiently high p. This fact explains their good performance in removing positive spikes. L_p and CH_p filters can be used for the removal of negative (black) impulses having mean $M = bm$ ($0 < b << 1$), if appropriately high p is chosen to satisfy the following relations, respectively:

$$q \leq \frac{1-a^p}{1-b^p} \qquad (7.5.13)$$

$$q \leq \frac{1-a}{1-a+b^p(a-b)} \qquad (7.5.14)$$

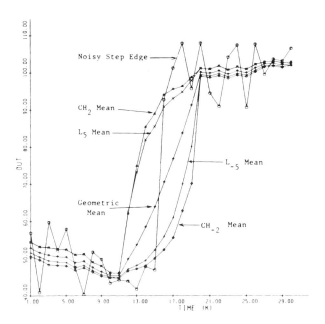

Figure 7.5.2: Response of nonlinear filters in the presence of a noisy edge.

The nonlinear mean filters have very good performance in the removal of positive or negative impulse noise. An example of positive noise removal is shown in Figure 7.5.1. The probability of occurrence of an impulse in Figure 7.5.1a is 30%. Median filter of 3×3 extent cannot remove all spikes, as seen in Figure 7.5.1b. The harmonic filter of 3×3 extent removes all impulses completely, as seen in Figure 7.5.1c. The L_{-2} filter has similar performance. However, nonlinear mean filters fail to remove salt-and-pepper noise. If such noise is present a cascade of an L_p mean filter followed by an L_{-p} mean filter can be used. Nonlinear mean filters tend to the maximum or to the minimum filters for sufficiently high p. This fact explains their behavior in the presence of edges shown in Figure 7.5.2. All nonlinear mean filters (except perhaps the geometric mean) tend to preserve the edge. However, they introduce bias. L_{-p} and CH_{-p} filters tend to enhance the black regions of the image, whereas L_p, CH_p filters tend to enhance the bright regions. The bias is not so strong and it is unnoticeable if p is low (e.g. $p=1,2$).

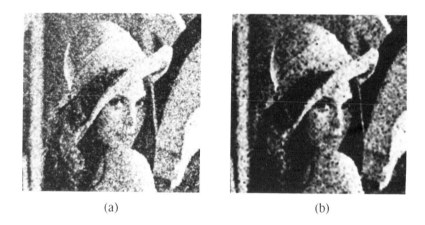

(a)   (b)

Figure 7.5.3: (a) Image corrupted by white additive Gaussian noise; (b) Output of a 3×3 harmonic mean filter.

An example of additive white noise filtering is shown in Figure 7.5.3. The noise distribution is Gaussian. Harmonic filter output, shown in Figure 7.5.3b, preserves the edges and suppresses the noise in the homogeneous image regions. The bias toward the black image regions is relatively small.

7.6 HOMOMORPHIC SYSTEMS FOR CONVOLUTION

There are many digital signal processing problems in which signals are combined by convolution:

$$x(i) = s(i) * h(i) = \sum_{k=-\infty}^{\infty} s(k)h(i-k) \tag{7.6.1}$$

Such situations occur in digital speech processing [1,15,20,21], seismic signal processing [1,2], digital signal processing in communication systems, and digital image restoration problems. $x(i)$ is the observed sequence, $s(i)$ is the desired sequence and $h(i)$ are the system impulse response, e.g., the impulse response of the earth, of the imaging system or of the communication channel. In all those cases a *deconvolution* of the observed signal is desired so that an estimate of sequence $s(i)$ is obtained. This can be done by using a homomorphic filter, whose canonic form is shown in Figure 7.2.2.

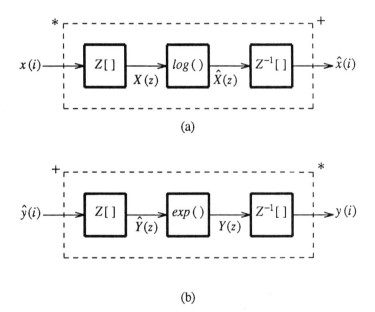

(a)

(b)

Figure 7.6.1: (a) Characteristic system D_* of the homomorphic system for convolution; (b) Representation of the inverse of the characteristic system.

In this case, the operations $*$, $\times$ on the input and the output signal space are the convolution. The characteristic system D_* transforms signal convolution to signal addition:

$$\hat{x}(i) = D_*[x(i)] = D_*[s(i)] + D_*[h(i)] = \hat{s}(i) + \hat{h}(i) \qquad (7.6.2)$$

The system D_*^{-1} is the inverse operation of D_*. The characteristic system D_* can be obtained by using the z transform property:

$$X(z) = S(z)H(z) \qquad (7.6.3)$$

The logarithm transforms z transform multiplication to addition:

$$\hat{X}(z) = log[X(z)] = log[S(z)] + log[H(z)] = \hat{S}(z) + \hat{H}(z) \qquad (7.6.4)$$

Therefore, the characteristic system D_* has the form shown in Figure 7.6.1a. The representation of the inverse system D_*^{-1} is shown in Figure 7.6.1b. The output of the characteristic system is the inverse z transform of the logarithm of $X(z)$:

$$\hat{x}(i) = \frac{1}{2\pi j} \int_C \hat{X}(z) z^{i-1} dz \qquad (7.6.5)$$

where C is a contour in the region of convergence of $\hat{X}(z)$ having the form $r_1 < |z| < r_2$. It is called *complex cepstrum* of the signal $x(i)$. If the sequence $x(i)$ is real, its cepstrum is a real sequence, too. If the cepstrum is a stable sequence, its region of convergence must contain the unit circle. In this case, (7.6.5) becomes:

$$\hat{x}(i) = \frac{1}{2\pi} \int_{-\pi}^{\pi} \hat{X}(e^{j\omega}) e^{j\omega i} d\omega \qquad (7.6.6)$$

where:

$$\hat{X}(e^{j\omega}) = log[X(e^{j\omega})] = log[\,|X(e^{j\omega})|\,] + jarg[X(e^{j\omega})] \qquad (7.6.7)$$

The cepstrum of of $x(i)$ is the sum of the the cepstra of $s(i)$, $h(i)$:

$$\hat{x}(i) = \hat{s}(i) + \hat{h}(i) \qquad (7.6.8)$$

The cepstra $\hat{s}(i)$, $\hat{h}(i)$ possess different properties, as will be shown in the next section. It is the linear filter L, shown in Figure 7.2.2 that can take advantage of this fact and separate the two cepstra. Therefore, the performance of a homomorphic system for convolution depends on the appropriate choice of L. The cepstra $\hat{x}(i)$, $\hat{s}(i), \hat{h}(i)$ are real sequences. Thus L must be a real linear system. In many cases the cepstra $\hat{s}(i)$, $\hat{h}(i)$ occupy different regions on the the axis i, as it will be seen later on. Therefore, the linear system L is chosen to be a window function of the form:

$$\hat{y}(i) = l(i)\hat{x}(i) \qquad (7.6.9)$$

In the frequency domain (7.6.9) becomes:

$$\hat{Y}(e^{j\omega}) = \frac{1}{2\pi} \int_{-\pi}^{\pi} \hat{X}(e^{j\theta}) \quad L(e^{j(\omega-\theta)})d\theta \tag{7.6.10}$$

The output of the whole homomorphic system is given by:

$$y(i) = \frac{1}{2\pi j} \int_C Y(z)z^{i-1} dz \tag{7.6.11}$$

$$Y(z) = exp[\hat{Y}(z)] \tag{7.6.12}$$

The properties of the cepstrum will be described in the next section.

7.7 PROPERTIES OF THE COMPLEX CEPSTRUM

The properties of the cepstrum of sequences having rational z transforms is of importance in many practical cases:

$$X(z) = \frac{Az^r \prod_{k=1}^{M_1}(1-a_k z^{-1})\prod_{k=1}^{M_2}(1-b_k z)}{\prod_{k=1}^{N_1}(1-c_k z^{-1})\prod_{k=1}^{N_2}(1-d_k z)} \tag{7.7.1}$$

where $|a_k|$, $|b_k|$, $|c_k|$, $|d_k|$ are less than one. The terms a_k, c_k correspond to zeroes and poles inside the unit circle, whereas $1/b_k$, $1/d_k$ correspond to zeroes and poles outside the unit circle. The logarithm of $X(z)$ has the form:

$$\hat{X}(z) = log[A] + log[z^r] + \sum_{k=1}^{M_1} log(1-a_k z^{-1}) + \sum_{k=1}^{M_2} log(1-b_k z) \tag{7.7.2}$$

$$- \sum_{k=1}^{N_1} log(1-c_k z^{-1}) - \sum_{k=1}^{N_2} log(1-d_k z)$$

The term $log[z^r]$ carries only time shift information and can be removed. All other logarithmic terms can be expanded as follows:

$$log(1-az^{-1}) = -\sum_{n=1}^{\infty} \frac{a^n}{n} z^{-n} \quad |z| > |a| \tag{7.7.3}$$

$$log(1-bz) = -\sum_{n=1}^{\infty} \frac{b^n}{n} z^n \quad |z| < |b^{-1}| \tag{7.7.4}$$

Thus the cepstrum $\hat{x}(i)$ can be obtained easily:

$$\hat{x}(i) = \begin{cases} log|A| & i=0 \\ -\sum_{k=1}^{M_1} \frac{a_k^i}{i} + \sum_{k=1}^{N_1} \frac{c_k^i}{i} & i>0 \\ \sum_{k=1}^{M_2} \frac{b_k^{-i}}{i} - \sum_{k=1}^{N_2} \frac{d_k^{-i}}{i} & i<0 \end{cases} \tag{7.7.5}$$

Important properties can be derived from (7.7.5): The complex cepstrum decays as $1/i$. Thus the cepstrum of signals having rational z transform occupies the low-time parts of the cepstrum (close to $i=0$). The cepstrum of minimum phase sequences (having no zeroes or poles outside the unit circle) is zero for negative times $i<0$. The cepstrum of maximum phase sequences (having no zeroes or poles inside the unit circle) is zero for positive times $i>0$.

Another important class of signals is the train of delta impulses of the form:

$$h(i) = \sum_{k=0}^{M} h_k \delta(i-kN) \tag{7.7.6}$$

Such signals are important because they can describe relatively well the impulse response of some physical systems, e.g., that of the earth and of the voice pitch. The z transform of such a signal is a polynomial given by:

$$H(z) = \sum_{k=0}^{M} h_k z^{-kN} \tag{7.7.7}$$

and can be factored in terms of the form:

$$H_k(z) = 1+a_k z^{-N} \tag{7.7.8}$$

The logarithm of (7.7.8) can be expanded in a series by using (7.7.3):

$$\hat{H}_k(z) = log\,(1+a_k z^{-N}) = -\sum_{i=1}^{\infty} \frac{(-a_k)^i}{i} z^{-iN} \tag{7.7.9}$$

Thus the complex cepstrum of each of the terms of $H_k(z)$ is given by:

$$\hat{h}_k(i) = \sum_{n=1}^{\infty} (-1)^{n+1} \frac{a_k^n}{n} \delta(i-nN) \tag{7.7.10}$$

It is a train of equally spaced impulses. Thus the cepstrum of the train of impulses is also a train of impulses. This fact is used in homomorphic speech processing and in homomorphic seismic deconvolution.

In the following the application of homomorphic filtering in digital speech analysis and in seismic deconvolution will be presented. A short speech segment can be modeled by the following equation [15]:

$$x(i) = g(i)*v(i)*r(i)*h(i) \tag{7.7.11}$$

where $g(i)$ represents the *glottal wave shape*. $v(i)$, $r(i)$ represent the impulse responses of the *vocal tract* and of the speech radiation, respectively. $h(i)$ is a quasi-periodic impulse train that excites the vocal tract. The spacing between the impulses corresponds to the *pitch period* of the speech. A reasonable model for the vocal tract system is a rational z transform function of the form (7.7.1).

The model of the glottal wave shape is of the form:

$$G(z) = A \prod_{k=1}^{M_1}(1-a_k z^{-1}) \prod_{k=1}^{M_2}(1-b_k z) \qquad (7.7.12)$$

Finally, the radiation effects can be reasonably modeled by a system of the form:

$$R(z) = 1-z^{-1} \qquad (7.7.13)$$

Thus the transfer function of the system:

$$s(i) = g(i)*v(i)*r(i) \qquad (7.7.14)$$

is rational. Its cepstrum $\hat{s}(i)$ decays rapidly and it occupies only the low-time parts of the cepstrum (close to $i=0$). The pitch cepstrum $\hat{h}(i)$ is of impulsive form and occupies the high-time parts of the cepstrum. Thus it can be assumed that the two cepstra do not overlap and can be separated by using suitable window functions. The following window separates the pitch impulses:

$$l(i) = \begin{cases} 1 & |i| > i_0 \\ 0 & |i| \le i_0 \end{cases} \qquad (7.7.15)$$

for appropriately chosen threshold i_0. A low-time window of the form:

$$l(i) = \begin{cases} 0 & |i| > i_0 \\ 1 & |i| \le i_0 \end{cases} \qquad (7.7.16)$$

can be used to recover the slowly varying signal $s(i)$. The interested reader can find more details on homomorphic speech processing in [15]. Applications of homomorphic speech processing in speaker verification and in speech recognition can be found in [22,23], respectively.

Seismic signals are also of the form:

$$x(i) = s(i)*h(i) \qquad (7.7.17)$$

where $x(i)$ is the received signal, $s(i)$ is the shape of the seismic wavelet, and $h(i)$ is the impulse response of the earth. This response comes from seismic signal reflections on the interfaces of the rock layers in the earth subsurface. Therefore, it has an impulsive form. Its cepstrum has also an impulsive form. The z transform of the seismic wavelet is rational and its cepstrum decays rapidly with time. Thus the earth impulse response can be separated from the seismic wavelet by using cepstral windows of the form (7.7.15-16). The whole procedure is called *homomorphic deconvolution*. It has found extensive application in the geophysical oil prospecting, although it has been proven that it is relatively noise-sensitive. The interested reader can find more details in the monography [2]. Finally, cepstra have been proposed as digital image processing tools [17, p.

259].

7.8 REALIZATIONS OF HOMOMORPHIC FILTERS FOR CONVOLUTION

The basic problem in the computer realization of homomorphic filters for convolution is the computation of the complex cepstrum:

$$X(e^{j\omega}) = \sum_{i=-\infty}^{\infty} x(i)e^{-j\omega i} \tag{7.8.1}$$

$$\hat{X}(e^{j\omega}) = log\,[X(e^{j\omega})] = log\,[\,|X(e^{j\omega})|\,] + jarg\,[X(e^{j\omega})] \tag{7.8.2}$$

$$\hat{x}(i) = \frac{1}{2\pi}\int_{-\pi}^{\pi} \hat{X}(e^{j\omega})e^{j\omega i}d\omega \tag{7.8.3}$$

The complex cepstrum can be realized by using the Discrete Fourier Transform:

$$X(k) = \sum_{n=0}^{N-1} x(n)e^{-j\frac{2\pi kn}{N}} \tag{7.8.4}$$

$$\hat{X}(k) = log\,[X(k)] + jarg\,[X(k)] \tag{7.8.5}$$

$$\hat{x}_p(n) = \frac{1}{N}\sum_{n=0}^{N-1}\hat{X}(k)e^{j\frac{2\pi kn}{N}} \tag{7.8.6}$$

and can be computed by employing Fast Fourier Transform algorithms. The computed cepstrum $\hat{x}_p(n)$ is related to the true cepstrum $\hat{x}(n)$ given by (7.8.3) as follows:

$$\hat{x}_p(n) = \sum_{k=-\infty}^{\infty} \hat{x}(n+kN) \tag{7.8.7}$$

due to the sampling theorem. Thus a sufficiently high number of points N must be chosen in order to avoid aliasing problems. Another problem that is encountered in the computation of the cepstrum is the computer realization of the function $arg\,[X(k)]$. Standard computer libraries produce an output in the range $[-\pi,\pi]$. Thus the function $arg\,[X(k)]$ becomes discontinuous and phase *unwarping* is required. Special algorithms have been developed for appropriate phase unwarping [2,16].

The use of the complex logarithm can be avoided if the following relation is used:

$$-i\hat{x}(i) = \frac{1}{2\pi j}\int_C \hat{X}'(z)z^{i-1}dz = \frac{1}{2\pi j}\int_C \frac{zX'(z)}{X(z)}z^{i-1}dz \tag{7.8.8}$$

$\hat{X}'(z)$, $X'(z)$ denote the derivatives of $\hat{X}(z)$, $X(z)$ with respect to z. Thus $\hat{x}(i)$ can be computed as follows:

$$\hat{x}(i) = -\frac{1}{2\pi ji} \int_{-\pi}^{\pi} \frac{X'(e^{j\omega})}{X(e^{j\omega})} e^{j\omega i} d\omega \qquad i \neq 0 \qquad (7.8.9)$$

$$X'(e^{j\omega}) = -j \sum_{i=-\infty}^{\infty} ix(i)e^{-j\omega i} \qquad (7.8.10)$$

The cepstrum at $i=0$ can be calculated by:

$$\hat{x}(0) = -\frac{1}{2\pi} \int_{-\pi}^{\pi} \hat{X}(e^{j\omega})d\omega = -\frac{1}{2\pi} \int_{-\pi}^{\pi} \hat{log}[|X(e^{j\omega})|]d\omega \qquad (7.8.11)$$

because the imaginary part of $\hat{X}(e^{j\omega})$ is an odd function and its integral over $[-\pi,\pi]$ vanishes. The DFT can be used for the calculation of (7.8.9-11). Thus the computation of the complex logarithm is avoided. However this algorithm introduces more severe aliasing [1].

7.9 DISCUSSION

Homomorphic filter theory and applications have been presented in this chapter. Their application has been focused on digital image processing, although their use in digital speech and seismic signal processing has also been described briefly. The use of the logarithmic homomorphic filters in image enhancement, multiplicative noise filtering, and cloud noise removal has been described in detail. The use of homomorphic filters having nonlinearities of the form x^γ in signal dependent noise filtering and in impulsive noise removal has been presented. Finally, the homomorphic systems for convolution and the complex cepstrum have been analyzed. Complex cepstrum has interesting properties that make it suitable for digital speech analysis and for seismic deconvolution. Its computation can be performed relatively fast, if Fast Fourier Transform algorithms are employed in its realization.

REFERENCES

[1]   A.V. Oppenheim, R.W. Schafer, *Digital signal processing*, Prentice-Hall, 1975.

[2]   J.M. Tribolet, *Seismic applications of homomorphic signal processing*, Prentice-Hall, 1979.

[3]   W.F. Schreiber, "Image processing for quality improvements", *Proc. of the IEEE*, vol. 66, no. 12, pp. 1640-1651, Dec. 1978.

[4]   R.W. Fries, J.W. Modestino, "Image enhancement by stochastic homomorphic filtering", *IEEE Transactions on Acoustics, Speech and Signal Processing,* vol. ASSP-27, no. 6, pp. 625-637, Dec. 1979.

[5]   R.J. Blackwell, W.A. Crisci, "Digital image processing technology and its application to forensic sciences", *Journal of Forensic Sciences,* vol. 20, no.2, pp. 288-304, 1975.

[6]   Z.K. Liu, B.R. Hunt, "A new approach to removing cloud cover from satellite imagery", *Computer Vision, Graphics and Image Processing* , vol. 25, pp. 252-256, 1984.

[7]   O.R. Mitchell, E. Delp, P.L. Chen, "Filtering to remove cloud cover in satellite imagery", *IEEE Transactions on Geoscience Electronics, vol. GE-15, pp. 137-141, 1977.*

[8]   T. Peli, T.F. Quatieri, "Homomorphic restoration of images degraded by light cloud cover", *Proc. IEEE Int. Conf. on Acoustics, Speech and Signal Processing*, 1984.

[9]   J.W. Goodman, "Statistical properties of laser speckle patterns" in *Laser speckle and related phenomena*, J.C. Dainty editor, Springer Verlag, 1976.

[10] H.H. Arsenault, M. Denis, "Image processing in signal-dependent noise", *Canadian Journal of Physics,* vol. 61, pp. 309-317, 1983.

[11] I. Pitas, A.N. Venetsanopoulos, "Nonlinear mean filters in image processing", *IEEE Transactions on Acoustics, Speech and Signal Processing,* vol. ASSP-34. no. 3, pp. 573-584, June 1986.

[12] I. Pitas, A.N. Venetsanopoulos, "Nonlinear order statistics filters for image filtering and edge detection", *Signal Processing*, vol. 10, pp. 395-413, 1986.

[13] A. Kundu, S.K. Mitra, P.P. Vaidyanathan, "Application of two-dimensional generalized mean filtering for the removal of impulse noises from images", *IEEE Transactions on Acoustics, Speech and Signal Processing,* vol. ASSP-32. no. 3, pp. 600-608, June 1984.

[14] A.V. Oppenheim, R.W. Schafer, T.G. Stockham, "Nonlinear filtering of multiplied and convolved signals", *Proc. of IEEE*, vol.56, pp. 1264-1291, Aug. 1968.

[15] L.R. Rabiner, R.W. Schafer, *Digital Processing of Speech Signals*, Prentice-Hall, 1978.

[16] J.M. Tribolet, "A new phase unwarping algorithm", *IEEE Transactions on Acoustics, Speech and Signal Processing,* vol. ASSP-25, no. 2, pp. 170-177, Apr. 1977.

[17] A.K. Jain, *Fundamentals of digital image processing*, Prentice-Hall, 1989.

[18] J.S. Lim, "Image enhancement", in *Digital Image Processing Techniques*, M.P. Ekstrom editor, Academic Press, 1984.

[19] A.K. Jain, C.R. Christensen, "Digital processing of images in speckle noise", *Proc. SPIE, Applications of Speckle Phenomena*, vol. 243, pp. 46-50, July 1980.

[20] S. Furui, *Digital Speech Processing, Synthesis and Recognition*, Marcel-Dekker, 1989.

[21] A.V. Oppenheim, R.W. Schafer, *Discrete-time signal processing*, Prentice-Hall, 1979.

[22] S. Furui, "Cepstral analysis techniques for automatic speaker verification", *IEEE Transactions on Acoustics, Speech and Signal Processing*, vol. ASSP-29, no. 2, pp. 254-272, Apr. 1981.

[23] S.B. Davis, P. Mermelstein, "Comparison of parametric representations of monosyllabic word recognition in continuously spoken sentence", *IEEE Transactions on Acoustics, Speech and Signal Processing*, vol. ASSP-28, no. 4, pp. 357-366, Apr. 1980.

CHAPTER 8

POLYNOMIAL FILTERS

8.1 INTRODUCTION

In many problems of digital signal processing it is necessary to introduce nonlinear systems. For example, it is well known that in detection and estimation problems, nonlinear filters arise in the case where the Gaussian assumption is not valid or the noise is not signal independent and/or additive. In the search for optimum signal processing systems, the task is to obtain general characterization procedures for nonlinear systems that retain at least a part of the simplicity that the impulse response method has for linear filters.

A possible way to describe the input-output relation in a nonlinear filter that is amenable to characterization, analysis, and synthesis is to use a discrete Volterra series representation. The Volterra series can be viewed as a Taylor series with memory [1]. Although the class of nonlinear systems that can be represented by the Volterra series expansion is limited, the Volterra series is one of a few representations that are easily amenable to characterization and analysis [1,2]. In many cases, a nonlinear system can be represented by a truncated version of the Volterra series, which results in a simpler representation and requires a limited amount of knowledge of higher order statistics.

Filters based on the Volterra series and another representation called Wiener series [1] will be referred to as polynomial filters. Such filters can be useful in many situations in which linear processors are known to be suboptimal, such as the equalization of nonlinear channels [3,4]. Polynomial filters can also be applied for compensating typical nonlinearities that affect adaptive echo cancellers [5-7]. This is, indeed, an important feature for echo cancellers, because in practical systems various kinds of nonlinearities are present in the echo path or in the echo replica, such as transmitted pulse asymmetry, saturation in transformers, nonlinearities of data converters, etc.

In optics, nonlinearities exist in almost every imaging system. Specifically, it has been shown [8] that the third term of the Volterra series, termed quadratic, describes many optical transformations. This is, in fact, due to the inherent bilinearity in optics, which results from the quadratic relation between the optical intensity and the optical field. Moreover, since we know that the nature of optical recording materials, rough surfaces, and grainy emulsions always contain a randomly fluctuating term interpreted as noise, determining the statistics of the response of a dynamic nonlinear system is of extreme importance. Understanding the propagation of such fluctuations through the system is of significance in correctly interpreting the results of optical experimental measurements.

In image processing a special class of polynomial filters have been used for image enhancement [9-11], edge extraction [12], nonlinear prediction [13-14], and nonlinear interpolation of TV image sequences [15]. Polynomial filters have also been used in modeling geophysical and sonar channels with nonlinearities.

Daily stock returns are another example of real data that exhibits nonlinearities in its behavior. Hinich and Patterson [16] tested fifteen randomly selected stock series for nonlinearity with a test based on the sample bispectrum. Linearity was rejected for all fifteen series, and the data were found to be suitably fitted to a second-order Volterra-type nonlinear model.

If the input-output relation is restricted only to the third term of a Volterra processor, the system becomes a quadratic filter. It is very common to reduce a nonlinear system to the quadratic part. In addition, quadratic filters appear in many problems of digital signal processing, such as in the optimal detection of a Gaussian signal in Gaussian noise [17] as well as texture discrimination. Quadratic filters are the simplest nonlinear, time-invariant systems, and their mathematical properties, harmonic representation, and efficient implementation have been considered [17,18].

8.2 DEFINITION OF POLYNOMIAL FILTERS

In this chapter, a class of nonlinear, 1-d, shift-invariant systems with memory based on the discrete-time Volterra series is considered [1,2]. The input-output relation satisfied by these systems, when the discrete independent variables are defined on a finite support, is then given by (8.2.1):

$$y(n) = h_o + \sum_{k=1}^{\infty} \overline{h}_k[x(n)] \tag{8.2.1}$$

where $x(n)$ and $y(n)$ denote the input and output, respectively, h_o is a constant term, and $\overline{h}_k[x(n)]$ is defined by (8.2.2):

$$\overline{h}_k[x(n)] = \sum_{i_1=0}^{N-1} \cdots \sum_{i_k=0}^{N-1} h_k(i_1,...,i_k)x(n-i_1)\cdots x(n-i_k) \tag{8.2.2}$$

Note that in (8.2.2) for $k=1$ the term $h_1(i_1)$ is the usual linear impulse response, and the term $h_k(i_1,...,i_k)$ can be considered as the finite extent k-th order impulse response, which characterizes the nonlinear behavior of the filter. In a Volterra filter of order K the upper limit of (8.2.1) is replaced by K [19]. As in the continuous case, researchers usually assume [2,19] that the operator $\bar{h}_k$ satisfies the symmetry conditions with respect to the discrete variables $i_1,...,i_k$. Therefore, $\bar{h}_k$ can be rewritten in the following form:

$$\bar{h}_k[x(n)] = \sum_{i_1=0}^{N-1} \sum_{i_2=i_1}^{N-1} \cdots \sum_{i_k=i_{k-1}}^{N-1} C(i_1,...,i_k) \cdot$$

$$\cdot h_k(i_1,...,i_k) x(n-i_1) \cdots x(n-i_k) \qquad (8.2.3)$$

where $C(i_1,...,i_k)$ are suitable constant values derived on a combinatorial basis [19]. Hence (8.2.1-3) allows us to define a class of 1-d polynomial filters, which play the same role as 1-d finite impulse response linear filters in the 1-d case. The quadratic filter is given by the third term of (8.2.1) or the second term of (8.2.2) and is described by (8.2.4):

$$y(n) = \sum_{i_1=0}^{N-1} \sum_{i_2=0}^{N-1} h_2(i_1,i_2) x(n-i_1) x(n-i_2) \qquad (8.2.4)$$

and can be written in compact form as (8.2.4a) indicates:

$$y_n = \sum_{i=1}^{N} \sum_{j=1}^{N} h_{i,j} x_{n-i+1} x_{n-j+1} \qquad (8.2.4a)$$

The quadratic kernel $h_{i,j}$ is usually assumed to be symmetric function of its indices [2,19]. In addition, we define the operation of (8.2.4a) as the *quadratic convolution* between the input x_n and the quadratic kernel $h_{i,j}$.

The input-output relationship of the quadratic filter can be written in a vectorial form:

$$y_n = \mathbf{x}^T \mathbf{H} \mathbf{x} = [x_n \; x_{n-1} \; \cdots \; x_{n-N+1}] \begin{bmatrix} h_{1,1} & \cdots & h_{1,N} \\ \vdots & & \vdots \\ h_{N,1} & \cdots & h_{N,N} \end{bmatrix} \begin{bmatrix} x_n \\ x_{n-1} \\ \vdots \\ x_{n-N+1} \end{bmatrix} \qquad (8.2.4b)$$

where $\mathbf{H}$ is a finite symmetric matrix, i.e. $\mathbf{H} = \mathbf{H}^T (h_{i,j} = h_{j,i})$. We call this matrix the kernel of the filter, although we sometimes refer to its elements as the kernel. In the case of a finite kernel quadratic filter, we can also say that the memory of the filter is finite.

For different applications, the quadratic kernel can assume different forms. We will now describe four special forms of the kernel matrix $\mathbf{H}$ that often arise in practical situations [18]:

1)   The kernel matrix $\mathbf{H}$ can be positive definite, which implies that for any input x_n, the output y_n is positive.

2)   The kernel matrix $\mathbf{H}$ can be diagonal, which means that the output y_n is obtained through a linear filtering of the square of the input x_n (Figure 8.2.1).

3)   The kernel matrix $\mathbf{H}$ is dyadic, i.e., $h_{i,j}=a_i a_j$, which means that the output is simply the square of the output of a linear filter with an impulse response a_i (Figure 8.2.2).

4)   The kernel matrix $\mathbf{H}$ is Toeplitz, which means that $h_{i,j}=h_{|i-j|}$.

The first case implies that $\mathbf{x}^T \mathbf{H} \mathbf{x} > 0$. This is true if the quadratic filter is used to estimate the power of a random process.

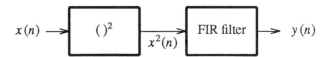

Figure 8.2.1: Special case of quadratic filter employing a square function followed by an FIR linear filter.

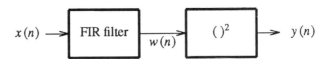

Figure 8.2.2: Special case of quadratic filter employing an FIR linear filter followed by a square function.

An example of the second case is a binary detection problem with multiple observations. Consider the following hypotheses:

$$H_0 : r_i = n_i$$
$$H_1 : r_i = s_i + n_i \quad i=1,2,...,N$$

where H_0 is a null hypothesis and H_1 corresponds to the alternate hypothesis; r_i, n_i and s_i are samples of the observation, noise, and signal, respectively. Consider the case when both s_i and n_i are independently and identically distributed, zero mean Gaussian random variables with variances σ_s^2 and σ_n^2, respectively. Thus, the probability densities of r_i under each hypothesis are

$$P_{r|H_0}(\mathbf{R}|H_0) = P_n(\mathbf{R}|H_0) = \prod_{i=1}^{N} \frac{1}{\sqrt{2\pi}\,\sigma_n} \exp\left[-\frac{R_i^2}{2\sigma_n^2}\right] \qquad (8.2.5a)$$

and

$$P_{\mathbf{r}|H_1}(\mathbf{R}|H_1) = \prod_{i=1}^{N} \frac{1}{\sqrt{2\pi}(\sigma_s+\sigma_n)} \exp\left[-\frac{R_i^2}{2(\sigma_s^2+\sigma_n^2)}\right] \qquad (8.2.5b)$$

then the likelihood ratio is given by

$$\Lambda(\mathbf{R}) = \frac{P_{\mathbf{r}|H_1}(\mathbf{R}|H_1)}{P_{\mathbf{r}|H_0}(\mathbf{R}|H_0)} \qquad (8.2.6)$$

Substituting (8.2.5a) and (8.2.5b) into (8.2.6) and taking the logarithm, we have

$$\Lambda(\mathbf{R}) = N \ln\left[\frac{\sigma_n}{\sigma_s+\sigma_n}\right] + \frac{1}{2}\left[\frac{1}{\sigma_n^2} - \frac{1}{\sigma_s^2+\sigma_n^2}\right] \sum_{i=1}^{N} R_i^2$$

Hence, the sufficient statistic is the sum of the square of the observations

$$l(\mathbf{R}) = \sum_{i=1}^{N} R_i^2 \qquad (8.2.7a)$$

and the likelihood ratio test reduces to

$$l(\mathbf{R}) \underset{H_0}{\overset{H_1}{\underset{<}{>}}} \gamma \qquad (8.2.7b)$$

where γ is the threshold, which depends on the statistics of the signal and noise as well as some predetermined cost. Hence, a decision can be made by passing the observation r_i through a quadratic filter and then a threshold device. It is quite obvious that the quadratic kernel matrix in this case is a diagonal matrix.

The fourth special case (Toeplitz) arises in some applications of adaptive detection [17]. Other special cases of interest of (8.2.1) can be shown to be a memoryless nonlinearity followed by an FIR linear filter and an FIR filter followed by a memoryless nonlinearity.

8.3 HARMONIC REPRESENTATION AND DESIGN OF QUADRATIC FILTERS

Linear filters, because of the convolution property, can be represented efficiently through Fourier, Laplace, and z-transforms. Polynomial filters can be also represented in such a way by means of multidimensional linear transforms, since their input-output relations are expressed by means of a sum of multidimensional convolutions [17].

It is possible to consider (8.2.4) as a reduction of a 2-d convolution (8.3.1):

$$y(n,m) = \sum_{i_1=0}^{N-1} \sum_{i_2=0}^{N-1} h_2(i_1,i_2)w(n-i_1,m-i_2) \tag{8.3.1}$$

for $w(n-i_1,m-i_2)=x(n-i_1)x(m-i_2)$ and $n=m$. Then it is possible to use instead of the kernel $h_2(i_1,i_2)$ its 2-d z-transform:

$$H_2(z_1,z_2) = \sum_{i_1=0}^{\infty} \sum_{i_2=0}^{\infty} h_2(i_1,i_2)z_1^{-i_1}z_2^{-i_2} \tag{8.3.2}$$

or its Fourier transform:

$$G_2(f_1,f_2) = H_2[e^{2\pi jf_1},e^{2\pi jf_2}] \tag{8.3.3}$$

By using this Fourier transform, we can write:

$$y(n) = \int_{-\frac{1}{2}}^{\frac{1}{2}} \int_{-\frac{1}{2}}^{\frac{1}{2}} G_2(f_1,f_2)\cdot X(f_1)X(f_2)e^{2\pi j(f_1+f_2)n} df_1 df_2 \tag{8.3.4}$$

where $X(f)$, is the Fourier transform of $x(n)$. Then the quadratic filter is represented through a double integral in the harmonic representation.

Such a relationship can be easily derived for every term of (8.2.1) and can represent the output of the general polynomial filter (8.2.1), as a sum of single, double, and multiple integrals, which contain the transforms of the kernels $h_k(i_1,...,i_k)$:

$$y(n) = h_o + \sum_{k=1}^{\infty} \overline{H}_k[X(f)] \tag{8.3.5a}$$

where

$$\overline{H}_k[X(f)] = \int_{-\frac{1}{2}}^{\frac{1}{2}} \cdots \int_{-\frac{1}{2}}^{\frac{1}{2}} G_k(f_1,f_2,...,f_k)X(f_1)\cdots X(f_k) \cdot$$

$$\cdot e^{2\pi j(f_1+\cdots +f_k)n} df_1 \cdots df_k \tag{8.3.5b}$$

where $X(f)$, is the Fourier transform of $x(n)$. Similar input output expressions can be easily derived involving m-d discrete Fourier transforms, which can be implemented efficiently by using fast algorithms. In fact, this is the basis for some fast implementations of polynomial filters which are based on similar implementations of linear m-d filters.

We now consider a few particular examples [17]. If the input is monochromatic, then $X(f)=a\delta(f-f_0)$ and the output described by (8.3.4) becomes:

$$y(n) = a^2 G_2(f_0,f_0)e^{2\pi j2f_0n} \tag{8.3.6}$$

which is still monochromatic, but with frequency $2n_1$. If the input is the sum of two monochromatic signals, then $X(f)=a_1\delta(f-f_0)+a_2\delta(f-f_0')$ and the output given by (8.3.4) becomes:

$$y(n) = a_1^2 G_2(f_0, f_0) e^{2\pi j 2 f_0 n} + a_2^2 G_2(f'_0, f'_0) e^{2\pi j 2 f'_0 n}$$
$$+ a_1 a_2 [G_2(f_0, f'_0) + G_2(f'_0, f_0)] e^{2\pi j (f_0 + f'_0) n} \qquad (8.3.7)$$

This output consists of the two terms, which would have resulted by superposition of the two outputs corresponding to the two monochromatic signals applied independently, plus a third term, which is due to the combination of the two inputs and reflects the nonlinear nature of the quadratic filter.

Limited results have been reported in the literature on the problem of quadratic filter design [20,21]. This is mainly due to the fact that quadratic filters do not have a simple characterization, such as the frequency response, that characterizes linear filters. For this reason attempts to design quadratic filters to date have been based on output considerations and have involved optimization techniques.

The approach introduced in [20,21] is based on the following statement of the problem [20]:

"Find the coefficients of a filter with a given structure and properties that optimally produces the desired output from any noisy input."

The resulting optimization problem is therefore non-analytical, i.e., the objective function values are obtained by calculating the filter response and unconstrained, i.e., symmetry conditions are used to reduce the number of independent variables and are not considered as constraints in the optimization procedure. Some designs of quadratic filters for image processing applications were reported in [20,21].

8.4 WIENER FILTERS

Many problems of digital signal processing require determination of a model matching the statistics of a given observation of a generally non-Gaussian random process. This has been attempted in a number of papers [22-30], which were devoted to utilizing the Volterra series for estimation and nonlinear system identification. More recently [2,31-37], discrete time filters of a similar form were studied with and without adaptive implementation.

Due to the convergence problem of Volterra series [1] and the difficulty associated with the measurements of its kernels, Wiener [1] proposed a general means of analyzing a wide range of nonlinear behaviors by subjecting a nonlinear system to a white Gaussian process input. He used a Gram-Schmidt orthogonalization procedure to form an orthogonal set of functionals from the Volterra functionals. For a discrete system with arbitrary input $x(n)$, the output

$y(n)$ is given by:

$$y(n) = \sum_{i=0}^{\infty} G_i[k_i; x(n)] \qquad (8.4.1)$$

where G_i, with a Gaussian input, is a complete set of orthogonal functionals and k_i are the kernels of the system. The first few terms of the discrete Wiener series are [1]:

$$G_0[k_0; x(n)] = k_0$$

$$G_1[k_1; x(n)] = \sum_{i_1} k_1(i_1) x(n-i_1)$$

$$G_2[k_2; x(n)] = \sum_{i_1} \sum_{i_2} k_2(i_1, i_2) x(n-i_1) x(n-i_2) - \beta \sum_{i_1} k_2(i_1, i_1)$$

$$G_3[k_3; x(n)] = \sum_{i_1} \sum_{i_2} \sum_{i_3} k_3(i_1, i_2, i_3) x(n-i_1) x(n-i_2) x(n-i_3)$$

$$-3\beta \sum_{i_1} \sum_{i_3} k_3(i_1, i_1, i_3) x(n-i_3) \qquad (8.4.2)$$

where β is the power spectral density of the white Gaussian input and by k_i we denote the Wiener kernel.

Based on Volterra and Wiener representations, several methods for characterizing a p-th order time-invariant nonlinear system have been developed, both in time/space and frequency domains. While the system identification problem deals with the determination of the optimum p-th order model of a given time-invariant system possibly by minimizing the mean-square difference between the model response and that of the given system, there are situations where this approach is not valid. If the available data are just a realization or the statistics of a non-Gaussian time series and the objective is to determine a model to fit the given observations, input-output based approaches cannot be used in general in estimating the parameters of the proposed model.

Two approaches have been suggested in literature to deal with such a problem. The first approach is to use linear model driven by a non-Gaussian process [38], while the other approach is to use nonlinear model driven by a Gaussian process [39,40]. In both cases, higher order statistics in conjunction with the autocorrelation function provide a powerful tool in estimating the parameters of the proposed model.

This and subsequent sections present a general analysis for the second approach where a nonlinear model is driven by white Gaussian noise. With the absence of prior knowledge required to select the model structure and the complicated numerical problems associated with multidimensional kernels, the second-order Wiener filter will be chosen as the model. Although this type of filters is considered as a special case of systems that can be represented by the

Wiener class, it leads to simplified estimation methods. Furthermore, the second-order Volterra and Wiener filters have the same mathematical structure, with the requirement imposed on their outputs to be zero-mean.

The input-output relationship of the discrete second-order Wiener filter can be obtained from (8.4.1) and (8.4.2):

$$y(n) = k_0 + \sum_{i_1} k_1(i_1)x(n-i_1) + \sum_{i_1}\sum_{i_2} k_2(i_1,i_2)x(n-i_1)x(n-i_2)$$

$$- \beta \sum_{i_1} k_2(i_1,i_1), \quad 0 \leq i_1, i_2 \leq N-1 \tag{8.4.3}$$

where k_0 is a constant whose value depends on the white Gaussian input $x(n)$, N denotes the filter length, and G_1 and G_2 are the first-order and second-order type operators. It is important here to notice that the second term is a linear shift-invariant operator, whose kernel $k_1(i_1)$ is the system impulse response. The last term, on the other hand, is a nonlinear shift-invariant operator, whose kernel $k_2(i_1,i_2)$ can be interpreted in terms of a two-dimensional impulse response.

By assuming that the input signal is a discrete, stationary, zero-mean, white Gaussian process and the output process is also a discrete, stationary, zero mean, non-Gaussian process, it is easy to show that $k_0=0$. In addition, if a second-order Volterra filter was used for system modeling of the form:

$$y(n) = h_0 + \sum_{i_1} h_1(i_1)x(n-i_1) + \sum_{i_1}\sum_{i_2} h_2(i_1,i_2)x(n-i_1)x(n-i_2) \tag{8.4.4}$$

this implies that:

$$h_0 = -\beta \sum_{i_1} k_2(i_1,i_1), \quad h_1(i_1) = k_1(i_1), \quad h_2(i_1,i_2) = k_2(i_1,i_2) \tag{8.4.5}$$

which shows that the second-order Volterra filter and the second-order Wiener filter have the same mathematical structure, with the requirement imposed on their outputs to be zero-mean [38-40].

8.5 POWER SPECTRUM ANALYSIS

In this section, the power spectrum of the output of a second-order Wiener filter will be considered. Since $y(n)$ is the output of the filter, the power spectrum of the zero mean process $y(n)$ is the Fourier transform of its autocorrelation function $r(k)$. Symbolically:

$$S(\omega) = \sum_{k} r(k)\exp(-j\omega k), \quad |\omega| \leq \pi \tag{8.5.1}$$

The autocorrelation function of the real process $y(n)$ is given by:

$$R(k) = E[y(n)y(n+k)] \tag{8.5.2}$$

It is clear from the definitions above that $R(k)$ and $S(\omega)$ contain the same information. The autocorrelation function $R(k)$ provides a time-domain description of the second-order statistics of the process, while the power spectrum $S(\omega)$ provides the frequency-domain description of the second-order statistics of the process. In cases where the process is Gaussian, the autocorrelation function and the power spectrum provide a complete statistical description. By virtue of (8.4.3) and for $k_0=0$, the autocorrelation function of the output $y(n)$ can be written as:

$$R(k) = \sum_{i_1} k_1(i_1)E[x(n-i_1)y(n+k)] +$$

$$+ \sum_{i_1}\sum_{i_2} k_2(i_1,i_2)E[x(n-i_1)x(n-i_2)y(n+k)]$$

$$- \beta \sum_{i_1} k_2(i_1,i_1)E[y(n+k)] \tag{8.5.3}$$

The first term of (8.5.3) involves averaging over the product of three Gaussian random variables, while the second term involves averaging over the product of four Gaussian random variables. It is well known in statistics that the average of the product of an odd number of zero-mean jointly Gaussian random variables is identically zero irrespective of their mutual correlation. And, the average of the product of an even number of zero-mean jointly Gaussian random variables is equal to the summation over all distinct ways of partitioning the random variables into products of averages of pairs [41]. For example, if x_1, x_2, x_3, x_4 are zero-mean jointly Gaussian random variables, then:

$$E[x_1x_2x_3] = 0 \tag{8.5.4}$$

$$E[x_1x_2x_3x_4] = E[x_1x_2] \cdot E[x_3x_4] + E[x_1x_3] \cdot E[x_2x_4] +$$

$$+ E[x_1x_4] \cdot E[x_2x_3] \tag{8.5.5}$$

With (8.5.4) and (8.5.5) in mind and the fact that $k_2(i_1,i_2)$ is a symmetric kernel, $R(k)$ reduces to the form:

$$R(k) = \beta \sum_{i_1} k_1(i_1)k_1(i_1+k) + 2\beta^2 \sum_{i_1}\sum_{i_2} k_2(i_1,i_2)k_2(i_1+k,i_2+k) \tag{8.5.6}$$

Now define $\phi_1(k)$ and $\phi_2(k,l)$ to be as follows:

$$\phi_1(k) = \sum_{i_1} k_1(i_1)k_1(i_1+k) \tag{8.5.7a}$$

$$\phi_2(k,l) = \sum_{i_1}\sum_{i_2} k_2(i_1,i_2)k_2(i_1+k,i_2+l) \tag{8.5.7b}$$

Then we can write (8.5.6) as:

$$R(k) = \beta\phi_1(k) + 2\beta^2\phi_2(k,k) \tag{8.5.8}$$

$\phi_1(k)$ and $\phi_2(k,k)$ are the autocorrelation functions of the kernels of the first-order and second-order operators, respectively. The Fourier transform of (8.5.8) is the power spectral density of the process $y(n)$. Therefore:

$$S(\omega) = \beta|K_1(\omega)|^2 + (\beta^2/\pi)\int_{-\pi}^{\pi}|K_2(\omega_1,\omega-\omega_1)|^2 d\omega_1 \tag{8.5.9}$$

where $K_1(\omega), K_2(\omega_1,\omega_2)$ are the Fourier transforms of the kernels $k_1(i_1), k_2(i_1,i_2)$. In system modeling the problem is to determine the optimum filter that can match the statistics of given data that are assumed to be generated by a second-order Wiener filter. The autocorrelation function given by (8.5.8) is not sufficient to solve the problem. This can simply be explained by two observations. First, the autocorrelation function is an even function of k; and hence, $R(-k)$ does not provide any additional information that can be incorporated in solving (8.5.8). Second, the number of unknowns present in (8.5.8) is much greater than the number of useful samples of $R(k)$. Since the kernels k_1 and k_2 have length N, the autocorrelation function $R(k)$ will have length $2N-1$; and hence, the number of equations provided is N, while the number of unknowns is $N(3+N)/2+1$. In the next section, additional information will be provided by looking at higher order statistics rather than restricting ourselves only to the second-order moment sequence.

8.6 BISPECTRAL ANALYSIS

The bispectrum of a zero-mean stationary process is defined as the two-dimensional Fourier transform of its third-order moment sequence. If we define $M(\tau,p)$ to be the third-order moment sequence of $y(n)$, then [38]:

$$M(\tau,p) = E[y(n)y(n+\tau)y(n+p)] \tag{8.6.1}$$

and, the corresponding bispectrum is:

$$B(\omega_1,\omega_2) = \sum_{\tau}\sum_{p} M(\tau,p)\exp(-j\omega_1\tau-j\omega_2 p), \quad |\omega_1|, |\omega_2|\leq\pi \tag{8.6.2}$$

In the following, the third-order moment sequence of the second-order Wiener filter will be derived. This can be achieved by substituting (8.4.3) for $k_0=0$ into (8.6.1) to obtain:

$$M(\tau,p) = E[(G[k_1;x(n)] + G[k_2;x(n)])$$
$$(G[k_1;x(n+\tau)] + G[k_2;x(n+\tau)])$$
$$(G[k_1;x(n+p)] + G[k_2;x(n+p)])] \tag{8.6.3}$$

For convenience, let us replace the first-order and second-order operators of (8.6.3) by the following notation:

$$B_1 = G[k_1; x(n)]$$

$$B_2 = G[k_2; x(n)]$$

$$B_{1\tau} = G[k_1; x(n+\tau)]$$

$$B_{2\tau} = G[k_2; x(n+\tau)]$$

$$B_{1p} = G[k_1; x(n+p)]$$

$$B_{2p} = G[k_2; x(n+p)] \tag{8.6.4}$$

Based on (8.6.4), one can easily expand (8.6.3) in the following compact form:

$$M(\tau,p) = E[G_1 \cdot G_{1\tau} \cdot G_{1p} + G_1 \cdot G_{1\tau} \cdot G_{2p}$$
$$+ G_1 \cdot G_{2\tau} \cdot G_{1p} + G_1 \cdot G_{2\tau} \cdot G_{2p}$$
$$+ G_2 \cdot G_{1\tau} \cdot G_{1p} + G_2 \cdot G_{1\tau} \cdot G_{2p}$$
$$+ G_2 \cdot G_{2\tau} \cdot G_{1p} + G_2 \cdot G_{2\tau} \cdot G_{2p}] \tag{8.6.5}$$

The first, fourth, sixth, and seventh terms of (8.6.5) involve averaging over an odd number of zero-mean jointly Gaussian random variables. Therefore, they are identically zero. Equation (8.6.5) then becomes:

$$M(\tau,p) = E[G_1 \cdot G_{1\tau} \cdot G_{2p} + G_1 \cdot G_{2\tau} \cdot G_{1p}$$
$$+ G_2 \cdot G_{1\tau} \cdot G_{1p} + G_2 \cdot G_{2\tau} \cdot G_{2p}] \tag{8.6.6}$$

Each term of (8.6.6) involves averaging over an even number of zero-mean jointly Gaussian random variables. Keeping in mind the procedure described in the previous section, one can decompose the average of the product of an even number of jointly Gaussian random variables into a product of averages of pairs. The first term of (8.6.6) can then be written as follows:

$$E[G_1 \cdot G_{1\tau} \cdot G_{2p}] = E[\sum_{i_1} k_1(i_1)x(n-i_1) \cdot \sum_{i_1} k_1(i_1)x(n-i_1+\tau)$$

$$(\sum_{i_1}\sum_{i_2} k_2(i_1,i_2)x(n-i_1+p)x(n-i_2+p) - \beta \sum_{i_1} k_2(i_1,i_2))]$$

$$= 2\beta^2 \sum_{i_1}\sum_{i_2} k_1(i_1)k_1(i_2)k_2(i_1+p,i_2+p-\tau) \tag{8.6.7}$$

Define $\phi_{12}(m,n)$ to be as follows:

$$\phi_{12}(m,n) = \sum_{i_1}\sum_{i_2} k_1(i_1)k_1(i_2)k_2(i_1+m,i_2+n) \tag{8.6.8}$$

Therefore:

$$E[G_1 \cdot G_{1\tau} \cdot G_{2p}] = 2\beta^2 \phi_{12}(p, p - \tau) \tag{8.6.9}$$

Similarly, one can show that:

$$E[G_1 \cdot G_{2\tau} \cdot G_{1p}] = 2\beta^2 \phi_{12}(\tau, \tau - p) \tag{8.6.10}$$

$$E[G_2 \cdot G_{1\tau} \cdot G_{1p}] = 2\beta^2 \phi_{12}(-\tau, -p) \tag{8.6.11}$$

The fourth term of (8.6.6) is quite different from the first three terms. It involves averaging over the product of four Gaussian random variables as well as averaging over the product of six Gaussian random variables. The latter can be broken into the sum of fifteen terms, where each term involves a product of averages of three distinct pairs of random variables. By doing so and defining:

$$\phi_{22}(x, y, z) = \sum_{i_1} \sum_{i_2} \sum_{m} k_2(i_1, i_2) k_2(i_1 + x, m + y) k_2(i_2 + z, m) \tag{8.6.12}$$

we obtain:

$$E[G_2 \cdot G_{2\tau} \cdot G_{2p}] = 8\beta^3 \phi_{22}(p, p - \tau, \tau) \tag{8.6.13}$$

By combining (8.6.6), (8.6.9), (8.6.10) and (8.6.11), (8.6.13) takes the final form:

$$M(\tau, p) = 2\beta^2 (\phi_{12}(p, p - \tau) + \phi_{12}(\tau, \tau - p) + \phi_{12}(-\tau, -p))$$
$$+ 8\beta^3 \phi_{22}(p, p - \tau, \tau) \tag{8.6.14}$$

The corresponding bispectrum of $M(\tau, p)$ is then:

$$B(\omega_1, \omega_2) = 2\beta^2 (2K_1(\omega_2) K_1(-\omega_2 - \omega_1) K_2(\omega_1 + \omega_2, -\omega_2) +$$
$$+ K_1(\omega_1) K_1(\omega_2) K_2(-\omega_1, -\omega_2)) +$$
$$+ 4\beta^3/\pi \int_{-\pi}^{\pi} K_2(-\omega_1 - \omega, -\omega_2 + \omega) K_2(\omega_1 + \omega, -\omega) K_2(\omega_2 - \omega, \omega) d\omega \tag{8.6.15}$$

Examining (8.6.14) shows that the third-order moment sequence $M(\tau, p)$ has been only related to the kernels of the second-order Wiener filter. This, in fact, will help in providing additional information to solve for the kernels k_1 and k_2, and the power of the driving process. Since we assume that the observed data are created by a second-order Wiener filter, their third-order moment should satisfy the conditions imposed by (8.6.14).

In fact, the third-order moment sequence of a zero-mean stationary process possesses some symmetry as is the case of the autocorrelation function. Specifically, if $s(n)$ is a discrete zero-mean stationary process whose third-order moment sequence is $R(m, n)$, then [38]:

$$R(m, n) = R(n, m)$$

$$= R(-n, m-n)$$

$$= R(n-m, -m)$$

$$= R(m-n, -n)$$

$$= R(-m, n-m) \qquad (8.6.16)$$

Based on (8.6.16), the $R(m,n)$ domain can be partitioned into six adjacent sectors. Knowing the samples of $R(m,n)$ in one sector will enable us to determine the samples of $R(m,n)$ in the other sectors. As a consequence, no more than one sector can provide samples that are independent of each other. The samples of the remaining sectors can be determined through the symmetry given by (8.6.16). The number of equations that can be drawn from (8.6.14) is equal to the number of samples contained in the working sector. Generally speaking, the sequence $R(m,n)$ is of an infinite extent. Hence, the number of samples is infinite. However, for a filter of symmetric kernels and length N, the number of independent samples that can be substituted in (8.6.14) is N^2. (8.5.8) in conjunction with (8.6.14) provides a sufficient number of equations required to solve for k_1 and k_2, and the parameter of the white Gaussian process β. The equations are nonlinear. So the least-squares method and other similar methods can be used to achieve the solution.

8.7 DISCUSSION AND APPLICATIONS

In data transmission over bandlimited channels, two types of distortions have been experienced to be a serious source of signal degradation: namely, linear and nonlinear distortions [42]. In low data rates one can ignore the effects of channel distortion and take into account the effect of additive noise only. However, as higher speeds are required, channel distortion becomes the dominant impairment on many voiceband telephone channels [43].

The effect of nonlinear distortion on linearly modulated data signals is to introduce nonlinear intersymbol interference and reduce the margin against noise [4]. Nonlinearities usually arise in telephone channels from nonlinear amplifiers, companders, and other elements. In this section, we address the problem of channel characterization in the presence of quadratic nonlinearities. Higher order nonlinearities can be taken into consideration, where the number of nonlinearities included represents the desired degree of accuracy in the characterization of the channel. However, quadratic and cubic nonlinearities often provide a fairly good model of the channel [42].

A block diagram of a data communication system is shown in Figure 8.7.1. The kernels $h_1(i)$ and $h_3(i)$ represent the impulse responses of the transmitter and receiver filters respectively, while the filter with kernel $h_2(i)$ accounts for the linear distortion introduced by the channel. The quadratic nonlinearity has been modeled as the third term of Volterra series where $h_2(i,j)$ represents its two-dimensional kernel. Since any real channel will contain a source of noise, the additive noise signal $w(n)$ is introduced in the model and assumed to be a zero-mean white Gaussian process, and also independent of the received signal $y(n)$. The overall system depicted in Figure 8.7.1a is equivalent to that of Figure 8.7.1b, where the latter is reduced to the compact form shown in Figure 8.7.1c. The kernels of Figure 8.7.1c can be expressed in terms those of Figure 8.7.1a as shown below [44]:

$$h(i) = h_1(i)*h_2(i)*h_3(i) \tag{8.7.1}$$

$$h(i,j) = \sum_m \sum_l \sum_k h_1(i-m-l)h_1(j-m-k)h_2(l,k)h_3(m)$$

where $*$ denotes convolution. The system shown in Figure 8.7.1c is, actually, a second-order Volterra filter whose first- and second-order kernels are given by (8.7.1).

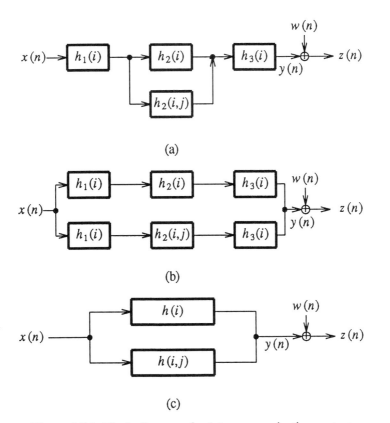

(a)

(b)

(c)

Figure 8.7.1: Block diagram of a data communication system.

Now, if we probed the channel by a discrete zero-mean stationary Gaussian process having a constant power spectral density β, the received signal $z(n)$ would be a noise corrupted version of the non-Gaussian process $y(n)$. To make use of the analysis presented in the previous sections, one should keep in mind the following:

1) The central second-order moment sequence (autocovariance) and the central third-order cumulant sequence of the second-order Volterra filter for $h_0=0$ are equal to the second-order and third-order moment sequences of the second-order Wiener filter for $k_0=0$, respectively.

2) Since $y(n)$ and $w(n)$ are two independent random processes, it follows that the central second-order moment sequence of $z(n)$ equals the central second-order moment sequence of $y(n)$ plus that of $w(n)$; and so does the cumulant.

3) Recognizing that the third-order cumulant of a stationary Gaussian random process is identically zero, the third-order cumulant of the received signal $z(n)$ is equal to the third-order cumulant of $y(n)$ alone.

4) Practically speaking, the additive noise transformed at the receiver output is generally a nonwhite random process. However, the autocorrelation function of $w(n)$ usually assumed to be so small at lags greater than zero, that it can be neglected. The main reason for that is to reduce the number of unknowns to the level of the number of equations.

Based on what is mentioned above, the mean, autocovariance and third-order cumulant sequences of $z(n)$ are given by:

$$\mu = -\beta \sum_{i_1} h_2(i_1, i_1) \tag{8.7.2}$$

$$C(\tau) = \beta\phi_1(\tau) + 2\beta^2\phi_2(\tau,\tau) + Q\delta(\tau) \tag{8.7.3}$$

$$M(\tau,p) = 2\beta^2(\phi_{12}(p,p-\tau) + \phi_{12}(\tau,\tau-p) + \phi_{12}(-\tau,-p))$$
$$+ 8\beta^3\phi_{22}(p,p-\tau,\tau) \tag{8.7.4}$$

where $Q\delta(\tau)$ is the autocorrelation function of the additive noise and $\phi_1, \phi_2, \phi_{12}, \phi_{22}$ are given by (8.5.7a), (8.5.7b), (8.6.8) and (8.6.12), respectively, with the replacement of $k_1(i_1), k_2(i_1,i_2)$ by the kernels of Figure 8.7.1c. (8.7.2)-(8.7.4) can be used when we have the true mean, autocovariance and third-order cumulant sequences of $z(n)$. If we are given finite samples of $z(n)$, we can use estimates of these quantities in place of the true values. We form these estimates by dividing up the received signal into k records of M samples each. For each record we compute the following estimates:

$$\mu' = (1/M)\sum_{n} z(n) \qquad 0 \le n \le M-1 \tag{8.7.5}$$

$$C(\tau)' = (1/M) \sum_n (z(n) - \mu')(z(n+\tau) - \mu') \qquad (8.7.6)$$

$$M(\tau,p)' = (1/M) \sum_n (z(n) - \mu')(z(n+\tau) - \mu')(z(n+p) - \mu') \qquad (8.7.7)$$

and then we average over all the k records. For a system of order two the parameter vector that should be estimated is:

$$[h(0) \quad h(1) \quad h(0,0) \quad h(0,1) \quad h(1,0) \quad h(1,1) \quad Q \quad \beta]^T \qquad (8.7.8)$$

such a vector can be estimated from the estimates of the mean, autocorrelation function (or spectral density) and third order moment (or bispectrum) of the output of such a system [44].

Recently, polynomial type filters have been extended to two and m-dimensions [45,46], adaptive polynomial filters have been considered [47], decision directed nonlinear filters for image processing have been studied [48], while considerable attention has been given to issues of fast and efficient implementation of polynomial filters [49]. This issue will be further considered in chapter 11.

In addition, the correspondence of classical Volterra filters described by (8.2.1) to FIR linear filters suggests possibilities for the study of nonlinear discrete systems, which would be analogous to IIR linear filters and would be described by rational Volterra theory [50]. The input output selection of such a filter is shown in (8.7.9):

$$y(n) = \frac{N(n)}{D(n)} = \frac{h_0^N + \sum_{k=1}^{K} \bar{h}_k[x(n)]}{h_0^D + \sum_{l=1}^{L} \bar{h}_l[x(n)]} \qquad (8.7.9)$$

here $\bar{h}_k[x(n)]$ and $\bar{h}_l[x(n)]$ are numerator and denominator Volterra kernels, which fully characterize the system. Such a method was introduced to estimate the kernel of a system in [50] in order to determine a model with a rational Volterra representation.

REFERENCES

[1]   M. Schetzen, *The Volterra and Wiener theories of nonlinear systems*, Wiley, 1980.

[2]   E. Biglieri, "Theory of Volterra processors and some applications", in *Proc. ICASSP-82*, Paris, pp. 294-297, 1982.

[3]   S. Benedetto, E. Biglieri, "Nonlinear equalization of digital satellite chan-
       nels", presented at the *9th AIAA Conf. Comm. Satellite Syst.*, San Diego,
       CA, March 1982.

[4]   D.D. Falconer, "Adaptive equalization of channel nonlinearities in QAM
       data transmission systems", *Bell Sys. Tech. J.*, vol. 57, pp. 2589-2611,
       Sept. 1978.

[5]   G.L. Sicuranza, A. Bucconi, P. Mitri, "Adaptive echo cancellation with
       nonlinear digital filters", in *Proc. IEEE Int. Conf. Acoust., Speech, Signal
       Processing*, San Diego, CA, pp. 3.10.1 - 3.10.4, 1984.

[6]   O. Agazzi, D.G. Messerschmitt, D.A. Hodges, "Nonlinear echo cancella-
       tion of data signals", *IEEE Trans. Commun.*, vol. COM-30, pp. 2421-
       2433, Nov. 1982.

[7]   G.L. Sicuranza, G. Ramponi, "Distributed arithmetic implementation of
       nonlinear echo cancellers", *Proc. IEEE Int. Conf. on Acoust., Speech and
       Signal Processing*, ICASSP-85, pp. 42.5.1 - 42.5.4, Tampa FL, 1985.

[8]   B.E.A. Saleh, "Optical bilinear transformation: General properties",
       Optica Acta, vol. 26, no. 6, pp. 777-799, 1979.

[9]   G. Ramponi, "Quadratic filters for image enhancement", *Proc. Fourth
       European Signal Processing Conf.*, EUSIPCO-88, Grenoble, France, pp.
       239-242, 1988.

[10]   G. Ramponi, "Enhancement of low-contrast images by nonlinear opera-
       tors", *Alta Frequenza*, vol. LVII, no. 7, pp. 451-455, Sept. 1988.

[11]   G. Ramponi, G. Sicuranza, "Quadratic digital filters for image process-
       ing", *IEEE Trans. on Acoust., Speech, Signal Processing*, vol. ASSP-36,
       no. 6, pp. 937-939, June 1988.

[12]   G. Ramponi, "Edge extraction by a class of second-order nonlinear
       filters", *Electronics Letters*, vol. 22, no. 9, April 24, 1986.

[13]   R. Glavina, G. Ramponi, S. Cucchi, G. Sicuranza, "Interframe image cod-
       ing with nonlinear prediction", *Tecnica Italiana*, no. 2, 1988.

[14]   G.L. Sicuranza, G. Ramponi, "Adaptive nonlinear prediction of TV image
       sequences", *Electronics Letters*, vol. 25, no. 8, pp. 526-527, April 13,
       1989.

[15]   R. Glavina, S. Cucchi, G. Sicuranza, "Nonlinear interpolation of TV
       image sequences", *Electronics Letters*, vol. 23, no. 15, pp. 778-780, July
       16, 1987.

[16]   M.J. Hinich, D.M. Patterson, "Evidence of nonlinearity in daily stock
       returns", *Journal of Business and Economic Statistics*, 3, pp. 69-77, 1985.

[17]   B. Picinbono, "Quadratic filters", *Proc. of IEEE ICASP*, Paris, France, pp.
       298-301, 1982.

[18] H.H. Chiang, C.L. Nikias, A.N. Venetsanopoulos, "Efficient implementations of quadratic filters", *IEEE Trans. on Acoustics, Speech and Signal Processing*, vol. ASSP-34, no. 6, pp. 1511-1528, Dec. 1986.

[19] G.L. Sicuranza, "Theory and realization of nonlinear digital filters", in *Proc. IEEE ISCAS-84*, Montreal, Canada, May 1984.

[20] G. Ramponi, G. Sicuranza, W. Ukovich, "An optimization approach to the design of nonlinear Volterra filters", *Proc. of EUSIPCO-86*, The Hague, The Netherlands, 1986.

[21] G. Ramponi, G. Sicuranza, W. Ukovich, "A computational method for the design of 2-D nonlinear Volterra filters", *IEEE Trans. on Circuits and Systems*, vol. CAS-35, no. 9, Sept. 1988.

[22] J. Katzenelson, L.A. Gould, "The design of nonlinear filters and control systems", *Inform. Contr.*, vol. 5, pp. 108-143, 1962.

[23] J.F. Barret, "The use of functionals in the analysis of nonlinear system", *J. Electron. Contr.*, vol. 15, no. 6, pp. 567-615, 1963.

[24] P. Eykhoff, "Some fundamental aspects of process-parameter estimation", *IEEE Trans. Automat. Contr.*, vol. AC-8, pp. 347-357, Oct. 1963.

[25] A.V. Barakrishnan, "A general theory of nonlinear estimation problems in control systems", *J. Math. Anal. Appl.*, vol. 8, pp. 4-30, Feb. 1964.

[26] M. Schetzen, "Nonlinear system modeling based on the Wiener theory", *Proc. IEEE*, vol. 69, pp. 1557-1573, Dec. 1981.

[27] A.S. French, E.G. Butz, "Measuring the Wiener kernels of a non-linear system using the fast Fourier transform algorithm", *Int. J. Contr.*, vol. 17, no. 3, pp. 529-539, 1973.

[28] Y.W. Lee, M. Schetzen, "Measurement of the Wiener kernels of a nonlinear system by cross-correlation", *Int. J. Contr.*, vol. 2, no. 3, pp. 237-254, 1965.

[29] A.V. Barakrishnan, "On a class of nonlinear estimation problems", *IEEE Trans. Inform. Theory*, vol. IT-10, pp. 314-320, Oct. 1964.

[30] D.R. Brillinger, "The identification of polynomial systems by means of higher order spectra", *J. Sound Vib.*, vol. 12, no. 3, pp. 301-313, 1970.

[31] M.J. Coker, D.N. Simkins, "A nonlinear adaptive noise canceller", in *Proc. 1980 IEEE Int. Conf. Acoust., Speech, Signal Processing*, pp. 470-473, 1980.

[32] D.M. Mansour, A.H. Gray Jr., "Frequency domain non-linear filter", in *Proc. 1981 IEEE Int. Conf. Acoust., Speech, Signal Processing*, pp. 550-553, 1981.

[33] S.R. Parker, F.A. Perry, "A discrete ARMA model for nonlinear system identification", *IEEE Trans. Circuits Syst.*, vol. CAS-28, pp. 224-233, March 1981.

[34] T. Koh, E.J. Powers, "An adaptive nonlinear digital filter with lattice orthogonalization", in *Proc. 1983 IEEE Int. Conf. Acoust., Speech, Signal Processing*, pp. 37-40, 1983.

[35] T. Koh, E.J. Powers, R.W. Miksad, F.J. Fischer, "Application of nonlinear digital filters to modeling low-frequency drift oscillations of moored vessels in random seas", in *Proc. 1984 Offshore Technol. Conf.*, pp. 309-314, 1984.

[36] S.R. Parker, "An autoregressive moving average (ARMA) discrete nonlinear model", in *Proc. IEEE Int. Symp. Circuits Syst.*, pp. 918-920, 1980.

[37] S.Y. Fakhouri, "Identification of the Volterra kernels of nonlinear systems", *Proc. IEE*, vol. 127, pt. D, pp. 296-304, Nov. 1980.

[38] C.M. Nikias, M.R. Raghuveer, "Bispectrum estimation: A digital signal processing framework", *Proc. IEEE*, vol. 75, pp. 869-891, July 1987.

[39] S.A. Diant, M.R. Raghuveer, "Estimation of the parameters of a second-order nonlinear system", *Int. Contr. on ACC*, Baton Rouge, Louisiana, Oct. 1988.

[40] V.Z. Marmarelis, D. Sheby, "Bispectrum analysis of weakly nonlinear quadratic systems", in *Proc. ASSP Spectrum Estimation and Modeling Workshop III*, Boston, MA, pp. 14-16, Nov. 1986.

[41] W.B. Davenport, W.L. Root, "An introduction to the theory of random signals and noise", McGraw-Hill, New York, 1958.

[42] W.J. Lawless, M. Schwartz, "Binary signaling over channels containing quadratic nonlinearities", *IEEE Trans.*, vol. COM-22, pp. 288-298, 1974.

[43] S. Benedetto, E. Biglieri, R. Daffara, "Performance of multilevel baseband digital systems in a nonlinear environment", *IEEE Trans. Commun.*, pp. 1166-1175, Oct. 1976.

[44] S.A. Alshebeili, *Linear and nonlinear modeling of non-Gaussian processes*, Ph.D. Thesis, Department of Electrical Engineering, University of Toronto, in preparation.

[45] G. Sicuranza, G. Ramponi, "Theory and realization of M-D nonlinear digital filters", *Proc. of IEEE Int. Conf. on Acoust., Speech, Signal Processing*, Tokyo, Japan, April 7-11, 1986.

[46] A.N. Venetsanopoulos, K.M. Ty, A.C.P. Loui, "High speed architectures for digital image processing", *IEEE Trans. on Circuits and Systems*, vol. CAS-34, no. 8, pp. 887-896, Aug. 1987.

[47] G. Sicuranza, G. Ramponi, "Adaptive nonlinear digital filters using distri-
      buted arithmetic", *IEEE Trans. on Acoust., Speech, Signal Processing*,
      vol. ASSP-34, no. 3, pp. 518-526, June 1986.

[48] G. Ramponi, G. Sicuranza, "Decision-directed nonlinear filter for image
      processing", *Electronics Letters*, vol. 23, no. 23, pp. 1218-1219, Nov. 5,
      1987.

[49] V.G. Mertzios, G.L. Sicuranza, A.N. Venetsanopoulos, "Efficient struc-
      tures for two-dimensional quadratic filters", *Photogrammetria*, vol. 43,
      pp. 157-166, 1989.

[50] E.V.D. Eijnd, J. Scfionkens, J. Renneboog, "Parameter estimation in
      rational Volterra models", *IEEE International Symposium of Circuits and
      Systems*, Philadelphia, PA, pp. 110-114, May 1987.

CHAPTER 9

ADAPTIVE NONLINEAR FILTERS

9.1 INTRODUCTION

The nonlinear filters described in the previous chapters are usually optimized for a specific type of noise and sometimes for a specific type of signal. However, this is not usually the case in many applications of nonlinear filtering, especially in image processing. Images can be modeled as two-dimensional stochastic processes, whose statistics vary in the various image regions. Images are nonstationary processes. Furthermore the noise statistics, e.g., the noise standard deviation and even the noise probability density function, vary from application to application, as was described in chapter 3. Sometimes, the noise characteristics vary in the same application from one image to the next. Such cases are the channel noise in image transmission and the atmospheric noise (e.g., the cloud noise) in satellite images. In these environments non-adaptive filters cannot perform well because their characteristics depend on the noise and signal characteristics, which are unknown. Therefore, adaptive filters are the natural choice in such cases. Their performance depends on the accuracy of the estimation of certain signal and noise statistics, namely the signal mean and standard deviation and the noise standard deviation. The estimation is usually local, i.e., relatively small windows are used to obtain the signal and noise characteristics. An important property of these estimators is their robustness to impulse noise, which is present in many image processing applications. Another reason for using adaptive filters is the fact that edge information is very important for the human eye and must be preserved. Certain filters, e.g., the moving average, perform well in homogeneous image regions but fail close to edges. The opposite is true for other filters, e.g., for the median. A combined filter which performs differently in the image edges than in the image plateaus can be used in such a case. These filters are also called *decision directed filters* because they employ an edge detector to decide if an edge is present or not. Decision directed filtering can also be used in the cases of mixed additive white noise and impulsive noise. Impulses can be detected and removed before the additive noise filtering is performed. Another approach related to decision directed filtering is the *two-component model* filtering. An image is assumed to consist of two components, the low-pass and the high-pass component. The first one is mainly related to homogeneous image regions, whereas the second one is related to edge information. These two components can be processed in different ways. The output of the two corresponding filters can be recombined to give the final filtered image. The two-component image processing model has been used both for noise removal and image enhancement applications.

The performance of most adaptive nonlinear filters is based on the reliable estimation of image and noise characteristics or on the robust edge detection in the image. Therefore, in subsequent sections robust edge detection and local statistics estimation will be described.

9.2 ROBUST ESTIMATORS OF SCALE

Standard deviation estimation of signal and/or of noise is very important for many adaptive nonlinear filters. It corresponds to the so called *scale estimation* in statistics. Perhaps the most widely known standard deviation estimator is the *sample standard deviation s :*

$$s = \left[\frac{1}{n-1} \sum_{i=1}^{n} (x_i - \overline{x})^2 \right]^{1/2} \tag{9.2.1}$$

where x_i, $i=1,..,n$ are the signal values in an image window and $\overline{x}$ is the arithmetic mean:

$$\overline{x} = \frac{1}{n} \sum_{i=1}^{n} x_i \tag{9.2.2}$$

If signal x is two-dimensional and the window size is $n \times n$ ($n=2v+1$), the local standard deviation is given by:

$$s_{ij} = \left[\frac{1}{n^2-1} \sum_{k=i-v}^{i+v} \sum_{l=j-v}^{j+v} (x_{kl} - \overline{x}_{ij})^2 \right]^{1/2} \tag{9.2.3}$$

$$\overline{x}_{ij} = \frac{1}{n^2} \sum_{k=i-v}^{i+v} \sum_{l=j-v}^{j+v} x_{kl} \tag{9.2.4}$$

The (biased) standard deviation is a maximum likelihood estimator of scale for the normal distribution, as was described in chapter 2. It is also an M-estimator of scale. Its main disadvantage is that it is not robust to outliers. Its breakdown point ε^* is equal to 0. Thus, even one outlier can destroy its performance. Therefore, more robust estimators of scale must be found. Robust M-estimators and L-estimators of scale have already been reviewed in chapter 2. The main disadvantage of the M-estimators of scale is their implicit definition which leads to iterative techniques for their computation. In contrast, L-estimators of scale are based on order statistics and are easily computed:

$$\sigma^* = \sum_{i=1}^{n} a_i x_{(i)} \tag{9.2.5}$$

where $x_{(i)}$ is the i-th order statistic. The coefficients a_i depend on the probability distribution $F(x)$ given by:

$$F(x) = F(\frac{x-\mu}{\sigma}), \qquad \sigma > 0 \tag{9.2.6}$$

A joint estimation of both the mean μ and the standard deviation σ can be made [1, pp.128-133], leading to the estimate of the form (9.2.5):

$$\sigma^* = e^T \frac{R^{-1}(em^T - me^T)R^{-1}}{|\Delta|} x \tag{9.2.7}$$

x is the vector of the ordered data $[x_{(1)},..,x_{(n)}]^T$, e is the unit vector, $m = [\mu_1,..,\mu_n]^T$ is the vector of the expected values of the ordered random variables $y_{(i)}$:

$$\mu_i = E[y_{(i)}] = E[\frac{x_{(i)} - \mu}{\sigma}] \tag{9.2.8}$$

R is the covariance matrix of the variables $y_{(i)}$:

$$R = \left[cov(y_{(i)}, y_{(j)}) \right] \tag{9.2.9}$$

The matrix Δ is given by:

$$\Delta = \begin{bmatrix} e^T R^{-1} e & e^T R^{-1} m \\ m^T R^{-1} e & m^T R^{-1} m \end{bmatrix} \tag{9.2.10}$$

The joint estimate of the mean μ is given by:

$$\mu^* = -m^T \frac{R^{-1}(em^T - me^T)R^{-1}}{|\Delta|} x \tag{9.2.11}$$

If the distribution $F(x)$ is symmetric, (9.2.5),(9.2.7) can be further simplified:

$$\sigma^* = \frac{m^T R^{-1} x}{m^T R^{-1} m} \tag{9.2.12}$$

Fortunately, there exist even simpler and faster scale estimates. One of the most known robust scale estimates is the *median of the absolute deviations from the median (MAD)*:

$$\sigma^* = 1.483 \, med\{|x_i - med(x_j)|\} \tag{9.2.13}$$

The constant $1/\Phi^{-1}(3/4) \simeq 1.483$ is just a scaling factor. It was noted in chapter 2 that the MAD is a robust M-estimator. Its breakdown point is $\varepsilon^* = 1/2$. Therefore, it can reject up to 50% of all outliers. Its influence function is bounded and its gross error sensitivity is $\gamma^* = 1.167$. Therefore, the MAD possesses excellent qualitative and quantitative robustness. Another well-known *L*-estimator of scale is the *t-quantile range* whose coefficients a_i are given by:

$$a_i = \begin{cases} -1 & i=tn \\ 1 & i=(1-t)n \\ 0 & i \neq tn, (1-t)n \end{cases} \qquad (9.2.14)$$

The *interquartile range* ($t = 1/4$) has the same influence function with $MAD(x_i)$ for the normal distribution. Therefore it is *B*-robust and it has gross-error sensitivity $\gamma^* \simeq 1.167$. However, its breakdown point is only $\varepsilon^* = 1/4$ instead of $\varepsilon^* = 1/2$ for the $MAD(x_i)$, i.e., the MAD has better global robustness than the interquantile range. The *t*-quantile range is equivalent to the *quasi range* :

$$W_{(i)} = x_{(n+1-i)} - x_{(i)} , \qquad 2 \leq i \leq [\frac{n}{2}] \qquad (9.2.15)$$

The quasi-range reduces to the *range W* for $t=0$:

$$W = x_{(n)} - x_{(1)} \qquad (9.2.16)$$

The range can be easily computed, but it is not robust. Even one outlier can destroy its performance completely.

9.3 NONLINEAR EDGE DETECTORS

As will be seen in subsequent sections, many adaptive filters employ edge information for their adaptation. Therefore, a description of some useful edge detectors is included in this section. Edges are basic features of images, which carry valuable information, useful both in image filtering and in image analysis. However, a precise and widely accepted mathematical definition of an edge is not available yet. In the following, we shall consider as an edge the border between two homogeneous image regions having different illumination intensities. This definition implies that an edge is also a local variation of illumination (but not vice versa).

Several edge detectors have been proposed in the literature [2,3]. Each of them has its advantages and disadvantages, according to the mathematical model of an edge that it uses. Edge detectors can be grouped in the following categories:

a)   Local techniques

b)   Filtering techniques

c)   Regional techniques

d)   Dynamic techniques

e)   Relaxation techniques.

An excellent survey of the most widely known edge detectors can be found in [2,3,5]. However, the comparison of the performance of the various edge detectors is usually qualitative, although some efforts have been made on the quantitative analysis of the performance of edge detectors [4]. In the following, some new edge detectors will be presented which are based on local techniques. These edge detectors are essentially measures of the local standard deviation of the image. Local standard deviation is small in the homogeneous image regions and larger close to the edges. Therefore, any local estimator of scale proposed in the previous section can be used as edge detector, if it is followed by a thresholder. The range W and the quasi range $W_{(i)}$ can be used as the *range edge detector* and the *quasi-range edge detectors*, respectively [6]. The *thickened edge* :

$$J_i = W + W_{(1)} + W_{(2)} + ... + W_{(i)} \tag{9.3.1}$$

can also be used as the so-called *dispersion edge detector*, which is more robust than the range and the quasi-range edge detectors [6]. All these edge detectors are essentially special cases of the L-filter, whose structure is shown in Figure 9.3.1, if the coefficients a_i, $i=1,..,n$ are chosen appropriately.

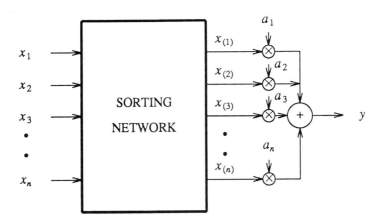

Figure 9.3.1: Filter structure for edge detection based on order statistics.

Three other measures of data dispersion [1] can also be used as edge detectors:

$$W_A = \frac{2\sqrt{\pi}}{n(n-1)} \sum_{i=1}^{n} [i - \frac{1}{2}(n+1)]x_{(i)} \tag{9.3.2}$$

$$W_A = \sum_{i=1}^{[n/2]} \frac{(n-2i+1)}{n(n-1)} W_{(i)} \tag{9.3.3}$$

$$W_A = \frac{2}{n(n-1)} \sum_{i=1}^{n-1} i(n-i)(x_{(i+1)} - x_{(i)}) \tag{9.3.4}$$

They can also be implemented by the structure shown in Figure 9.3.1.

The range edge detector can be considered as the difference of two nonlinear filters maximum and minimum. Other nonlinear filters can replace the max/min operators. The resulting edge detector filter is shown in Figure 9.3.2.

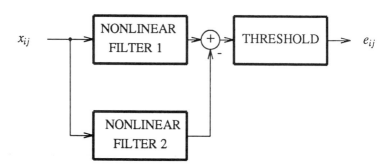

Figure 9.3.2: Edge detector based on the difference of two nonlinear filters.

Filters which approximate maximum and the minimum filters are the L_p filter and the L_{-p} filter, respectively, which have been described in chapter 5:

$$y_{L_p} = \left(\frac{\sum_{i=1}^{n} a_i x_i^p}{\sum_{i=1}^{n} a_i} \right)^{1/p} \tag{9.3.5}$$

$$y_{L_{-p}} = \left(\frac{\sum_{i=1}^{n} a_i x_i^{-p}}{\sum_{i=1}^{n} a_i} \right)^{-1/p} \tag{9.3.6}$$

The difference of the L_p filter and the L_{-p} filter is the L_p *edge detector*. Another set of filters approximating the maximum and the minimum are the CH_p and CH_{-p} filters:

$$y_{CH_p} = \frac{\sum_{i=1}^{n} x_i^{p+1}}{\sum_{i=1}^{n} x_i^p} \tag{9.3.7}$$

$$y_{CH_p} = \frac{\sum\limits_{i=1}^{n} x_i^{-p+1}}{\sum\limits_{i=1}^{n} x_i^{-p}} \tag{9.3.8}$$

Their difference is the CH_p *edge detector* . Their performance in the presence of a noisy edge is shown in Figure 9.3.3. An example of the performance of the described nonlinear edge detectors is shown in Figure 9.3.4. The range edge detector detects even low contrast edges, but its output has high background noise. Thus a large threshold is required. L_p and CH_p edge detectors and dispersion edge detector do not detect some low-contrast edges, but have much less background noise.

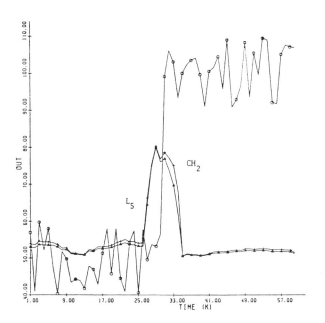

Figure 9.3.3: Output of the L_5 and CH_2 edge detectors in the presence of a noisy edge.

Another set of edge detectors is based on the local estimates $\hat{\sigma}_x$ and $\hat{\sigma}_n$ of image and background noise:

$$e_{ij} = \begin{cases} (\hat{\sigma}_x/\hat{\sigma}_n)(\hat{\sigma}_x/\hat{\sigma}_n - 1) & \hat{\sigma}_x \geq \hat{\sigma}_n \\ 0 & \hat{\sigma}_x < \hat{\sigma}_n \end{cases} \tag{9.3.9}$$

(a)   (b)

(c)   (d)

Figure 9.3.4: (a) Original image; (b) Range edge detector output; (c) Dispersion edge detector output; (d) L_4 edge detector output.

At homogeneous image regions the signal standard deviation estimate is approximately equal to noise standard deviation estimate and the output of the edge detector is 0. At edge regions the signal standard deviation is much greater than the standard deviation of the noise and the output of the edge detector (9.3.9) is large. Any estimator of signal and noise standard deviation can be used in (9.3.9). The sample standard deviation and the MAD can be such estimates [7]. An example of the performance of this edge detector is given in Figure 9.3.5. Its output has low output noise compared to the output of the Sobel edge detector

[2]. However, it tends to produce gaps in the detected edges.

(a)   (b)

(c)

Figure 9.3.5: (a) Original image; (b) Sobel edge detector output; (c) Output of the edge detector based on the local standard deviation estimate.

A completely different approach in using order statistics for edge detection is followed in [8]. Let A_L, A_R be two neighboring windows, shown in Figure 9.3.6. Let also $\hat{x}_L$, $\hat{x}_R$ be the mean image intensities in the corresponding windows. If their difference is larger than a threshold:

$$|\hat{x}_L - \hat{x}_R| > \tau \qquad (9.3.10)$$

an edge is declared. The local intensity mean estimates can be the local arithmetic means or the local medians [8]. The use of local arithmetic mean differences leads to an edge detection scheme similar to the well known Prewitt edge detectors, if A_L, A_R correspond to the three nearest pixels to the left or to the right of the center of a 3×3 window. This edge detector is very susceptible to impulsive noise as, it is shown in Figure 9.3.7. In contrast, the absolute difference of medians is much more robust to the existence of impulses.

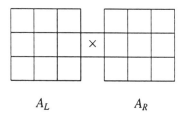

$$A_L \qquad\qquad A_R$$

Figure 9.3.6: Two neighboring windows used in edge detection.

A modification of the above-mentioned edge detector is the following. Let x_i, $i=1,..,n$ be the pixels belonging to A_L and x_i, $i=n+1,..,2n$ the pixels belonging to the window A_R. Two hypotheses can be formed: The first one (H_0) assumes that all pixels have the same mean m. The second one (H_1) assumes that the mean in the two windows have different means $m_1 \neq m_2$. If the second hypothesis is correct, an edge is present. The hypotheses test can be done by the *rank tests* [9]. Let

$$x_{(1)} \le \cdots \le x_{(n)} \le \cdots x_{(2n)} \qquad\qquad (9.3.11)$$

be the order statistics of x_i, $i=1,..,2n$. Let also R_i, $i=1,..,2n$ be the ranks of x_i, $i=1,..,2n$ in the order statistics (9.3.11). The *Wilcoxon test* uses the following statistic:

$$S = \sum_{i=1}^{n} R_i \qquad\qquad (9.3.12)$$

For the hypothesis H_1: $m_1 > m_2$ against H_0 it is expected that the observations $x_1,..,x_n$ are statistically larger then the observations $x_{n+1},..,x_{2n}$. Therefore, the sum of their ranks (9.3.12) is expected to be larger than a constant denoted by $(S | H_0)$:

$$(S | H_1 : m_1 > m_2) \ge (S | H_0) \qquad\qquad (9.3.13)$$

The opposite is true for the hypothesis that $m_1 < m_2$:

$$(S | H_1 : m_1 < m_2) \le (S | H_0) \qquad\qquad (9.3.14)$$

Therefore, an edge is present when S is larger or lower than $(S \mid H_0)$ by a certain threshold.

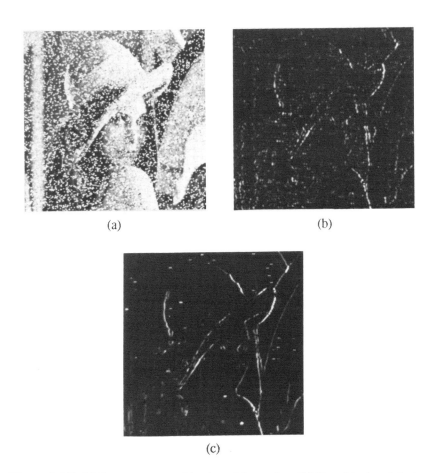

(a)   (b)

(c)

Figure 9.3.7: (a) Image corrupted by impulsive noise; (b) Prewitt edge detector output; (c) Absolute difference of the medians edge detector output.

Another useful test is the *median test*, which is based on the following statistic:

$$M = \sum_{i=n+1}^{2n} Z(i) \qquad (9.3.15)$$

$$Z(i) = \begin{cases} 1 & \text{if } x_{(i)} \in \{x_1, \ldots, x_n\} \\ 0 & \text{if } x_{(i)} \in \{x_{n+1}, \ldots, x_{2n}\} \end{cases} \qquad (9.3.16)$$

Again if $m_1 > m_2$, the larger order statistics $x_{(i)}$, $i = n+1, .., 2n$ are expected to belong to $\{x_1, .., x_n\}$ and the statistic M is expected to be greater than a constant denoted by $(M \mid H_0)$:

$$(M \mid H_1 : m_1 > m_2) \geq (M \mid H_0) \tag{9.3.17}$$

The opposite is true for the hypothesis $m_1 < m_2$:

$$(M \mid H_1 : m_1 < m_2) \leq (M \mid H_0) \tag{9.3.18}$$

Therefore, an edge exists if (9.3.17) or (9.3.18) are satisfied, i.e. if local statistic M is greater or lesser than $(M \mid H_0)$ by a threshold value. Modifications of both the Wilcoxon and the median test have been used for edge detection [10]. A major problem with both of these edge detectors is the choice of the appropriate thresholds.

9.4 REJECTION OF IMPULSES BY ORDER STATISTICS

Sometimes, the observed image obviously contains impulses. Adaptive filters must remove these impulses. A simple remedy in this case would be the following: detect the impulses (outliers), reject them, and filter the image based on the rest of the image pixels. Several tests for outlier detection have been proposed in robust estimation theory and they can by used in image filtering, too. In the following such outlier tests will be described.

Let $x_1, .., x_n$ be the observed data, which are assumed to follow a normal distribution having mean μ and standard deviation σ. This is the "null" situation. If one or more of the data follow another distribution or they have a shift of the mean and (possibly) of the standard deviation, they are called outliers. Usually the distribution mean and standard deviation and the number of outliers k ($k < n$) is unknown. Also the direction of the shift of the mean is unknown. These two facts justify the existence of many tests for outliers [1, p.221].

If the mean μ and the standard deviation σ are known and only one outlier exists with a shift to the right (to greater values), the following test is a natural choice:

$$A_1 = \frac{x_{(n)} - \mu}{\sigma} \tag{9.4.1}$$

The outlier is rejected if:

$$A_1 > c \tag{9.4.2}$$

When c is large, only the "wild" outliers are rejected. When c is smaller, even "mild" outliers are rejected. If the direction of the shift of the outlier is unknown, the following test is appropriate:

$$A_2 = \frac{\max |x_i - \mu|}{\sigma} \qquad (9.4.3)$$

When the standard deviation σ is known and the mean μ is unknown, an estimate of the mean can be used instead. If arithmetic mean $\bar{x}$ is used as an estimate, the tests A_1, A_2 take the form:

$$B_1 = \frac{x_{(n)} - \bar{x}}{\sigma} \qquad (9.4.4)$$

$$B_2 = \frac{\max |x_i - \bar{x}|}{\sigma} \qquad (9.4.5)$$

However, $\bar{x}$ is not a robust estimator of the mean. Therefore, the median can be used instead leading to the following tests:

$$B'_1 = \frac{x_{(n)} - med(x_j)}{\sigma} \qquad (9.4.6)$$

$$B'_2 = \frac{\max |x_i - med(x_j)|}{\sigma} \qquad (9.4.7)$$

If both mean and standard deviation are unknown, the arithmetic mean $\bar{x}$ and the sample standard deviation s can be used. The following tests are obtained:

$$D_1 = \frac{x_{(n)} - \bar{x}}{s} \qquad (9.4.8)$$

$$D_2 = \frac{\max |x_i - \bar{x}|}{s} \qquad (9.4.9)$$

The median and the maximum absolute deviation from the median MAD can be used instead of the arithmetic mean and the standard deviation. The resulting tests are much more robust:

$$D_1' = \frac{x_{(n)} - med(x_j)}{1.483 MAD(x_j)} \qquad (9.4.10)$$

$$D_2' = \frac{\max |x_i - med(x_j)|}{1.483 MAD(x_j)} \qquad (9.4.11)$$

If more than one outlier is present in the data, an iterative technique can be applied. First the most extreme outlier is removed by using one of the proposed tests (e.g., D_2'). Then the same procedure can be applied to the remaining $n-1$ data, adjusting its threshold to the data size $n-1$. This procedure is repeated until all the existing outliers are rejected.

Having discussed the detection of outliers, we come to the problem of the estimation of the data mean. Let the observation x_M be an outlier detected by using the test B_2. Let us also suppose that the arithmetic mean $\bar{x}$ is used as the estimator of the mean. A reasonable approach is to reject x_M and to exclude it from the calculation of the arithmetic mean. This approach leads to the following estimator of the mean:

$$T_n = \begin{cases} \bar{x} & \dfrac{|x_M - \bar{x}|}{\sigma} < c \\ \bar{x} - \dfrac{x_M - \bar{x}}{n-1} & elsewhere \end{cases} \tag{9.4.12}$$

Another approach is not to eliminate the outlier, but to reduce its contribution. Such an approach is called *Windsorization* :

$$T_n = \begin{cases} \bar{x} & \text{if } B_2 = \dfrac{|x_M - \bar{x}|}{\sigma} < c \\ \bar{x} - \dfrac{x_{(n)} - x_{(n-1)}}{n} & \text{if } B_2 > c \text{ and } x_M = x_{(n)} \\ \bar{x} - \dfrac{x_{(2)} - x_{(1)}}{n} & \text{if } B_2 > c \text{ and } x_M = x_{(1)} \end{cases} \tag{9.4.13}$$

The estimator (9.4.12) is devised for one outlier. Iterative versions can be used, when multiple outliers are present.

9.5 ADAPTIVE FILTERS BASED ON LOCAL STATISTICS

Let the observed image x_{ij} be corrupted by additive white noise:

$$x_{ij} = s_{ij} + n_{ij} \tag{9.5.1}$$

The linear Minimal Mean Square Error Estimate (MMSE) of s_{ij} is given by the following formula:

$$\hat{s}_{ij} = (1 - \frac{\sigma_n^2}{\sigma_x^2})x_{ij} + \frac{\sigma_n^2}{\sigma_x^2}\hat{m}_x \tag{9.5.2}$$

where σ_n, σ_x, $\hat{m}_x$ are the local estimates of the noise standard deviation, the signal standard deviation, and the signal mean, respectively. In the homogeneous image regions the noise standard deviation is approximately equal to the signal standard deviation. Therefore, for these regions the adaptive MMSE filter (9.5.2) is reduced to the local estimate of the signal mean $\hat{s}_{ij} \simeq \hat{m}_x$. At edge regions, the signal standard deviation is much greater than the noise signal deviation ($\sigma_n \ll \sigma_x$). In those regions, no filtering is performed at all ($\hat{s}_{ij} = x_{ij}$). Thus the adaptive MMSE filter preserves the edges, although it does not filter the noise in the edge regions. The performance of the adaptive MMSE filter depends on the choice of the local measures of signal mean and standard deviation and of the noise standard deviation.

(a)   (b)

(c)

Figure 9.5.1: (a) Original image; (b) Image corrupted by white additive Gaussian noise; (c) Output of the adaptive filter.

Several of the already discussed local estimators can be used. The local arithmetic mean and sample standard deviation have been used in [11,12] for the estimation of the signal mean and standard deviation, respectively. The local median has been proposed in [13,14] for the estimation of the signal mean. The

midpoint (MP) statistic can also be used if the noise distribution is short-tailed
[13]. The interquantile distance [14] and the range W [13] have been proposed
as measures of the local signal standard deviation. The noise standard deviation
σ_n either has to be known or to be evaluated at a homogeneous image region,
where the signal standard deviation is approximately equal to noise standard
deviation. It is difficult to evaluate the local noise standard deviation adaptively
if it changes spatially. A modification of the MMSE filter is the adaptive filter
described by the following equation [13]:

$$y_{ij} = \begin{cases} (1-a)x_{ij} + a\hat{m}_x & a \geq c \\ med(x_{ij}) & a < c \end{cases}$$ (9.5.3)

$$a = \begin{cases} b + \hat{\sigma}_n/\hat{\sigma}_x & b + \hat{\sigma}_n/\hat{\sigma}_x \leq 1 \\ 1 & otherwise \end{cases}$$ (9.5.4)

where $b \in [0.3, 0.5]$, $c \in [0.5, 0.6]$. This modification is based on experimentation
with different images. The performance of this filter in the removal of additive
white Gaussian noise noise is shown in Figure 9.5.1. The arithmetic mean and
the sample standard deviation have been used as the local estimates of the mean
and signal standard deviation.

Another filter which uses local statistics for adaptation is the so-called
adaptive double window modified trimmed mean (DW MTM) filter [27]. This
filter is the adaptive version of the DW MTM filter described in section 5.5 of
chapter 5. It adapts the threshold value q in such a way that can be used for the
filtering of signal-dependent noise mixed with impulses and additive noise of the
form:

$$x = \begin{cases} s + (k_1 f(s) + k_2)n & \text{with probability } 1 - (p_1 + p_2) \\ x_{max} & \text{with probability } p_1 \\ x_{min} & \text{with probability } p_2 \end{cases}$$ (9.5.5)

where x_{max}, x_{min} are the values of positive and negative impulses having probabilities of occurrence p_1, p_2, respectively. k_1, k_2 are scalar constants. The noise
term $k_1 f(s)n$ is signal-dependent, whereas the term $k_2 n$ is additive. The noise
process n has zero mean unit variance Gaussian distribution. The noise term
$(k_1 f(s) + k_2)n$ has the following standard deviation:

$$\sigma_s = k_1 f(s) + k_2$$ (9.5.6)

If m_{ij} is a local mean estimate, most of the noisy samples are in the range
$[m_{ij} - q, m_{ij} + q]$ with q given by:

$$q = c\sigma_s = c\left[k_1 f(m_{ij}) + k_2\right] \tag{9.5.7}$$

The constant c is chosen [27] to be in the range [1.5, 2.5]. Almost all noisy samples outside $[m_{ij}-q, m_{ij}+q]$ correspond to impulses.

(a)   (b)

(c)

Figure 9.5.2: (a) Original image; (b) Image corrupted by signal-dependent noise mixed with positive and negative impulses and white additive Gaussian noise; (c) Output of the adaptive DW MTM filter.

The adaptive DW MTM filter operates as follows. It employs a median filter of small window size $n \times n$ to find an estimate of the local mean m_{ij}. Then it employs a larger window of size $m \times m$ ($m > n$) and calculates the arithmetic mean of all points lying in this window and having values in the range

$[m_{ij}-q,m_{ij}+q]$, with q given by (9.5.7). This mean is the output of the adaptive DW MTM filter. Thus, the averaging range of the adaptive DW MTM filter varies with the local signal intensity. It becomes small when the signal level is low and it increases when the signal level is high. An example of adaptive DW MTM filter in the presence of noise of the form (9.5.5) is shown in Figure 9.5.2. The probability of impulse occurrence in Figure 9.5.2b was 12%. The signal-dependent and the additive Gaussian noise terms are given by:

$$x = (s^{0.5} + 7)n \tag{9.5.8}$$

The adaptive DW MTM filter employed two windows of size 3×3 and 5×5 respectively. The threshold value q was chosen as follows:

$$q = 2(m_{ij}^{0.5} + 10) \tag{9.5.9}$$

The output of the adaptive DW MTM filter is shown in Figure 9.5.2c. The impulses have been almost completely removed. The edges have been preserved sufficiently well and the additive and the signal-dependent noise have been suppressed.

9.6 DECISION DIRECTED FILTERS

Another kind of adaptive filter is based on decisions on the presence of impulses, edges, or homogeneous image regions in the neighborhood of an image pixel, and they modify their operation according to the outcome of these decisions [14-18]. Thus these filters can be called *decision directed filters* . The local decisions are usually based on simple edge detectors or impulse detectors described in the previous sections. The filters used are special cases of order statistic filters (e.g., medians, L-filters, α-trimmed filters) or the moving average filter. Such a decision directed filter is shown in Figure 9.6.1. It consists of two L-filters [16]. One is performing as edge detector (e.g. range edge detector) and the other is performing as a usual L-filter. The output of the edge detector controls the coefficients of the second L-filter. If no edge is detected, the L-filter operates as a moving average filter. If an edge is detected, the L-filter operates as a median. Thus, if e denotes the output of the edge detector, the coefficients a_i, $i=1,..,n$ of the L-filter are chosen in the following way:

$$a_i = \frac{1}{n} , \qquad i=1,..,n, \qquad e < threshold \tag{9.6.1}$$

$$a_{\lfloor n/2 \rfloor} = 1 , \quad a_i = 0 \quad i \neq \lfloor n/2 \rfloor , \qquad e \geq threshold \tag{9.6.2}$$

The threshold used in (9.6.1-2) determines the filter edge preservation properties. A small threshold is used when the preservation of edges of small contrast is desirable. High threshold values are used when noise removal is more important then edge preservation. The choice of the threshold can be done by using the histogram of the edge detector output. The threshold can be chosen in such a way, so that only a desired percentage of the edge detector output pixels are declared as edge pixels.

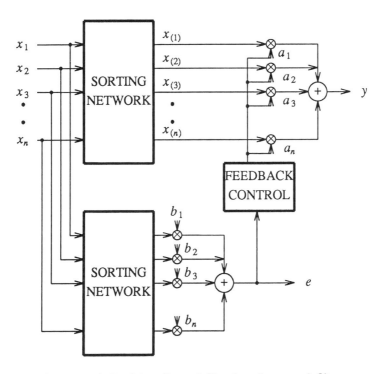

Figure 9.6.1: Decision directed filter based on two L-filters.

The performance of the decision directed filter is shown in Figure 9.6.2. The noisy image is shown in Figure 9.6.2a. The output of the decision directed filter is shown in Figures 9.6.2b. The edge detector output is shown in Figure 9.6.2c. The decision on the appropriate threshold is based on the image histogram shown in Figure 9.6.2d. Decision directed filters can take into account both edge information and impulsive noise information. Impulses, when detected, can be removed from the estimation of the local mean, median, and standard deviation. Furthermore, when an edge is detected, the window of the filter can become smaller so that the edge blurring is minimized. Such an impulse sensitive filter, called the *adaptive window edge detection (AWED)* filter, is shown in Figure 9.6.3 [17].

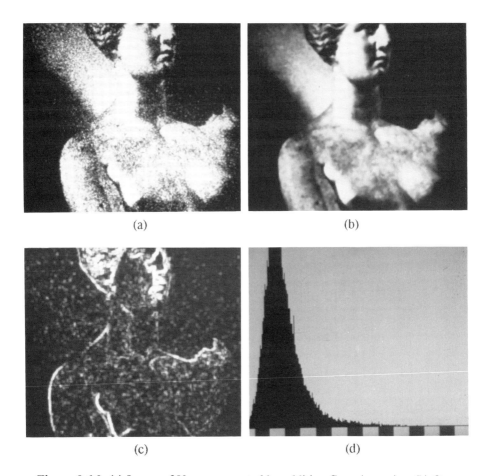

Figure 9.6.2: (a) Image of Venus corrupted by additive Gaussian noise; (b) Output of the decision directed filter; (c) Edge detector output; (d) Histogram of the edge detector output.

The filter initially starts with a 7×7 or 5×5 window. The local image histogram in the filter window is examined. If impulses are detected, they are rejected and the local image standard deviation calculation is based on the rest of the pixels in the window. If the local standard deviation is low enough, a homogeneous image region is assumed and the moving average filter is used. If the local standard deviation is large (above a certain threshold) an edge is declared. If the window size is 3×3, the median filter is used for image filtering. If the window size is greater than 3×3, it is reduced and the whole procedure is repeated. The window size is increased at each pixel, if no edge has been detected. The performance of the impulse sensitive decision directed filter has proven to be superior than the non-adaptive median and moving average filter in several simulations. It has also been proven to be superior than the previously mentioned decision directed filter and other adaptive filters (see section 9.10).

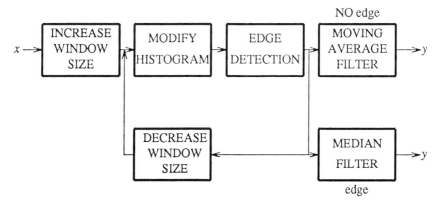

Figure 9.6.3: Adaptive window edge detection (AWED) filter.

An example of the preformance of the AWED filter in the presence of mixed additive Gaussian and impulsive noise is shown in Figure 9.6.4. The filter suppresses the impulses effectively and preserves the edges.

Finally, an adaptive version of the α-trimmed mean filter has been proposed [18]. The α-trimmed mean filter:

$$y_i(\alpha) = \frac{1}{n(1-2\alpha)} \sum_{j=\alpha n+1}^{n-\alpha n} x_{(j)} \qquad (9.6.3)$$

ranges from the moving average filter to the median for appropriate choices of the coefficient α. Therefore, its performance is good for long-tailed noise distributions (when it tends to the median) up to the Gaussian distribution, when it tends to the moving average filter (for $\alpha=0$). The complementary α-trimmed mean filter:

$$y'_i(\alpha) = \frac{1}{2\alpha n}[\sum_{j=1}^{\alpha n} x_{(j)} + \sum_{j=n-\alpha n+1}^{n} x_{(j)}] \qquad (9.6.4)$$

ranges from the midpoint to the moving average filter filter, for appropriate choices of α. Therefore, it is suitable for short-tailed and medium-tailed distributions. Thus the α-trimmed mean filter and the complementary α-trimmed mean filter form a filter pair, which can be easily adapted to the tail behavior of the noise probability distribution. A traditional measure of the distribution tail behavior is the *kurtosis*. However, more robust measures of the probability tail behavior have been found [19]. Such a measure is the statistic $V_i(\alpha)$ [18,19]:

$$V_i(\alpha) = \frac{U_i(\alpha) - L_i(\alpha)}{U_i(0.5) - L_i(0.5)} \tag{9.6.5}$$

(a)   (b)

Figure 9.6.4: (a) Original image; (b) Image corrupted by mixed additive Gaussian and impulsive noise; (c) Output of the AWED filter.

The statistics $U_i(\alpha)$, $L_i(\alpha)$ are L-statistics and they can be calculated by L-filters having coefficients given by:

$$a_j = \begin{cases} 1/(n\alpha) & 2v+2-\lfloor n\alpha \rfloor \leq i \leq 2v+1 \\ 1-\lfloor n\alpha \rfloor/(n\alpha) & i=2v+1-\lfloor n\alpha \rfloor \\ 0 & \textit{otherwise} \end{cases} \tag{9.6.6}$$

$$a_j = \begin{cases} 1/(n\alpha) & 1 \leq i \leq \lfloor n\alpha \rfloor \\ 1-\lfloor n\alpha \rfloor/(n\alpha) & i = \lfloor n\alpha \rfloor + 1 \\ 0 & otherwise \end{cases} \qquad (9.6.7)$$

respectively, where $n = 2v+1$. The denominator of (9.6.5) is a very good robust estimator of the scale for the double-exponentially distributed data. If α is small, (e.g., $\alpha=0.05$), the numerator of (9.6.5) is a very good estimator of the scale of uniformly distributed data. If α is moderate (e.g., $\alpha=0.2$), the numerator of (9.6.5) is a good estimator of the scale of Gaussian data. The statistic $V_i(\alpha)$ is expected to take large values for long-tailed distributions and small values for short-tailed distributions. It has been observed that the exact value of α is not critical for the performance of $V_i(\alpha)$ [18]. Therefore, the following filter can be used, which adapts to the noise probability distribution:

$$y_i = \begin{cases} MP(x_i) & 0 \leq V_i(\alpha) < \tau_1 \\ y'_i(\alpha) & \tau_1 \leq V_i(\alpha) < \tau_2 \\ \bar{x}_i & \tau_2 \leq V_i(\alpha) < \tau_3 \\ y_i(\alpha) & \tau_3 \leq V_i(\alpha) < \tau_4 \\ med(x_i) & \tau_4 \leq V_i(\alpha) < \infty \end{cases} \qquad (9.6.8)$$

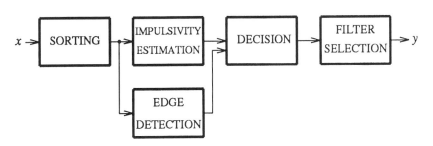

Figure 9.6.5: Structure of the adaptive α-trimmed mean filter.

Monte Carlo techniques can be used for the derivation of the thresholds τ_i, $i=1,..,4$ [18]. Quasi-range edge detectors can also be used to provide edge information to the filter. If the quasi-edge detector output e_i is greater than a threshold, an edge is declared and median filtering is used. Otherwise the filtering scheme (9.6.8) is used. The structure of the decision directed filter based on (9.6.8) is shown in Figure 9.6.5. Simulation examples have shown that this filter performs well in a variety of input noise distributions [18].

9.7 TWO-COMPONENT IMAGE FILTERING

An image $\mathbf{x}$ can be considered to consist of two parts: a low-frequency part $\mathbf{x}_L$ and a high-frequency part $\mathbf{x}_H$:

$$\mathbf{x} = \mathbf{x}_L + \mathbf{x}_H \tag{9.7.1}$$

The low-frequency part is dominant in the homogeneous image regions, whereas the high-frequency part is dominant in the edge regions. The two-component image model allows different treatment of its components. Therefore, it can be used for adaptive image filtering and enhancement, provided that the two components can be separated. A low-pass and a high-pass filter can be used for the separation of the two components. In most of the cases the moving average filter or the median filter are used as estimators of the low-frequency component, whereas the high-frequency component is given by:

$$x_{Hij} = x_{ij} - x_{Lij} \tag{9.7.2}$$

Let us suppose that the observed image x_{ij} is corrupted by mixed additive and signal dependent noise:

$$x_{ij} = u_{1ij} + u_{2ij} n_{1ij} + n_{2ij} \tag{9.7.3}$$

where $u_{1ij} = r(s_{ij})$, $u_{2ij} = t(s_{ij})$ are nonlinearly related to the original image s_{ij}. This type of noise can be filtered by a two-component filter shown in Figure 9.7.1 [20].

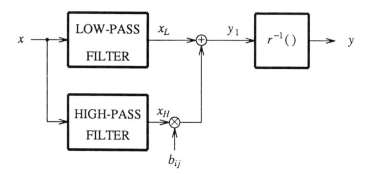

Figure 9.7.1: Two-component image filtering.

Its output signal is given by:

$$y_{1ij} = \bar{x}_{ij} + b_{ij}(x_{ij} - \bar{x}_{ij}) \tag{9.7.4}$$

$$y_{ij} = r^{-1}(y_{1ij}) \tag{9.7.5}$$

The performance of the adaptive filter (9.7.4) depends of the choice of the coefficient b_{ij}. This coefficient can be chosen to minimize the mean square error:

$$minimize \quad E[(u_{1ij} - y_{1ij})^2] \tag{9.7.6}$$

This minimization leads to the following choice of b_{ij} [20]:

$$b_{ij} = \frac{\sigma_{u_1}^2}{\sigma_x^2} = 1 - \frac{\overline{u_2}^2 \sigma_{n_1}^2 + \sigma_{n_2}^2}{\sigma_x^2} \tag{9.7.7}$$

where $\overline{u_2^2} = E[u_{2ij}^2]$. The coefficient b_{ij} takes the following form for purely multiplicative noise:

$$b_{ij} \simeq 1 - \frac{\bar{x}_{ij}^2 \sigma_{n_1}^2 + \sigma_{n_2}^2}{\sigma_x^2} \tag{9.7.8}$$

For signal dependent-noise of the form (9.7.3), with $t(s) = \sqrt{r(s)}$, coefficient b_{ij} takes the form:

$$b_{ij} = 1 - \frac{\bar{x}_{ij}\sigma_{n_1}^2 + \sigma_{n_2}^2}{\sigma_x^2} \tag{9.7.9}$$

A generalization of the form (9.7.7) is also possible:

$$b_{ij} = \begin{cases} (1 - \alpha\sigma_n^2/\sigma_x^2)^\beta & \alpha\sigma_n^2 < \sigma_x^2 \\ 0 & else \end{cases} \tag{9.7.10}$$

where the noise variance is given by:

$$\sigma_n^2 = g(m_x)\sigma_{n_1}^2 + \sigma_{n_2}^2 \tag{9.7.11}$$

m_x is the local mean estimate of the observed image x and $g(.)$ is a function that depends on the specific noise type, as it is seen in (9.7.8-9). The coefficient b_{ij} takes values between 0 and 1. When it is close to 0 it suppresses the high-frequency component. When it is close to 1 it allows all high frequencies to pass. The parameter α controls the threshold of the local signal to noise ratio up to which the high-frequency components are entirely suppressed. The parameter β controls the suppression of noise close to edges. The smaller it is, the larger is b_{ij} for a specific signal-to-noise ratio. Thus small values of β make the filter sensitive to edge information. The coefficient b_{ij} can also be used for the adaptation of the filter window. We can start filtering by using initial window size 5×5 or 7×7. If the coefficient b_{ij} becomes greater than an appropriate threshold b_t, the window size is decreased until the coefficient becomes less then the threshold or until the window reaches the size 3×3. Otherwise the window size is increased to its maximum size. The local estimates for signal mean and standard deviation can be the local arithmetic mean and sample standard deviation. If impulsive noise is present, the impulses can be detected and removed from the filter window. Then the local mean and scale estimates are calculated by using the rest of the pixel in the filter window. Also the local median of the remaining pixels can be used as an estimate of the signal mean. The window size of the median can also be adapted, according to the coefficient b_{ij}, as it has already been described. The *signal-adaptive median (SAM)* filter is shown in Figure 9.7.2.

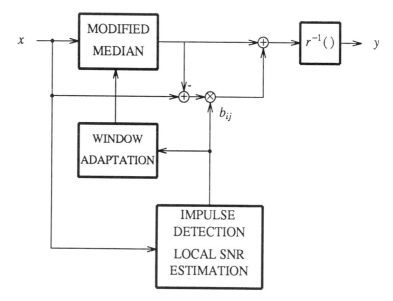

Figure 9.7.2: Signal-adaptive median (SAM) filter structure.

If S_{max} and S_{min} are the amplitudes of the positive and of the negative impulses, the coefficient b_{ij} of this filter is given by:

$$b_{ij} = \begin{cases} 0 & \text{If } \alpha\sigma_n^2 \geq \sigma_x^2 \\ & \text{or } x_{ij}-med(x_{ij})<c(S_{min}-med(x_{ij})) \\ & \text{or } x_{ij}-med(x_{ij})>c(S_{max}-med(x_{ij})) \\ (1-\alpha\sigma_n^2/\sigma_x^2)^\beta & \text{else} \end{cases} \qquad (9.7.12)$$

with $c=5/6$. The performance of the SAM filter in mixed additive Gaussian and impulsive noise is shown in Figure 9.7.3. The SAM filter removes impulses effectively, preserves edges, and smooths the Gaussian noise.

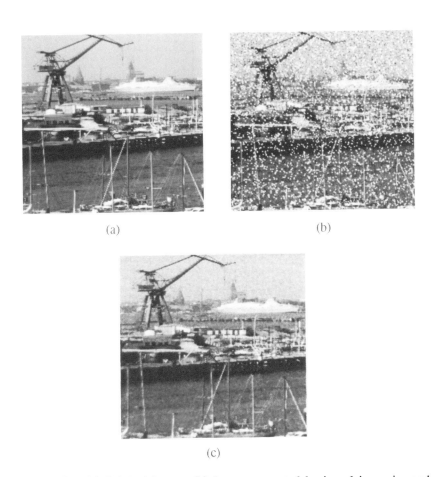

(a)

(b)

(c)

Figure 9.7.3: (a) Original image; (b) Image corrupted by impulsive noise and additive Gaussian noise; (c) Output of the SAM filter.

Another form of two-component image filtering, which has originally been proposed for cloud noise removal [21,22], is shown in Figure 9.7.4. The low-frequency component is obtained by low-pass filtering. The high-frequency component is obtained by subtracing the low-frequency component from the original image. The high frequency component is directly related to the local image contrast. It can be modified if it is multiplied by a scalar $k(x_L)$, which depends on the low-contrast component. If $k(x_L)$ is greater then one, the local contrast is increased, whereas if it is smaller than one, the local contrast decreases. The low-frequency component, which is the local luminance can be modified by a point-wise nonlinearity. The modified local contrast x_H' and the modified local luminance x_L' are combined to produce the output image. $k(x_L)$ is chosen to be larger than one in order to increase the local contrast, especially for high local luminance. If a decrease in the local luminance is desired, especially when it is high, the nonlinearity in Figure 9.7.4 is chosen in such a way so that it is saturated for large local input luminances.

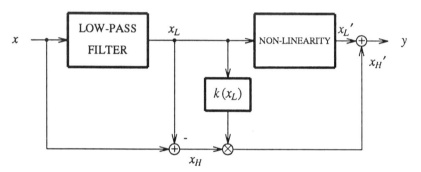

Figure 9.7.4: Filter for local luminance and local contrast modification.

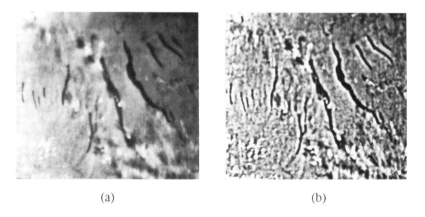

(a)   (b)

Figure 9.7.5: (a) Cloudy satellite image; (b) Output of the two-component enhancement filter.

This adaptive filter can be used for cloud noise removal. The regions covered by cloud have high local luminance and low local contrast, as it is shown in Figure 9.7.5a. The output of the filter, shown in Figure 9.7.5b, has much larger local contrast and much less luminance in the regions covered by clouds. Thus the cloud noise is removed fairly well.

9.8 ADAPTIVE L-FILTERS

Most adaptive filters presented in the previous sections are based on rather ad-hoc or heuristic techniques to derive the filter parameters and, sometimes, also the filter structure itself. In contrast, a rigid mathematical theory has been developed for linear adaptive filters [23,24]. Therefore, an effort has been made recently to extend this mathematical approach to nonlinear filters based on order statistics [25], [50-52].

Let $d(i)$ be the uncorrupted zero-mean signal, which will also be used as a reference signal for the derivation of the adaptive L-filter. The observed signal $x(i)$ is given by:

$$x(i) = d(i) + n(i) \qquad (9.8.1)$$

The adaptive L-filter is defined as follows:

$$\hat{d}(i) = y(i) = \sum_{i=1}^{n} a_j(i)x_{(j)}(i) = \mathbf{a}^T(i)\mathbf{x}(i) \qquad (9.8.2)$$

The coefficient vector $\mathbf{a}(i) = [a_1(i),..,a_n(i)]^T$ must be updated at each step i in such a way, so that the mean square error J is minimized:

$$J = E[e(i)^2] = E[|d(i) - \sum_{i=1}^{n} a_j(i)x_{(j)}(i)|^2] \qquad (9.8.3)$$

$$= \sigma^2_{d_i} - 2\sum_{j=1}^{n} a_j(i)r^{(i)}_{dx}(j) + \sum_{j=1}^{n}\sum_{l=1}^{n} a_j(i)a_l(i)r^{(i)}_{xx}(j,l)$$

where $\sigma^2_{d_i}$ is the variance of the reference signal and

$$r^{(i)}_{xx}(j,l) = E[x_{(j)}(i)x_{(l)}(i)] \qquad (9.8.4)$$

$$r^{(i)}_{dx}(j) = E[d(i)x_{(j)}(i)]$$

If the gradient of J with respect to the filter coefficients is set equal to zero, the following set of normal equations results:

$$\sum_{l=1}^{n} a_l(i)r^{(i)}_{xx}(k,l) = r^{(i)}_{dx}(k) \qquad k=1,2,..,n \qquad (9.8.5)$$

It can be written in the following matrix format:

$$\mathbf{R}(i)\mathbf{a}(i) = \mathbf{d}(i) \tag{9.8.6}$$

where $\mathbf{R}(i) = [r_{xx}^{(i)}(k,l)]$ and $\mathbf{d}(i) = [r_{dx}^{(i)}(j)]$. Therefore, the optimal coefficient vector at step i is given by:

$$\mathbf{a}(i) = \mathbf{R}(i)^{-1}\mathbf{d}(i) \tag{9.8.7}$$

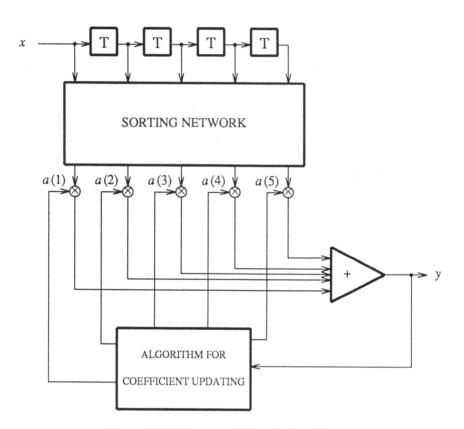

Figure 9.8.1: Structure of the adaptive L-filter.

However, (9.8.7) cannot be easily used, especially in real time applications, because the autocorrelation matrix $\mathbf{R}(i)$ of the ordered observations and the crosscorrelation vector $\mathbf{d}(i)$ must be estimated and then the autocorrelation matrix must be inverted. A much simpler iterative algorithm for the minimization of J can be found in the following way.

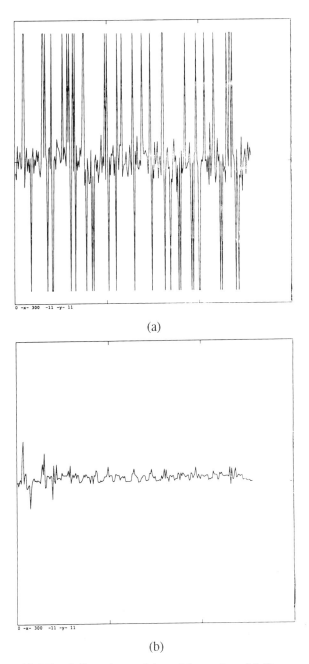

(a)

(b)

Figure 9.8.2: (a) Mixed Gaussian and impulsive noise; (b) Output of the LMS
L-filter.

The partial derivative of J with respect to the coefficient $a_k(i)$ is given by:

$$\frac{\partial J}{\partial a_k(i)} = -2[r_{dx}^{(i)}(k) - \sum_{l=1}^{n} a_l(i)r_{xx}^{(i)}(k,l)] \tag{9.8.8}$$

$$= -2E\,[(d(i) - \sum_{l=1}^{n} a_l(i)x_{(l)}(i))x_{(k)}(i)] = -2E\,[e(i)x_{(k)}(i)]$$

A noisy estimate of the gradient can be found by dropping the expectation operator in (9.8.8):

$$\frac{\partial \hat{J}}{\partial a_k(i)} = -2e(i)x_{(k)}(i) \tag{9.8.9}$$

By using this noisy gradient, the steepest descent method can be used for the minimization of J:

$$\mathbf{a}(i+1) = \mathbf{a}(i) + \mu e(i)\mathbf{x}(i) \tag{9.8.10}$$

where μ denotes the step size. Algorithm (9.8.10) is equivalent to the *least mean square (LMS) algorithm* (also called *Widrow algorithm*), which is very popular in linear adaptive filtering [23,24]. The only difference is that (9.8.10) uses the vector of ordered observations $\mathbf{x}(i)$ to update the adaptive L-filter coefficients, whereas LMS algorithm uses the vector $[x(i) \cdots x(i-n+1)]^T$ to update the coefficients of the adaptive FIR filter. Therefore, (9.8.10) will be called *LMS L-filter algorithm*. The structure of the adaptive L-filter is shown in Figure 9.8.1. The performance of the adaptive LMS L-filter in the presence of mixed Gaussian and impulsive noise is shown in Figure 9.8.2. After a short adaptation time, the adaptive L-filter removes the impulses completely, as is seen in Figure 9.8.2b. The adaptive L-filter can adapt well to the probability distribution of the noise. Therefore, it tends to the midpoint, moving average and median filters for short-tailed, medium-tailed, and long-tailed distributions, respectively. It can also adapt fairly well to changing noise or signal statistics. Other algorithms for faster adaptation are also presented in [25].

9.9 ADAPTIVE VOLTERRA FILTERS

Volterra filters are an interesting class of nonlinear filters that has found many applications in digital signal processing, as was described in chapter 8. Many of these applications demand filter adaptation to changing signal or noise statistics. Thus, adaptive Volterra filters have been developed recently for echo cancellation [36,38,44], interference, and noise reduction [32,42,48]. In the following, the theory of adaptive Volterra filters will be reviewed [24,49].

As was described in chapter 8, the output $y(i)$ of a nonlinear system can be expanded into a Volterra series in terms of the input signal $x(i)$:

$$y(i) = \sum_{n=0}^{\infty} y_n(i)$$

$$= h_0 + \sum_{k_1=0}^{\infty} h_1(k_1)x(i-k_1) + \sum_{k_1=0}^{\infty} \sum_{k_2=0}^{\infty} h_2(k_1,k_2)x(i-k_1)x(i-k_2) + \cdots$$

$$+ \sum_{k_1=0}^{\infty} \cdots \sum_{k_n=0}^{\infty} h_n(k_1,k_2,\cdots,k_n)x(i-k_1)x(i-k_2)\cdots x(i-k_n) + \cdots \qquad (9.9.1)$$

where h_0, h_1, ... , h_n, ... , are called the kernels of the system, and y_n indicates the $n-th$ order term, which has the kernel h_n. The first-order term, y_1, is the convolution for a linear system which has the impulse response $h_1(i)$. The second-order (quadratic) term, y_2, together with the higher order terms y_n, models the nonlinearity.

In practice, a Volterra series can be truncated into a Volterra polynomial with the first few most significant terms. Many of the adaptive Volterra filters previously presented were developed based on the quadratic Volterra series, namely, only with terms of h_0, h_1, h_2 [28,31,35,41,47].

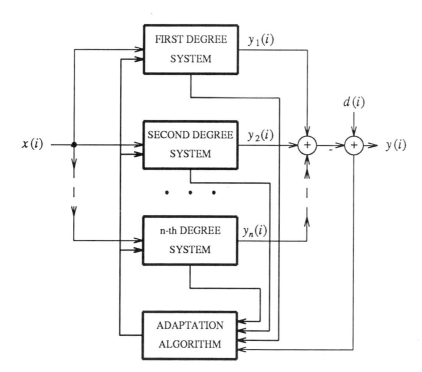

Figure 9.9.1: Structure of the adaptive Volterra filter.

The general structure of the adaptive Volterra filters is shown in Figure 9.9.1. Its coefficients are updated by an adaptation algorithm as a new signal arrives. Several adaptation algorithms have been developed recently. An adaptive Volterra filter using the LMS criterion has been developed in [48] for adaptive noise cancellation. The nonlinear adaptive filter was assumed to have finite memory and be characterized by a nonlinear function $f(\mathbf{X}(i))$, where $\mathbf{X}(i)$ is the vector of delayed input values in the delay line at time i, and f is a (generally nonlinear) scalar function of $\mathbf{X}(i)$. The function f was expanded into a Taylor series about the vector $\mathbf{0}$. It is quite interesting to observe that this expansion is the Volterra series expansion of a nonlinear finite-memory system. This Taylor series expansion was truncated to p terms for approximation. The updating formula for the coefficients of the expansion was derived using the LMS concept. Although this algorithm was not developed based on the theory of Volterra series, the results obtained are exactly of the same form as the ones derived from the Volterra series. In fact, the Volterra series can be introduced in this way, i.e., by using the Taylor series truncation of a finite-memory system [24].

Another second-order adaptive Volterra filter with minimum mean square error criterion was presented in [28,41,47]. It was assumed that the filter has finite memory, thus the second-order Volterra series can be written as:

$$y(i) = \mathbf{H}_1^T \mathbf{X}(i) + tr\{\mathbf{H}_2[\mathbf{X}(i)\mathbf{X}^T(i) - \mathbf{R}_x]\} \qquad (9.9.2)$$

where

$$\mathbf{H}_1 = [\, h_1(0)\, h_1(1)\, \cdots\, h_1(N-1)\,]^T$$
$$\mathbf{X}(i) = [\, x(i)\, x(i-1)\, \cdots\, x(i-N+1)\,]^T$$

$$\mathbf{H}_2 = \begin{bmatrix} h_2(0,0) & . & . & h_2(0,N-1) \\ . & . & . & . \\ . & . & . & . \\ h_2(N-1,0) & . & . & h_2(N-1,N-1) \end{bmatrix}$$

$$\mathbf{R}_x = \begin{bmatrix} r(0) & . & . & r(N-1) \\ . & . & . & . \\ . & . & . & . \\ r(N-1) & . & . & r(0) \end{bmatrix}$$

where r indicates the autocorrelation of the input signal x. The term $-tr(\mathbf{H}_2\mathbf{R}_x)$ is the dc component, namely, the term h_0 in equation (9.9.1). It is included to maintain the unbiasness of estimate. When the input signal is Gaussian, the

LMS solution of the second-order Volterra filter is as following:

$$\mathbf{H}_1 = \mathbf{R}_x^{-1}\mathbf{R}_{dx} \tag{9.9.3}$$

$$\mathbf{H}_2 = \frac{1}{2}\mathbf{R}_x^{-1}\mathbf{T}_{dx}\mathbf{R}_x^{-1} \tag{9.9.4}$$

where $\mathbf{R}_{dx}$ and $\mathbf{T}_{dx}$ are matrices associated with cross-covariance and cross-bicovariance between d and x, respectively. $d(i)$ is the desired signal. Direct solution of the matrices $\mathbf{H}_1$ and $\mathbf{H}_2$ based on equations (9.9.3-4) is called batch-processing method [47]. Another proposed adaptation algorithm is stochastic gradient method. The updating formula based on the approximate gradient for the linear part is the same as Widrow's formula:

$$\mathbf{H}_1(i+1) = \mathbf{H}_1(i) + 2\mu_1[d(i)-y(i)]\mathbf{X}(i) \tag{9.9.5}$$

where y is the output of the filtering system. The updating formula based on the approximate gradient for the second-order kernels is [28,41,48]:

$$\mathbf{H}_2(i+1) = \mathbf{H}_2(i) + 2\mu_2\mathbf{X}(i)\mathbf{X}^T(i)[d(i)-y(i)] \tag{9.9.6}$$

It was shown that, when the reference signal is Gaussian, the stochastic-gradient-based second-order Volterra filter converges to the optimum solution asymptotically. The convergence speed depends on the squared ratio of maximum to minimum eigenvalues of the input autocovariance matrix. The lattice orthogonalization was considered for improving convergence [28].

A sign algorithm was developed in [35] on the basis of LMS concept. This algorithm updates the filter coefficients using only the sign of the gradient. The gradient-based adaptive algorithms use a step size to control the convergence. The step size is fixed in most existing nonlinear adaptive algorithms. A technique for adjusting the step size in the sign algorithm was presented in [35] in order to obtain faster convergence and reduce final errors without excessively affecting the advantage of the sign algorithm in the implementation complexity. From the standpoint of the realization complexity, the structure resulting from the modified sign algorithm still remains much simpler than that related to the stochastic iteration algorithm. The advantage becomes greater if the input signal wordlength or the order of the nonlinearity increases. Adaptive nonlinear IIR Volterra filters have been proposed recently [39]. Most adaptive Volterra filters work in the time domain. An efficient frequency domain adaptive Volterra filter was presented in [30]. For a finite memory of length N, the filter converges to the equivalent time domain nonlinear adaptive filter presented previously. For the second-order Volterra series, the proposed filter requires $O(N^2)$ multiplications-additions as opposed to an $O(N^3)$ for the time domain. This filter was demonstrated successfully on simulation of microphone acoustic noise reduction. The algorithms discussed above are based on the LMS concept. An algorithm based on a fast Kalman filter algorithm was presented in [31]. The algorithm developed is for a second-order Volterra filter. It uses a stochastic approximation in the least squares recursion equation for the quadratic weight matrix. The convergence of the algorithm for the quadratic weights was established and was found to be quadratic for the quadratic filter. In a simulation experiment, the algorithm proposed converged faster than an LMS-based adaptation algorithm.

In recent years, adaptive Volterra filters have found a variety of applications. Adaptive echo cancellation has been extensively investigated in connection with the new services introduced in the telephone network because it allows for full-duplex transmission on two-wire circuits with full bandwidth in both directions. Application of this concept can be found in digital subscriber loop modems and voiceband data modems. Adaptive nonlinear echo cancellation methods based on Volterra filters are described in [32,36,38,44]. Adaptive interference cancellers based on the Volterra series were presented in [42], for removing intersymbol interference due to channel nonlinearities. An investigation of a QAM receiver was reported in [40]. The receiver incorporates adaptive equalization of nonlinearities as well as adaptive decision feedback equalization and data-aided carrier recovery for mitigation of linear distortion and phase jitter, respectively. Although most of the applications of the adaptive filters are in communications systems, some work has been done to apply these filters to solve problems in other fields. An LMS algorithm for noise canceler was developed in [48], which was used in an ECG cancellation simulation. The nonlinear canceller achieved 22.3 dB of noise suppression, whereas the linear canceller was capable of suppressing the interference by only 6.4 dB. An adaptive quadratic Volterra filter was used in [41,47] to model and forecast the sway motion response of a moored vessel to random sea waves.

9.10 COMPARISONS OF ADAPTIVE NONLINEAR FILTERS

The comparison of the performance of various adaptive nonlinear filters for image processing applications is very difficult due to the lack of a well-defined mathematical criterion that reflects consistently the subjective human criteria for image quality. Several mathematical criteria have been proposed e.g., the normalized mean square error and the perceptual mean square error, as described in chapter 3. However, none of them is proven to be consistent with human quality criteria. Therefore, a set of criteria has been used to compare the nonlinear filters [26]. The measure of the overall performance of a filter was chosen to be the *peak-to-peak signal to noise ratio (PSNR)* defined as follows:

$$PSNR = \left[\frac{NM\,(x_{max}-x_{min})^2}{\sum\limits_{i=1}^{N}\sum\limits_{j=1}^{M}(x_{ij}-y_{ij})^2} \right]^{\frac{1}{2}} \tag{9.10.1}$$

where x_{max}, x_{min} are the maximum and minimum intensity on the uncorrupted image and x, y are the uncorrupted and the filtered images, respectively. The performance of the filter in the homogeneous image regions has been measured similarly:

$$PSNR_h = \left[\frac{N_h(x_{max}-x_{min})^2}{\sum_i \sum_j (x_{ij}-y_{ij})^2} \right]^{\frac{1}{2}} \qquad (i,j) \in \text{homogeneous area} \qquad (9.10.2)$$

where N_h is the number of pixels belonging to the homogeneous areas of the uncorrupted image.

(a)

(b)

(c)

(d)

Figure 9.10.1: Test images: (a) Lenna; (b) Geometrical; (c) Harbor; (d) Face.

The measure of the performance of the filter close to edges is the following:

$$PSNR_e = \left[\frac{N_e(x_{max}-x_{min})^2}{\sum_i\sum_j(x_{ij}-y_{ij})^2}\right]^{\frac{1}{2}} \qquad (i,j) \in edge \qquad (9.10.3)$$

where N_e is the number of pixels belonging to the edges of the uncorrupted image. A simple edge detector is used to identify the edges and the homogeneous regions on the uncorrupted image. The performance of the filter in the presence of impulses is measured by:

$$PSNR_i = \left[\frac{N_i(x_{max}-x_{min})^2}{\sum_i\sum_j(x_{ij}-y_{ij})^2}\right]^{\frac{1}{2}} \qquad (i,j) \in impulse\ in\ y \qquad (9.10.4)$$

where N_i is the number of impulses in the filtered image y. Finally, the computation time on a SUN 3/260 workstation using UNIX and Fortran 77 is the measure of the computational load of the filter algorithm. The following filters have been compared: (a) non-adaptive moving average filter; (b) non-adaptive median filter; (c) SAM filter; (d) AWED filter; and (e) AWED filter without window size adaptation (called AED filter). The test images used are shown in Figure 9.10.1. They have chosen in such a way so that they are representatives of geometric images and of images with/without fine details. Each of these images has been corrupted by three different types of noise: (a) impulsive noise; (b) additive noise; and (c) mixed additive and impulsive noise. The resulting 12 images have been filtered by the filters under comparison. The results are illustrated in Figure 9.10.2.

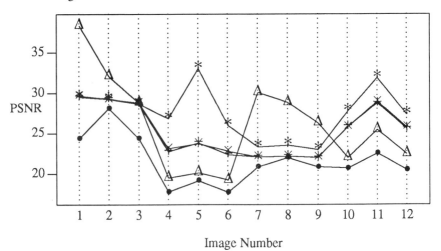

Image Number

(a)

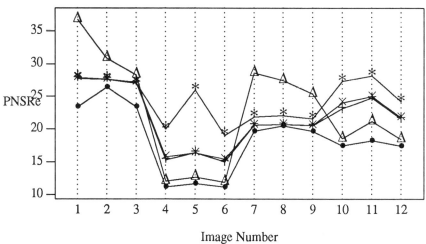

(b)

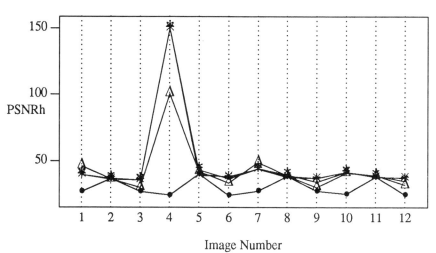

(c)

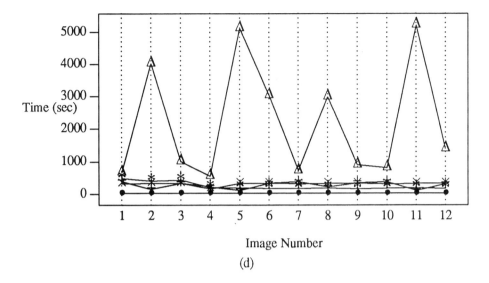

Image Number

(d)

Figure 9.10.2: Performance of the SAM filter (Δ), AWED filter (*), AED filter (+), median filter ($\times$) and moving average filter ($\bullet$) in the presence of various images and noise types: (a) *PSNR* measure; (b) $PSNR_e$ measure; (c) $PSNR_h$ measure; (d) Computation time.

Image numbers 1-3, 4-6, 7-9 and 10-12 correspond to images "Lenna", "Geometrical", "Harbor", and "Face" corrupted by impulsive, additive, and mixed noise respectively. It is observed that all filters perform similarly in the homogeneous image regions and they have approximately equal $PSNR_h$. Thus, it is proven experimentally that most of the errors occur close to the edges. SAM filter performs best on images having fine details (e.g., in "Harbor" and "Lenna"), but it is inferior to the AWED filter on images having large homogeneous regions. This fact correlates well with the sharp edges produced by the SAM filter in the images "Lenna" and "Harbor". The performance of the median filter is somewhere in between the performance of the above-mentioned filters. As expected, the performance of the moving average filter is the worst of all. The SAM filter is computationally very expensive, as can be seen in Figure 9.10.2d. The fastest of all is the moving average filter.

9.11 DISCUSSION

Several adaptive nonlinear filters have been analyzed in this chapter. Most of them belong to two main classes: adaptive Volterra filters and adaptive nonlinear filters based on order statistics. The first class has found extensive applications in telecommunications, e.g., in echo cancellation. It uses similar adaptation criteria with the adaptive linear filters. The second class has mainly been used in image processing applications. It does not possess a concrete mathematical foundation, and several filters of this class employ heuristics and have a rather ad-hoc structure. Some of the filters of this class adapt to the local image and noise statistics. Others are decision directed filters, i.e., their operation is based on local decisions on the existence of an edge or of impulses. Therefore they are usually driven by edge detectors or by impulse detectors. A variety of such detectors can be implemented by using order statistics. Another class of adaptive nonlinear image processing filters is based on the two-component image model. An image is assumed to consist of a low-frequency and a high-frequency component. The adaptive filters of this class operate differently on the two components and adapt locally to the image and noise characteristics as well. Finally, adaptive L-filters based on the mean square error criterion have been proposed recently. Their adaptation algorithms are similar to the ones used in adaptive linear filters. These filters have a rigid mathematical foundation and promising performance in a variety of noise characteristics. Similar adaptation techniques can be used for other nonlinear filters based on order statistics.

REFERENCES

[1]   H.A. David, *Robust statistics*, Wiley, 1981.

[2]   W.K. Pratt, *Digital image processing*, Wiley, 1978.

[3]   M.D. Levine, *Vision in Man and Machine*, McGraw-Hill, 1985.

[4]   I.E. Abdou, W.K. Pratt, "Quantitative design and evaluation of enhancement/ thresholding edge detectors", *Proc. IEEE*, vol. 67, May 1979.

[5]   D.H. Ballard, C.M. Brown, *Computer vision*, Prentice-Hall, 1982.

[6]   I. Pitas, A.N. Venetsanopoulos, "Edge detectors based on order statistics", *IEEE Transactions on Pattern Analysis and Machine Intelligence* , vol. PAMI-8, no.4, pp.538-550, July 1986.

[7]   X.Z. Sun, A.N. Venetsanopoulos, "Adaptive schemes for noise filtering and edge detection by the use of local statistics", *IEEE Transactions on Circuits and Systems*, vol. CAS-35, no.1, pp. 57-69, Jan. 1988.

[8]   A.C. Bovik, D.C. Munson, "Edge detection using median comparisons", *Computer Vision Graphics and Image Processing*, vol. 33, pp. 377-389, 1986.

[9]   J. Hajek, Z. Sidak, *Theory of rank tests*, Academic Press, 1967.

[10] A.C. Bovik, T.S. Huang, D.C. Munson, "Nonparametric tests for edge detection in noise", *Pattern recognition*, vol. 19, no. 3, pp. 209-219, 1986.

[11] J.S. Lee, "Digital image enhancement and noise filtering by local statistics", *IEEE Transactions on Pattern Analysis and Machine Intelligence* , vol. PAMI-2, no. 2, pp. 165-168, March 1980.

[12] J.S. Lee, "Refined filtering of image noise using local statistics", *Computer Graphics and Image processing*, vol. 15, pp. 380-389, 1981.

[13] X.Z. Sun, A.N. Venetsanopoulos, "Adaptive schemes for noise filtering and edge detection by use of local statistics", *IEEE Transactions on Circuits and Systems*, vol. CAS-35, no. 1, pp. 57-69, Jan. 1988.

[14] I. Scollar, B. Weidner, T.S. Huang, "Image filtering using the interquartile distance", *Computer Vision, Graphics and Image processing*, vol. 25, pp.236-251, 1984.

[15] N.F. Nahi, A. Habibi, "Decision directed recursive image enhancement", *IEEE Transactions on Circuits and Systems*, vol. CAS-6, pp. 286-293, March 1975.

[16] I. Pitas, A.N. Venetsanopoulos, "Nonlinear order statistic filters for image filtering and edge detection" *Signal Processing*, vol. 10, pp. 395-413, 1986.

[17] R.L.J. Martens, *Adaptive nonlinear order statistic filters: analysis and comparison*, M.A.Sc. Thesis, University of Toronto, 1988.

[18] A. Restrepo, A.C. Bovik, "Adaptive trimmed mean filters for image restoration", *IEEE Transactions on Acoustics, Speech and Signal Processing*, vol. 36, no.8, pp. 1326-1337, 1988.

[19] R.V. Hogg, "More light on the kurtosis and related statistics", *Journal Amer. Statist. Assoc.*, vol. 67, pp. 422-424, 1972.

[20] R. Bernstein, "Adaptive nonlinear filters for simultaneous removal of different kinds of noise in images", *IEEE Transactions on Circuits and Systems*, vol. CAS-34, no. 11, pp. 1275-1291, Nov. 1987.

[21] T. Peli, J.S. Lim, "Adaptive filtering for image enhancement", *Journal Opt. Eng.*,vol. 21, pp.108-112, 1982.

[22] J.S. Lim, "Image enhancement", in *Digital image processing techniques*, M.P. Ekstrom editor, Academic Press, 1984.

[23] T. Alexander, *Adaptive signal processing*, Springer Verlag, 1986.

[24] M. Bellanger, *Adaptive digital filters and signal analysis*, Marcel Dekker, 1987.

[25] I. Pitas, A.N. Venetsanopoulos, "Adaptive filters based on order statis-
      tics", *IEEE Transactions on Acoustics, Speech and Signal Processing,*
      under review.

[26] R. Martens, A.N. Venetsanopoulos, "A comparison of adaptive nonlinear
      filters using different images", *Proc. IEEE Int. Conf. on Systems
      Engineering,* Dayton, Ohio, 1989.

[27] R. Ding, A.N. Venetsanopoulos, "Generalized homomorphic and adaptive
      order statistic filters for the removal of impulsive and signal dependent
      noise", *IEEE Transactions on Circuits and Systems,* vol. CAS-34, no.8,
      pp. 948-955, Aug. 1987.

[28] T. Koh, E.J. Powers, "An adaptive nonlinear digital filter with lattice
      orthogonalization", *Proc. IEEE ICASSP-83,* pp. 37-40, 1983.

[29] H.H. Chiang, C.L. Nikias, A.N. Venetsanopoulos, "Efficient implementa-
      tions of quadratic digital filters" *IEEE Transactions on Acoustics, Speech,
      and Signal Processing,* vol.ASSP-34, pp.1511-1528, Dec. 1986.

[30] D. Mansour, A.H. Gray, "Frequency domain non-linear adaptive filter",
      1981 IEEE Int. Conf. Acoustics, Speech and Signal Processing, pp.550-
      553, 1981.

[31] C.E. Davila, A.J. Welch, H.G. Rylander, "A second-order adaptive Vol-
      terra filter with rapid convergence", *IEEE Transactions on Acoustics,
      Speech, and Signal Processing,* vol.ASSP-34, pp.1259-1263, Sept. 1987.

[32] J.C. Stapleton, S.C. Bass, "Adaptive noise cancellation for a class of non-
      linear, dynamic reference channels", *1984 Int. Symp. on Circuits and Sys-
      tems,* pp.268-271, 1984.

[33] G.L. Sicuranza, G. Ramponi, "A variable-step adaptation algorithm for
      memory-oriented Volterra filters", *IEEE Transactions on Acoustics,
      Speech, and Signal Processing,* vol.ASSP-35, pp.1492-1494, Oct. 1987.

[34] G.L. Sicuranza, G. Ramponi, "Adaptive nonlinear digital filters using dis-
      tributed arithmetic", *IEEE Transactions on Acoustics, Speech, and Signal
      Processing,* vol.ASSP-34, pp.518-526, June 1986.

[35] Y. Lou, C.L. Nikias, A.N. Venetsanopoulos, "VLSI array processing
      structure of quadratic digital filters with LMS algorithm", *1987 IEEE Int.
      Conf. Acoustics, Speech and Signal Processing,* pp.1394-1397, 1987.

[36] E.J. Thomas, "Some considerations on the application of the Volterra
      representation of nonlinear networks to adaptive echo cancelers", *Bell
      System Technical Journal,* vol.50, pp. 2797-2805, Oct. 1971.

[37] H.H. Chiang, C.L. Nikias, A.N. Venetsanopoulos, "Reconfigurable sys-
      tolic array implementation of quadratic digital filters", *IEEE Transactions
      on Circuits and Systems,* vol. CAS-33, pp.845-848, Aug. 1986.

[38] G. L. Sicuranza, A. Bucconi, P. Mitri, "Adaptive echo cancellation with nonlinear digital filters", *1984 IEEE Int. Conf. Acoustics, Speech and Signal Processing,* pp.3.10.1-3.10.4, 1984.

[39] X.Y. Gao, W.M. Snelgrove, D.A. Johns, "Nonlinear IIR adaptive filtering using a bilinear structure", *1989 Int. Symp. on Circuits and Systems,* Portland Oregon, 1989.

[40] D.D. Falconer, "Adaptive equalization of channel nonlinearities in QAM data transmission systems", *Bell System Technical Journal,* vol.57, pp. 2589-2611, Sept. 1978.

[41] T. Koh, E.J. Powers, "Second-order Volterra filtering and its application to nonlinear system identification", *IEEE Transactions on Acoustics, Speech, and Signal Processing,* vol.ASSP-33, pp.1445-1455, Dec. 1985.

[42] E. Biglieri, A. Gersho, R.D. Gitlin, T.L. Lim, "Adaptive cancellation of nonlinear intersymbol interference for voiceband data transmission", *IEEE J. Selected Areas in Communications,* vol.SAC-2, pp.765-777, Sept. 1984.

[43] G.L. Sicuranza, "Nonlinear digital filter realization by distributed arithmetic", *IEEE Transactions on Acoustics, Speech, and Signal Processing,* vol. ASSP-33, pp.939-945, Aug. 1985.

[44] O. Agazzi, D.G. Messerschmitt, D.A. Hodges, " Nonlinear echo cancellation of data signals", *IEEE Transactions on Communications,* vol. COM-30, pp.2421-2433, Nov. 1982.

[45] C.F.N. Cowan, P.F. Adams, "Nonlinear system modeling: concept and application", *1984 IEEE Int. Conf. Acoustics, Speech and Signal Processing,* Mar. 1984.

[46] H.H. Chiang, C.L. Nikias, A.N. Venetsanopoulos, "Efficient implementations of digital Volterra filters", *1986 IEEE Int. Conf. Acoustics, Speech and Signal Processing,* pp.857-860, 1986.

[47] T. Koh, E.J. Powers, R.W. Miksad, F.J. Fischer, "Application of nonlinear digital filters to modeling low-frequency, nonlinear drift oscillations of moored vessels in random seas", *The 16th Annual Offshore Technology Conference,* pp.309-314, Houston, Texas, May 1984.

[48] M.J. Coker and D.N. Simkins, "A nonlinear adaptive noise canceler", *1980 IEEE Int. Conf. Acoustics, Speech and Signal Processing,* pp.470-473, 1980.

[49] X.Y. Gao, *Adaptive Volterra filters: algorithms and applications,* Technical report, University of Toronto, 1988.

[50] F. Palmieri, C.G. Boncelet Jr., "A class of nonlinear adaptive filters", *Proc. IEEE International Conference on Acoustics, Speech and Signal*

Processing, pp. 1483-1486, New York, 1988.

[51] I. Pitas, A.N. Venetsanopoulos, "Adaptive *L*-filters", *Proc. European Conference on Circuit Theory and Design*, Brighton, England, 1989.

[52] F. Palmieri "A backpropagation algorithm for multilayer hybrid order statistics filters", *Proc. IEEE International Conference on Acoustics, Speech and Signal Processing*, Glasgow, Scotland, 1989.

CHAPTER 10

GENERALIZATIONS AND NEW TRENDS

10.1 INTRODUCTION

Nonlinear digital filters have had an impressive growth in the past two decades. This growth continues nowadays and gives new theoretical results, new filtering tools, and interesting applications. In the following, generalizations and current trends in nonlinear filtering will be described.

One of the main problems usually encountered in digital image filtering applications is the bulk of the image data and the large amount of operations per image pixel. Many linear and nonlinear image processing operators are applied to the neighborhood of each image pixel. They usually involve several additions, multiplications, comparisons, and nonlinear function evaluations per image pixel. Thus, the computational complexity required is relatively high. In certain image processing applications (e.g., in video image processing, automatic inspection) real-time image processing is required. In such cases, the common Single Instruction Single Data (SISD) computers fall short in speed requirements. Therefore, parallel digital filter implementation is the solution in this case. In general, parallel computation is an important implementation trend, as will be seen in the next chapter.

In most image filtering routines, operations are identical in each image pixel. Therefore, it is desirable to have one processor per image pixel. Since all processors execute the same operations, they are easily controllable from a central processor. This architecture is the Single Instruction Multiple Data (SIMD) one, which has already been used in image processing. The main problem with such a solution is the massive parallelism required, which is high even by today's standards. A second problem is the structure of the pixel processor itself. The solution is to use a standard processor structure capable of performing all basic image processing operations. This processor may be optimized for certain frequent image processing operations (e.g., moving average filtering, median filtering). However even in the case of a fairly simple processor structure, the amount of hardware required is relatively large. Another solution to the real-time image processing is to use one or more relatively sophisticated coprocessors on an image processing board that is driven by a host computer. Each coprocessor may have a structure which is optimized for a specific image processing task. Such processors have already been implemented in VLSI for certain operations (e.g., median filtering, edge detection) [2,4]. Another more reasonable solution is to construct filter structures that can realize a variety of image processing operations relatively fast [3]. The main problem in this approach is that most image filters result from different theoretical considerations and solve different problems. Thus several different families of image processing filters have been formed (e.g., order statistic filters, homomorphic filters, morphological filters, to name a few). Therefore, a unifying image structure that has all these classes as subcases is desirable. Such a structure has also various interesting theoretical properties and can be used as an analytical tool in the study of the performance of nonlinear filters. Furthermore, this structure can be versatile and can be used to design new filters that perform well in the presence of certain kinds of noise. Such a general structure will be described in the next section.

Another important current trend is nonlinear color image processing. It is driven by the urgent need to develop digital color image processing techniques for digital video and for High Definition TV (HDTV) application. An effort is currently being made to extend black and white (BW) image processing techniques to color image processing. Although the research results are not sufficient to form an opinion about the complexity of such an extension, we think that it is not trivial.

Finally, a trend that is well under way is nonlinear filtering based on neural networks. Neural nets (NN) have already found extensive applications in pattern recognition and in image analysis, mainly for binary images. They have also been proposed as digital signal and image processing schemes. An effort will be made in this chapter to analyze their performance as digital signal and image processing tools.

10.2 A GENERAL NONLINEAR FILTER STRUCTURE

A general nonlinear filter structure is shown in Figure 10.2.1.

Figure 10.2.1: General nonlinear filter structure.

This structure employs two nonlinear monotonic pointwise functions $g(.), f(.)$, two linear filters L_1, L_2, and a sorting network. The following families of nonlinear filters are special cases of the general nonlinear filter structure, if the functions $g(x), f(x)$ and the linear filters are chosen appropriately [3,4]:

a)   Filters based on order statistics

b)   Homomorphic filters

c)   Nonlinear mean filters

d)   Morphological filters.

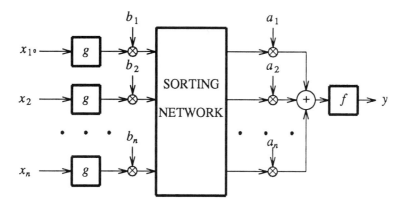

Figure 10.2.2: General nonlinear filter module.

The general nonlinear filter structure is a very versatile and powerful one. Its versatility lies in the two linear filters and the two nonlinear functions it employs. Their choice determines the characteristics of the nonlinear filter, which can be adapted for a variety of applications. In the following, we shall concentrate to a special case of the general nonlinear filter structure called *general nonlinear filter module*, shown in Figure 10.2.2. This module can be used as a building block for a variety of linear and nonlinear filters. The implementation of some nonlinear filters belonging to those classes by the nonlinear filter module is summarized in Table 10.2.1. It has also been proven that the nonlinear filter module can be used for the implementation of quadratic filters [9]. Furthermore, certain classes of edge detectors and pattern recognition algorithms are special cases of this structure [3]. The implementation of the range edge detector, the quasi-range edge detector, and the dispersion edge detector, described in chapter 8, can be implemented by the general nonlinear filter module, as is summarized in Table 10.2.1.

Table 10.2.1: Implementation of various nonlinear filters by the nonlinear filter module.

Filter	b_i	a_i	$g(x)$	$f(x)$
Linear	*	$a_i=1, i=1,..,n$	x	x
Homomorphic	*	$a_i=1, i=1,..,n$	*	$f(x)=g^{-1}(x)$
Harmonic	*	$a_i=1, i=1,..,n$	$1/x$	$f(x)=g^{-1}(x)$
Geometric	*	$a_i=1, i=1,..,n$	$ln(x)$	$f(x)=g^{-1}(x)$
L_p mean	*	$a_i=1, i=1,..,n$	x^p	$f(x)=g^{-1}(x)$
CH_p mean	$x_i^p/\sum_{i=1}^n x_i^p$	$a_i=1, i=1,..,n$	x	x
Median	$b_i=1, i=1,..,n$	$a_{v+1}=1, a_i=0, i\neq v+1$	x	x
i-th order	$b_i=1, i=1,..,n$	$a_i=1, a_j=0, j\neq i$	x	x
α-trimmed mean	$b_i=1, i=1,..,n$	$a_i=\dfrac{1}{n(1-2\alpha)}, i=\alpha n+1,..,n-\alpha n, a_i=0$ elsewhere	x	x
Erosion	$b_i=1, i\in B^s$ $b_i=0, i\notin B^s$	$a_1=1, a_i=0, i\neq 1$	x	x
Dilation	$b_i=1, i\in B^s$ $b_i=0, i\notin B^s$	$a_n=1, a_i=0, i\neq n$	x	x
Range	$b_i=1, i=1,..,n$	$a_1=-1, a_n=1, a_i=0, i\neq 1,n$	x	x
Quasi-range	$b_i=1, i=1,..,n$	$a_i=-1, a_{n+1-i}=1, a_j=0, j\neq i,n+1-i$	x	x
Dispersion	$b_i=1, i=1,..,n$	$a_i=\dfrac{-2}{n}, i=1,..,v, a_i=\dfrac{2}{n}, i=v+1,..,n, a_{v+1}=0$	x	x

* any choice

It has been mentioned that the general nonlinear filter structure can be used for image filtering in a variety of applications. In the following section an example of the use of the general nonlinear filter structure for signal-dependent noise filtering is given.

10.3 SIGNAL-DEPENDENT NOISE FILTERING

Signal-dependent noise is described by the following equation, as has already been described in chapter 3:

$$x = t(s)+r(s)\mathrm{n} \qquad (10.3.1)$$

A decoupling of the noise n from signal s can be obtained by passing the observed image x from a nonlinearity $g(x)$ as has already been described in chapter 7. If $g(x)$ is chosen as follows:

$$g^{(1)}[t(s)] = \frac{1}{r(s)} \tag{10.3.2}$$

the noise becomes additive:

$$g(x) \simeq g[t(s)] + n \tag{10.3.3}$$

The additive noise term n can be removed by any convenient linear technique, i.e., by a moving average filter, a low-pass linear filter, or a Wiener filter. However, linear filters tend to smooth edges. Thus, they are not desirable. Nonlinear filters can be used instead. L-filters are highly desirable in this case because their coefficients can adapt to the probability distribution of noise n:

$$y_i = \sum_{j=1}^{n} a_j \tilde{x}_{(j)} \tag{10.3.4}$$

where $\tilde{x}_i = g(x_i)$. The following equation gives the optimal coefficients of the L-filter section, as has already been described in chapter 5:

$$a = \frac{R^{-1}e}{e^T R^{-1} e} \tag{10.3.5}$$

R is the autocorrelation matrix of the ordered variates $[\tilde{x}_{(1)}, ..., \tilde{x}_{(n)}]^T$. After additive noise filtering, the distorted image $g[t(s)]$ can be restored by using another nonlinear function f(.) such that:

$$f\{g[t(s)]\} = s \tag{10.3.6}$$

The resulting filter is called *Nonlinear Order Statistic Filter* and it is shown in Figure 10.3.1. It is clearly a special case of the general nonlinear filter structure.

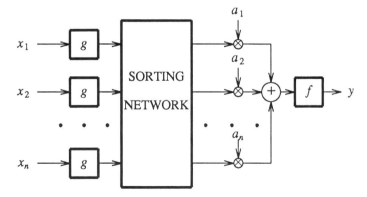

Figure 10.3.1: Nonlinear order statistics filter structure.

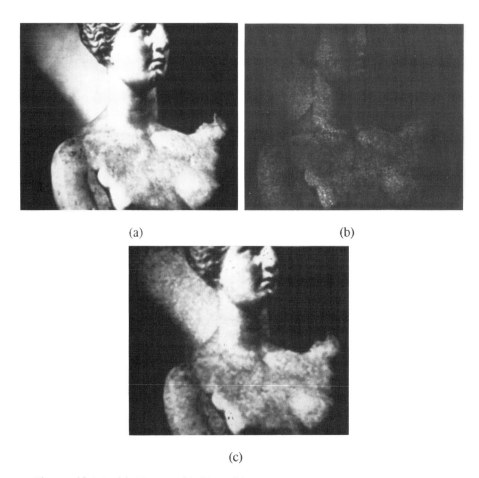

(a)   (b)

(c)

Figure 10.3.2: (a) Venus of Milos; (b) Image corrupted by signal-dependent noise; (c) Output of the NLOS filter.

An example of NLOS filtering of signal dependent noise is shown in Figure 10.3.2. The image formation model used is [5]:

$$x = s^{0.7} + s^{0.35}n \tag{10.3.7}$$

where s is the original image, shown in Figure 10.3.2a, n is additive Gaussian noise independent of s, and x is the observed image shown in Figure 10.3.2b. The nonlinear functions $g(x)$, $f(x)$ have been calculated according to (10.3.2),(10.3.6), respectively:

$$g(x) = 2x^{0.5} \tag{10.3.8}$$

$$f(x) = (0.5x)^{2.857} \tag{10.3.9}$$

The L-filter used is the NLOS filter has coefficients chosen in such a way that the order statistics close to the median are weighted. The filter is chosen in this way so that a compromise between the edge preservation of the median and the filtering of the additive Gaussian noise in the homogeneous image regions is obtained. The output of the NLOS filter is shown in Figure 10.3.2c. The signal-dependent noise has been removed and the edges have been preserved fairly well.

NLOS filters have also been used in the filtering of signal-dependent noise mixed with additive and salt-pepper noise [8]. The noise model employed was the following:

$$
x = \begin{cases}
s+(k_1 f(s)+k_2)n & \textit{with probability } 1-(p_1+p_2) \\
x_{max} & \textit{with probability } p_1 \\
x_{min} & \textit{with probability } p_2
\end{cases} \tag{10.3.10}
$$

where x_{max}, x_{min} are the values of positive and negative impulses having probabilities of occurrence p_1, p_2, respectively. k_1, k_2 are scalar constants. The noise term $k_1 f(s)n$ is signal-dependent, whereas the term $k_2 n$ is additive. The noise $(k_1 f(s)+k_2)n$ in (10.3.10) can be transformed to additive noise by using a pointwise nonlinearity $g(x)$ given by the following relation, according to (10.3.2):

$$
g^{(1)}(s) = \frac{1}{k_1 f(s) + k_2} \tag{10.3.11}
$$

$$
g(s) = \int \frac{ds}{k_1 f(s) + k_2} \tag{10.3.12}
$$

The transformed image $g(x)$ takes the following form:

$$
g(x) = \begin{cases}
g(s)+n & \textit{with probability } 1-(p_1+p_2) \\
g(x_{max}) & \textit{with probability } p_1 \\
g(x_{min}) & \textit{with probability } p_2
\end{cases} \tag{10.3.13}
$$

The noise encountered in the transformed image $g(x)$ is additive noise mixed with positive and negative impulses. Any suitable filter based on order statistics (e.g., median filter, L-filter, α-trimmed mean filter) can be used to remove this type of noise. The use of double window modified trimmed mean (DW MTM) filter, described in chapter 5, is proposed in [8]. The resulting filter is the so-called *homomorphic DW MTM* filter. The performance of this filter in removing signal-dependent noise mixed with additive noise and impulses is shown in Figure 10.3.3. The original signal, shown in Figure 10.3.3a consists of a ramp edge, step edges, and a narrow pulse. This signal corrupted by positive and negative impulses and by additive and signal-dependent noise of the form:

$$x = (s^{0.5} + 7.0)n \qquad\qquad (10.3.14)$$

is shown in Figure 10.3.3b. The noise process n in (10.3.14) is white Gaussian zero mean unit variance noise. The pointwise nonlinearity $g(x)$ has the form:

$$g(s) = \int \frac{ds}{s^{0.5} + 7} \qquad\qquad (10.3.15)$$

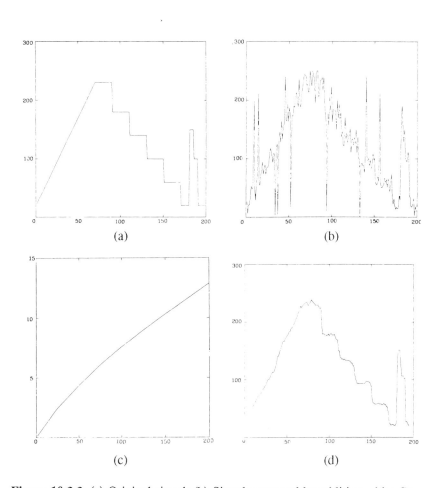

(a)   (b)

(c)   (d)

Figure 10.3.3: (a) Original signal; (b) Signal corrupted by additive white Gaussian noise, signal-dependent noise and salt-pepper noise; (c) Linear piecewise approximation of the nonlinear function $g(x)$; (d) Output of the homomorphic DW MTM filter.

Its linear piecewise approximation is shown in Figure 10.3.3c. The homomorphic DW MTM filter used employs two windows of size 11 and 7, respectively. The threshold of the averaging range used was $q=40$. The result of the homomorphic DW MTM filter is shown in Figure 11.3.3d. The impulses have been removed completely. The ramp edge, the narrow pulse, and the step edges have been preserved fairly well.

10.4 COMPUTATIONAL COMPLEXITY OF THE GENERAL NONLINEAR FILTER MODULE

If the computational complexity study of the general nonlinear filter module is to be done, the exact form of the sorting network, of the evaluation of the nonlinear functions, and the way of the execution of additions (serial or parallel) must be known. The evaluation of the nonlinear functions $g(.)$, $f(.)$ can be easily performed by using look-up tables. There are several sorting networks, as it will be discussed in the next chapter [6]. The *odd-even transportation* sorting network, shown in Figure 10.4.1, will be used because it has small throughput delay and a very uniform structure. The same sorting network has been used also for the development of VLSI chips for median filtering [4].

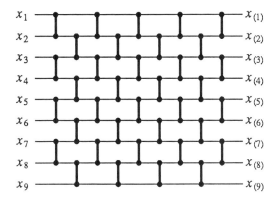

Figure 10.4.1: Odd/even transportation network. The vertical bars denote comparators.

The absolutely minimum parallel sorting time for this sorting network (also called *critical time*) is given by:

$$T_c = nT_{comp} \qquad (10.4.1)$$

where T_{comp} denotes the time required for a single comparison and $n=2v+1$

denotes the filter length. The number of comparators needed for the sorting network is given by:

$$n_c = n(n-1)/2 \qquad (10.4.2)$$

The additions in the filter module are proposed to be done by a tree network shown in Figure 10.4.2. This implementation has the advantage of high parallelism and the disadvantage of the high number of adders required.

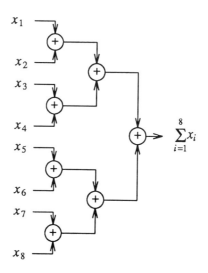

Figure 10.4.2: A tree network for additions.

The critical time for the computation of n additions by this network is:

$$T_c \simeq \lfloor \log_2 n \rfloor T_{add} \qquad (10.4.3)$$

where T_{add} denotes the time required for a real addition and the symbol $\lfloor \log_2 n \rfloor$ denotes the smallest integer which is larger than or equal to $\log_2 n$. The formula (10.4.3) is exact when n is a power of two. The number of adders needed for the addition network is:

$$N_{add} = n-1 \qquad (10.4.4)$$

The structure of the general filter module that employs the sorting and addition networks shown in Figures 10.4.1-2 is presented in Figure 10.4.3. The critical time of the filter (i.e., the least possible time required for parallel calculation) is:

$$T_c = nT_{comp}+2T_m+\log_2 n \, T_{add}+2T_n \qquad (10.4.5)$$

where T_m, T_n denote the time required for multiplication and nonlinear function evaluation, respectively.

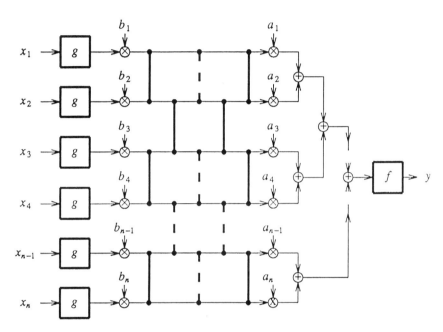

Figure 10.4.3: General nonlinear filtering module.

The time required for a completely serial computation is given by:

$$T_s = n(n-1)T_{comp}/2 + (n-1)T_{add} + 2nT_m + 2T_n \qquad (10.4.6)$$

The serial computation time is not of great importance because we are interested in parallel computation.

In the following the general nonlinear filter module will be compared with structures for the implementations of the following filters:

a)   moving average filter

b)   median filter

c)   erosion/dilation filter.

These filters are chosen because they are the most commonly used filters in image processing and represent the families of linear filters, order statistics filters and morphological filters respectively. The aim of the comparison is to study the increase in the hardware demands and the decrease in the filter throughput rate, which are paid for the flexibility of the proposed module. The median is computed by the sorting network shown in Figure 10.4.1. The moving average filter is calculated by the last part of the implementation of Figure 10.4.3. A separate structure for the implementation of the erosion/dilation filter, which is essentially a maximum/minimum filter, is shown in Figure 10.4.4.

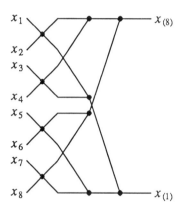

Figure 10.4.4: Erosion/dilation filter.

Its critical time is given by:

$$T_c \simeq \left\lfloor \log_2 n \right\rfloor T_{comp} \tag{10.4.7}$$

The number of comparators used in the erosion/dilation filter (for n even) is:

$$N_c = \frac{3n}{2} - 2 \tag{10.4.8}$$

The figure of merit used are the following:

a) number of adders N_{add}

b) number of multipliers N_m

c) number of comparators N_c

d) number of nonlinear function evaluators N_n

e) critical time T_c.

The results of the comparison are summarized in Table 10.4.1. The general filter module requires almost the same amount of hardware as a linear filter together with a median filter. Its critical time is also the sum of the critical times of the linear and the median filters. This is the price that is paid for the versatility and the flexibility of the general filter structure.

Table 10.4.1: Comparison of various image filters.

	N_{add}	N_m	N_c	N_n	T_c
Nonlinear module	$n-1$	$2n$	$\dfrac{n(n-1)}{2}$	2	$2T_m+2T_n+nT_{comp}+\lfloor \log_2 n \rfloor T_{add}$
Moving average	$n-1$	n	0	0	$T_m+\lfloor \log_2 n \rfloor T_{add}$
Median	0	0	$\dfrac{n(n-1)}{2}$	0	nT_{comp}
Max-Min	0	0	$\dfrac{3n}{2}-2$	0	$\lfloor \log_2 n \rfloor T_{comp}$

10.5 OVERVIEW OF THE PERFORMANCE OF VARIOUS NONLINEAR FILTERS

In the previous chapters a multitude of nonlinear filter structures have been discussed. Most of them, with the exception of the polynomial filters, can be implemented by the general filter structure. However, the multitude of filters poses some difficulties to the design/ applications engineer. Most of the filters have their advantages and disadvantages. Therefore, it is not clear which kind of filters are suitable for a specific application. This problem arises from the fact that different kinds of noise are present in different applications. The most commonly encountered types of noise can be summarized in the following categories:

a)   Short-tailed additive white noise

b)   Additive white Gaussian noise

c)   Long-tailed additive white noise

e)   Positive impulse noise

f)   Negative impulse noise

g)   Salt-pepper noise

h)   Multiplicative noise

i)   Signal-dependent noise.

j)   Mixed noise consisting of additive white noise, signal-dependent noise, and salt-pepper noise.

It is well known that most filters are designed to perform well in the presence of some types of noise. Usually their performance deteriorates rapidly in the presence of different types of noise. Furthermore, the performance measures depend greatly on the application itself. The most commonly used figures of merit for filter performance are the following:

a)   Noise filtering characteristics for different types of noise

b)   Edge preservation

c)   Fine detail preservation

d)   Unbiasedness

e)   Computational complexity.

The noise filtering properties of a filter are usually measured by its output variance or by the rate of success in the impulsive noise removal. Other performance measures are the Normalized Mean Square Error (NMSE) and the Perceptual Mean Square Error (PMSE) described in chapter 3. The main disadvantage of all these measures (except perhaps the rate of success in impulse noise removal) is that they are relatively poorly correlated to the subjective human criteria. The main reason for this is that human vision is rather complicated and its properties cannot be described by a single performance formula.

The edge preservation is an important property of a filter and it refers to its capability to preserve edges. The fine detail preservation properties refer to its ability to preserve image lines, sharp corners, and other fine image details. The bias of a filter refers to its directional or illuminational bias. It is known that certain filters tend to enhance image along certain directions (e.g., along the horizontal and the vertical direction for separable filters). Other filters (e.g., the erosion or dilation filters) tend to enhance image regions having certain illumination characteristics (e.g., low or high illumination). All figures of merit (b-d) are rather qualitative and they have not been described yet by quantitative criteria. Therefore, the performance of the filter according to the figures of merit (b-d) is relatively subjective. Most researchers demonstrate the performance of their filters by performing simulations of noisy images and by comparing the filtered images this the original images.

The computational complexity of a filter usually refers to the number of algebraic operations (multiplications, comparisons, additions) required per output pixel. In the case of the parallel computation it refers to the number of hardware resources (adders, multipliers, comparators) required and to the throughput delay per output pixel. Throughput delay is also connected to the least possible parallel computation time (called critical time).

Table 10.5.1: Overview of the performance of various nonlinear filters.

Filter	Figures of merit											
	a	b	c	d	e	f	g	h	i	j	k	l
Moving average	1	2	0	0	0	0	0	0	0	0	2	1
Median	0	1	2	2	2	2	0	0	2	0	2	1
Separable median	0	1	2	2	2	2	0	0	2	0	2	2
Recursive median	0	1	2	2	2	2	0	0	2	0	2	1
Max-median	0	1	1	1	2	1	0	0	2	1	0	1
Multistage median	0	1	2	2	2	2	0	0	2	2	2	1
Median hybrid	1	1	1	1	1	1	0	0	2	2	2	2
Low Ranked order	1	1	0	2	0	0	0	0	2	1	0	1
High Ranked order	1	1	0	0	2	0	0	0	2	1	0	1
L_p mean	1	1	0	0	2	0	0	0	1	1	1	0
CH_p mean	1	1	0	0	2	0	0	0	2	1	0	0
L_{-p} mean	1	1	0	2	0	0	0	0	1	1	1	0
CH_{-p} mean	1	1	0	2	0	0	0	0	2	1	0	0
Harmonic mean	1	1	0	2	0	0	0	0	1	1	1	0
Geometric mean	1	1	1	0	0	0	2	0	0	0	2	0
α-trimmed mean	0	1	2	2	2	2	0	0	2	1	2	0
compl α-trimmed mean	2	1	0	0	0	0	0	0	0	0	2	0
Midpoint	2	1	0	0	0	0	0	0	0	0	2	2
L-filter	2	2	2	2	2	2	0	0	2	1	2	0
NLOS filter	2	2	2	2	2	2	2	2	2	1	2	0
Wilcoxon	1	1	1	1	1	1	0	0	0	1	2	0
Modified Wilcoxon	1	1	1	1	1	1	0	0	1	1	2	0
STM-filter	1	2	2	2	2	2	0	0	2	1	2	0
Dilation (Maximum)	1	1	0	0	2	0	0	0	2	0	0	2
Erosion (Minimum)	1	1	0	2	0	0	0	0	2	0	0	2
Closing	1	1	0	0	2	0	0	0	2	0	2	1
Opening	1	1	0	2	0	0	0	0	2	0	2	1
Close-opening	1	1	1	1	2	1	0	0	2	0	2	0
Open-closing	1	1	1	2	1	1	0	0	2	0	2	0

a. Short-tailed additive white noise
b. Gaussian additive white noise
c. Long-tailed additive white noise
d. Positive impulses
e. Negative impulses
f. salt-pepper noise
g. Multiplicative noise
h. Additive signal-dependent noise
i. Edge preservation
j. Detail preservation
k. Bias
l. Computational complexity

Performance	0: poor	1: average	2: good
Bias	0: strong	1: average	2: low
Computational complexity	0: high	1: average	2: low

For completely serial computation on a general purpose computer, the computational complexity of each filter can be measured by the speed of its execution. This speed depends, of course, on the algorithm used, on the programming language, and on the computer characteristics. Although computational complexity has well-defined figures of merit, it depends greatly on the processor architecture used and on the amount of parallelism required. Therefore, the computational complexity of a filter depends on many factors and cannot be described by a single number. In the following chapter a thorough analysis of the algorithms, the processor architectures, and the computational complexity of the most commonly used nonlinear filters is given.

Until now, no comprehensive comparison of all known nonlinear filters has appeared in the literature. Most comparisons include at most 3-4 known nonlinear filters. Only a recent comparison includes a wide variety of nonlinear filters [7]. The results of the comparison are both quantitative and qualitative. Taking into account all the above-mentioned facts, we have decided to use the following crude ranking of the performance of a filter:

0:    Poor performance

1:    Average performance

2:    Good performance.

       For the bias, the ranking is the following:

0:    Strong bias

1:    Average bias

2:    Low bias.

       The following ranking has been used for computational complexity:

0:    High computational complexity

1:    Average computational complexity

2:    Low computational complexity.

       The filter comparison is summarized in Table 10.5.1. The entries to this table have been filled according to the characteristics of the filters given by the various researchers who have done comparisons (e.g., [7]) and according to the experience of the authors of this book. This table is intented to be a qualitative guide for design and application engineers who want to choose a filter for a specific application fast and without much experimentation.

10.6 COLOR IMAGE PROCESSING

       Color perception is a physiopsychological phenomenon that has been studied intensively in the past [10,12] and that is still subject of ongoing research. The colors that are perceived in an object basically depend on the light reflected by the object. Visible light occupies a relatively narrow band in the electromagnetic spectrum. Objects reflecting light that is distributed in all the visible spectrum appear white. Objects reflecting light in specific parts of the visible spectrum appear colored. This fact is explained by the anatomy of the human eye. The human retina has three types of cones that are sensitive to different parts of the visible spectrum, one for each of the three principal colors: red (R), green (G), and blue (B). The CIE (Commission Internationale de l'Eclairage) designated the following wavelengths to the three primary colors: 700 nm (red), 546.1 nm (green), 435.8 nm (blue). Different primary colors can be used instead of the RGB, as will be seen later. In general, any color can be matched in appearance with the additive mixture of three primaries P_1, P_2, P_3 [13]:

$$C = \alpha P_1 + \beta P_2 + \gamma P_3 \qquad (10.6.1)$$

If the primaries are the RGB colors, $E(\lambda)$ is the light spectrum and $f_r(\lambda), f_g(\lambda), f_g(\lambda)$ are the color matching factors, the coefficients α, β, γ are given by:

$$\alpha = \int_\lambda E(\lambda) f_r(\lambda) d\lambda \qquad (10.6.2)$$

$$\beta = \int_\lambda E(\lambda) f_g(\lambda) d\lambda$$

$$\gamma = \int_\lambda E(\lambda) f_b(\lambda) d\lambda$$

The coefficients α, β, γ are the tristimulus values of the given light. Each color can be represented by a vector in a three-dimensional space defined by the basis (R, G, B), as is shown in Figure 10.6.1.

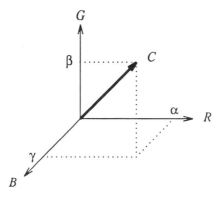

Figure 10.6.1: Representation of a color vector in the RGB space.

The perceptual attributes of the color are its *brightness, hue*, and *saturation*. Brightness represents the perceived luminance. Hue is associated with the dominant wavelength in a mixture of light wavelengths and represents the dominant color of the light. Saturation refers to the amount of the white light mixed with a hue. Hue and saturation constitute the *chromacity* of the light. They are relatively independent from the brightness. Therefore, brightness can be factored out of the primary system:

$$r = \frac{R}{R+G+B} \qquad\qquad (10.6.3)$$

$$g = \frac{G}{R+G+B}$$

$$b = \frac{B}{R+G+B}$$

where the coefficients α, β, γ have been replaced by R,G,B for notation consistency with the existing literature. The coefficients r,g,b are called chromatic coordinates and satisfy:

$$r+g+b = 1 \qquad\qquad (10.6.4)$$

Therefore, only two of them are sufficient to determine the chromacity. Several other primary color systems have been proposed and used, whose description is beyond the scope of this book. A relatively complete description can be found in [10, pp.66-70]. The only system that will be described briefly is the *XYZ* system, accepted by CIE in 1931. It is linearly connected to the *RGB* system:

$$
\begin{bmatrix} X \\ Y \\ Z \end{bmatrix} = \begin{bmatrix} 0.490 & 0.310 & 0.200 \\ 0.177 & 0.813 & 0.011 \\ 0.000 & 0.010 & 0.990 \end{bmatrix} \begin{bmatrix} R \\ G \\ B \end{bmatrix} \qquad\qquad (10.6.5)
$$

The chromacity coordinates in the *XYZ* system are given by:

$$x = \frac{X}{X+Y+Z} \qquad\qquad (10.6.6)$$

$$y = \frac{Y}{X+Y+Z}$$

$$z = \frac{Z}{X+Y+Z}$$

where:

$$x+y+z = 1 \qquad\qquad (10.6.7)$$

The white color corresponds to $x=y=z=0.33$. In any color coordinate system, each color is represented by a vector in a three-dimensional space. In the RGB and XYZ systems the chromacity can be represented as a vector (r,g) or (x,y) in the two-dimensional space $[0,1] \times [0,1]$. Any color image can be considered as a two-dimensional vector sequence or, in other words, as a two-dimensional three-channel sequence. In cases of constant brightness, the chromacity of a color image can be considered as a two-dimensional two-channel sequence. Therefore, color image processing is essentially multichannel signal processing. Several scientists working in color image processing did not pay sufficient attention to the multichannel nature of color images. Some of them simply applied black and white (BW) image processing techniques to each channel (e.g., RGB) separately. Some other scientists have used transformation techniques (e.g., Karhunen-Loeve transformation) to decorrelate the three channels and to apply BW image processing techniques afterwards (e.g., in [13]). Although we think that this approach is valid, we believe that the application of multichannel signal processing techniques (e.g., in [14-17]) is a more natural approach in color image processing.

In the following, the use of ordering and order statistics [18] in multichannel signal and image processing will be investigated. Order statistics filtering and especially median filtering have found extensive applications in BW image filtering, as we have seen in the previous chapters. Therefore, there is a natural interest in extending them to multichannel signals.

10.7 ORDERING OF MULTIVARIATE DATA

Let us denote by $\mathbf{X}$ a p-dimensional random variable, i.e., a p-dimensional vector of random variables $\mathbf{X} = [X_1, \ldots, X_p]^T$. We shall denote by $f(\mathbf{X})$, $F(\mathbf{X})$ the probability density function and the cumulative density function of this p-dimensional random variable respectively. One of the most commonly used multivariate distributions is the Gaussian distribution [19, p.18]:

$$f(\mathbf{x}) = (2\pi)^{-p/2} |\mathbf{C}|^{-1/2} \exp\{-\frac{1}{2}(\mathbf{x}-\mathbf{m})^T \mathbf{C}^{-1}(\mathbf{x}-\mathbf{m})\} \qquad (10.7.1)$$

where $\mathbf{C}$ is a positive definite matrix. The expected value of $\mathbf{X}$ is $E[\mathbf{X}]=\mathbf{m}$ and the dispersion matrix of $\mathbf{X}$ is given by:

$$E[(\mathbf{X}-\mathbf{m})(\mathbf{X}-\mathbf{m})^T] = \mathbf{C} \qquad (10.7.2)$$

In the case of the Gaussian distribution we say that $\mathbf{X}$ is distributed as $N(\mathbf{m}, \mathbf{C})$. If $\mathbf{X}$ is a two-dimensional Gaussian random vector, its distribution is a special case of (10.7.1) given by [20, p.126]:

$$f(x_1,x_2) = \frac{1}{2\pi\sigma_1\sigma_2\sqrt{1-r^2}} \cdot \tag{10.7.3}$$

$$\cdot \exp\left\{-\frac{1}{2(1-r^2)}\left[\frac{(x_1-m_1)^2}{\sigma^2_1} - 2r\frac{(x_1-m_1)(x_2-m_2)}{\sigma_1\sigma_2} + \frac{(x_2-m_2)^2}{\sigma^2_2}\right]\right\}$$

with $|r|<1$. Another useful distribution is the uniform distribution:

$$f(\mathbf{x}) = \begin{cases} 1 & \mathbf{x} \in [-1/2,1/2] \times .. \times [-1/2,1/2] \\ 0 & elsewhere \end{cases} \tag{10.7.4}$$

The uniform distribution is a separable one:

$$f(\mathbf{x}) = \prod_{i=1}^{p} f(x_i) \tag{10.7.5}$$

where $f(x_i)$ is the one-dimensional uniform distribution in $[-1/2, 1/2]$. There are several other multivariate distributions, e.g., the multivariate beta distribution and the Wishardt distribution [19]. A useful family of elliptically symmetric distributions is described in [21,22] and can give long- and short-tailed alternatives to the multivariate normal distribution. More information on multivariate distributions can be found in [19,22].

Let $\mathbf{x}_1, \ldots, \mathbf{x}_n$ be n random samples from a p-dimensional distribution having mean $\mathbf{m}$ and dispersion matrix $\mathbf{C}$. The sample arithmetic mean $\bar{\mathbf{x}}$ and the sample unbiased estimate $\mathbf{S}$ of the dispersion matrix are given by:

$$\bar{\mathbf{x}} = \frac{1}{n}\sum_{i=1}^{n}\mathbf{x}_i \tag{10.7.6}$$

$$\mathbf{S} = \frac{1}{n-1}\sum_{i=1}^{n}(\mathbf{x}_i-\bar{\mathbf{x}})(\mathbf{x}_i-\bar{\mathbf{x}})^T \tag{10.7.7}$$

Both $\bar{\mathbf{x}}$ and $\mathbf{S}$ are straightforward extensions of the univariate case. Their expected values and dispersions are given by:

$$E[\bar{\mathbf{x}}] = \mathbf{m} \tag{10.7.8}$$

$$E[(\bar{\mathbf{x}}-\mathbf{m})(\bar{\mathbf{x}}-\mathbf{m})^T] = \frac{1}{n}\mathbf{C} \tag{10.7.9}$$

The arithmetic mean reduces the data dispersion, as can be seen in (10.7.9), and can be used for noise reduction in color image filtering. However, at the same time, it blurs the color image edges, as it does in the BW image processing. Furthermore, its performance will be poor in the presence of impulsive noise. Therefore, different techniques must be used for noise filtering in color images.

The simplest approach is to try to extend the notion of median filtering for the color images. Median filtering uses the notion of data ordering, which is very natural in the one- dimensional case $(p=1)$. This notion can not be extended in a straightforward way in the case of multivariate data. An excellent treatment of the ordering of multivariate data can be found in [23]. It is shown that there are several ways to order multivariate data. There is no unambiguous, universally agreeable total ordering of the n multivariate samples $\mathbf{x}_1,...,\mathbf{x}_n$. The following so-called sub-ordering principles are discussed in [23]: *marginal ordering, reduced (aggregate) ordering, partial ordering,* and *conditional (sequential) ordering.*

In marginal ordering, the multivariate samples are ordered along each one of the p-dimensions:

$$x_{1(1)} \leq x_{1(2)} \leq \cdots \leq x_{1(n)} \tag{10.7.10}$$

$$x_{2(1)} \leq x_{2(2)} \leq \cdots \leq x_{2(n)}$$

$$x_{p(1)} \leq x_{p(2)} \leq \cdots \leq x_{p(n)}$$

i.e., ordering is performed in each channel of the multichannel signal. $x_{1(1)}, x_{2(1)}, \ldots, x_{p(1)}$ are the minimal elements in each dimension. $x_{1(n)}, x_{2(n)}, \ldots, x_{p(n)}$ are the maximal elements in each dimension. $x_{1(v+1)}, x_{2(v+1)}, \ldots, x_{p(v+1)}$ is the (marginal) median of the multivariate data for $n=2v+1$. The i-th marginal order statistic is the vector $\mathbf{x}_{(i)}=[x_{1(i)}, x_{2(i)}, \ldots, x_{p(i)}]^T$. Needless to say, the median or any i-th marginal order statistic may not correspond to any of the samples $\mathbf{x}_1, \ldots, \mathbf{x}_n$. In contrast, in the one-dimensional case there exists an one-to-one correspondence between the samples $x_1, \ldots, x_n$ and the order statistics $x_{(1)}, \ldots, x_{(n)}$.

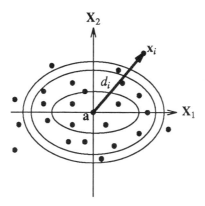

Figure 10.7.1: Generalized distance and R-ordering of multivariate data.

The reduced ordering (R-ordering) is based on the generalized distance:

$$d = (\mathbf{x}-\mathbf{a})^T \Gamma^{-1}(\mathbf{x}-\mathbf{a}) \qquad (10.7.11)$$

of a sample $\mathbf{x}$ from a point $\mathbf{a}$ which may be either the origin or the sample arithmetic mean $\bar{\mathbf{x}}$ or the marginal median $\mathbf{x}_{(v+1)}$. Γ may be the identity matrix, the dispersion matrix $\mathbf{C}$ or the sample dispersion matrix $\mathbf{S}$. The various data $\mathbf{x}_i$ are ordered according to their distances d_i from $\mathbf{a}$, as it is shown in Figure 10.7.1. Thus multivariate ordering is reduced to one-dimensional ordering.

Partial ordering (P-ordering) is based on the notion of the convex hull of the points $\mathbf{x}_1, \ldots, \mathbf{x}_n$, which is the minimum convex set which encloses all n samples [24]. Some of the samples lie on the perimeter of the convex set. These samples are denoted c-order group 1 and are discarded. The process is repeated with the rest of points and produces the c-order group 2 and so on. The resulting P-ordering is shown in Figure 10.7.2.

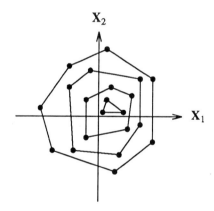

Figure 10.7.2: Convex hulls and P-ordering of multivariate data.

The conditional ordering (C-ordering) is conducted on one of the marginal sets of observations conditional on ordering within the data in terms of other marginal sets of observations. The samples $\mathbf{x}_1, \ldots, \mathbf{x}_n$ can be ordered in the following way [25]:

$$x_{1(1)} \le x_{1(2)} \le \cdots \le x_{1(n)} \qquad (10.7.12)$$

$$x_{2[1]} \le x_{2[2]} \le \cdots \le x_{2[n]}$$

$$x_{p[1]} \le x_{p[2]} \le \cdots \le x_{p[n]}$$

$x_{1(i)}$, $i=1,...,n$ are the marginal order statistics of the first dimension. $x_{j[i]}$, $j=2,...,p$, $i-1,...,n$ are the quasi-ordered samples in the dimensions $j=2,...,p$ conditional on the marginal ordering of the first dimension. In the two-dimensional case ($p=2$) the statistics $x_{2[i]}$, $i=1,...,n$ are called concomitants of the order statistics of X_1 [26].

In the following we shall concentrate on the probability distributions of the marginal order statistics. The study of the probability distribution of marginal order statistics started relatively early in [27], where the probability density function of the marginal median is investigated. Recurrence relations for the probability distribution of marginal order statistics are given in [28]. The probability distribution of the p-dimensional marginal order statistics is described in [29]. In the following the derivation of the cumulative distribution function and the probability distribution function of the two-dimensional marginal order statistics will be described. Let us denote by $F_{(r_1,r_2)}(x_1,x_2)$ the cdf:

$$F_{(r_1,r_2)}(x_1,x_2) = P\{X_{1(r_1)} \leq x_1, X_{2(r_2)} \leq x_2\} \qquad (10.7.13)$$

of the marginal order statistic $X_{1(r_1)}$, $X_{2(r_2)}$ when n data samples are available. Let us also denote by $F_i(x_1,x_2)$ $i=0,...,3$ the probability masses on the four regions of the plane defined by (x_1,x_2):

$$F_0(x_1,x_2) = P\{X_1 \leq x_1, X_2 \leq x_2\} = F(x_1,x_2) \qquad (10.7.14)$$

$$F_1(x_1,x_2) = P\{X_1 > x_1, X_2 \leq x_2\}$$

$$F_2(x_1,x_2) = P\{X_1 \leq x_1, X_2 > x_2\}$$

$$F_3(x_1,x_2) = P\{X_1 > x_1, X_2 > x_2\}$$

where $F(x_1,x_2)$ is the joint cdf of the random vector $\mathbf{X} = [X_1, X_2]^T$. The cdf $F_{(r_1,r_2)}(x_1,x_2)$: is given by [18, p.25]:

$$F_{(r_1,r_2)}(x_1,x_2) = \sum_{i_1=r_1}^{n} \sum_{i_2=r_2}^{n} P\{i_1 \text{ of } X_{1i} \leq x_1, i_2 \text{ of } X_{2i} \leq x_2\} = \qquad (10.7.15)$$

$$\sum_{i_1=r_1}^{n} \sum_{i_2=r_2}^{n} \sum_{n_0=max(0,i_1+i_2-n)}^{min(i_1,i_2)} \frac{n!}{n_0!(i_1-n_0)!(i_2-n_0)!(n-i_1-i_2+n_0)!}$$

$$\cdot F_0^{n_0}(x_1,x_2) F_1^{i_2-n_0}(x_1,x_2) F_2^{i_1-n_0}(x_1,x_2) F_3^{n-i_1-i_2+n_0}(x_1,x_2)$$

(10.7.15) is relatively complicated and does not give analytic expressions of $F_{(r_1,r_2)}(x_1,x_2)$ for arbitrary cdf $F(x_1,x_2)$. However, it can be easily computed numerically. The probability density function $f_{(r_1,r_2)}(x_1,x_2)$ of the marginal order statistics

$$f_{(r_1,r_2)}(x_1,x_2) = \frac{\partial^2 F_{(r_1,r_2)}(x_1,x_2)}{\partial x_1 \partial x_2} \qquad (10.7.16)$$

can easily be calculated from (10.7.15) by numerical differentiation. The cdf of the marginal median $r_1 = r_2 = v+1$ is a special case of (10.7.15):

$$F_{(v+1,v+1)}(x_1,x_2) = \sum_{i_1=v+1}^{n} \sum_{i_2=v+1}^{n} \sum_{n_*=1}^{min(i_1,i_2)} \frac{n!}{n_0!(i_1-n_0)!(i_2-n_0)!(n-i_1-i_2+n_0)!} \qquad (10.7.17)$$

$$\cdot F_0^{n_*}(x_1,x_2) F_1^{i_1-n_*}(x_1,x_2) F_2^{i_2-n_*}(x_1,x_2) F_3^{n-i_1-i_2+n_*}(x_1,x_2)$$

In the case of the marginal maximum $(r_1 = r_2 = n)$ (10.7.15) reduces to:

$$F_{(n,n)}(x_1,x_2) = F^n(x_1,x_2) \qquad (10.7.18)$$

10.8 MARGINAL ORDER STATISTICS AS ESTIMATORS OF THE MULTIDIMENSIONAL LOCATION

The definitions of L-estimators can be easily extended to the p-dimensional case by using marginal order statistics. The following estimator will be called p-dimensional marginal L-estimator:

$$\mathbf{T}_n = \sum_{i_1=1}^{n} \cdots \sum_{i_p=1}^{n} \mathbf{A}_{i_1,\ldots,i_p} \mathbf{x}_{(i_1,\ldots,i_p)} \qquad (10.8.1)$$

where $\mathbf{x}_{(i_1,\ldots,i_p)} = [x_{1(i_1)},\ldots,x_{p(i_p)}]^T$ are the marginal order statistics and $\mathbf{A}_{i_1,\ldots,i_p}$ are $p \times p$ matrices. The performance of the marginal L-estimator depends on the choice of those matrices. The marginal median, maximum, and minimum are special cases of (10.8.1) for appropriate choices of the matrices $\mathbf{A}_{i_1,\ldots,i_p}$. The p-dimensional marginal α-trimmed mean [32]:

$$\mathbf{T}_n = \begin{bmatrix} \dfrac{1}{n(1-2\alpha_1)} \sum_{i=\alpha_1 n+1}^{n-\alpha_1 n} x_{1(i)} \\ \cdots \cdots \\ \dfrac{1}{n(1-2\alpha_p)} \sum_{i=\alpha_p n+1}^{n-\alpha_p n} x_{p(i)} \end{bmatrix} \qquad (10.8.2)$$

is another special case of the p-dimensional L-estimator.

L-estimators can be used in multichannel image filtering in the case of additive white noise:

$$\mathbf{x} = \mathbf{s} + \mathbf{n} \qquad (10.8.3)$$

where $\mathbf{s}$ is the desired signal, $\mathbf{n}$ is a white noise process and $\mathbf{x}$ is the corrupted signal. L-estimators can be used in estimating $\mathbf{s}$, which is a multidimensional location estimation problem. In certain cases, e.g., in the homogeneous image regions, $\mathbf{s}$ is approximately constant. In this case, the performance of the estimator is measured by the dispersion matrix of its output:

$$\mathbf{D}(\mathbf{T}_n) = E\left[(\mathbf{T}_n - \mathbf{m}_T)(\mathbf{T}_n - \mathbf{m}_T)^T\right] \tag{10.8.4}$$

$$\mathbf{m}_T = E\left[\mathbf{T}_n\right] \tag{10.8.5}$$

The smaller the elements of $\mathbf{D}(\mathbf{T}_n)$ are, the better the performance of $\mathbf{T}_n$ is. The dispersion matrix of the two dimensional marginal median $\mathbf{x}_{(3)}$ for $n=5$ has been calculated numerically by using (10.7.16-17) for the uniform distribution in $[-1/2, 1/2] \times [-1/2, 1/2]$ and for the Gaussian distribution:

$$f(x_1, x_2) = \frac{1}{2\pi\sqrt{0.75}} \exp\{-\frac{2}{3}(x_1^2 - x_1 x_2 + x_2^2)\}$$

It is tabulated in Table 10.8.1 and compared with the dispersion matrix of the arithmetic mean for $n=5$ given by (10.7.7).

Table 10.8.1: Dispersion matrices of the arithmetic mean and of the marginal median for uniform and Gaussian distributions.

	Uniform noise		Gaussian noise	
Marginal	0.034615	0.000000	0.326553	0.104926
median ($n=5$)	0.000000	0.034615	0.104926	0.326553
Arithmetic	0.016	0.0	0.2	0.1
mean ($n=5$)	0.0	0.016	0.1	0.2

It is seen that the arithmetic mean performs better in both cases. However, the difference in the performance is much greater in the case of the uniform noise. The performance of the marginal median is expected to be superior in the case of impulsive noise of the form:

$$\mathbf{x} = \begin{cases} \mathbf{s} & \text{with probability } 1-p \\ \mathbf{n} & \text{with probability } p \end{cases} \tag{10.8.6}$$

where $\mathbf{n}$ is noise whose distribution function is much different from the distribution of $\mathbf{s}$. However, the performance of the marginal median filtering on multichannel impulses is not as efficient as in the single channel case [34].

Another definition of the multichannel median has been proposed for color image filtering [33]. It is the vector $\mathbf{x}_{med}$ that minimizes the L_1 error norm:

$$\sum_{i=1}^{n} |\mathbf{x}_i - \mathbf{x}_{med}| \rightarrow \min \tag{10.8.7}$$

This definition of the multichannel median is a direct extension of the corresponding single channel median definition. Generally, it gives different results than the marginal median. The vector $\mathbf{x}_{med}$ may or may not be one of the data $\mathbf{x}_i$, $i=1,..,n$. This implicit definition of the median makes difficult its application in color image filtering because the use of an iterative optimization algorithm is required to produce each output image pixel. The computational load is reduced, if the median $\mathbf{x}_{med}$ is forced to belong to the set $\{\mathbf{x}_i, i=1,..,n\}$. In this case, the L_1 norm (10.8.7) is computed for every $\mathbf{x}_{med} = \mathbf{x}_i$, $i=1,..,n$. The median is the one that minimizes (10.8.7). The situation is much simpler for $n=3$. In this case the median is given by:

$$\mathbf{x}_{med} = \begin{cases} \mathbf{x}_1 & \text{if } |\mathbf{x}_2-\mathbf{x}_3|\geq|\mathbf{x}_1-\mathbf{x}_2| \text{ and } |\mathbf{x}_2-\mathbf{x}_3|\geq|\mathbf{x}_1-\mathbf{x}_3| \\ \mathbf{x}_2 & \text{if } |\mathbf{x}_1-\mathbf{x}_3|\geq|\mathbf{x}_2-\mathbf{x}_1| \text{ and } |\mathbf{x}_1-\mathbf{x}_3|\geq|\mathbf{x}_2-\mathbf{x}_3| \quad (10.8.8) \\ \mathbf{x}_3 & \text{if } |\mathbf{x}_1-\mathbf{x}_2|\geq|\mathbf{x}_3-\mathbf{x}_1| \text{ and } |\mathbf{x}_1-\mathbf{x}_2|\geq|\mathbf{x}_3-\mathbf{x}_2| \end{cases}$$

The 3-point median can be combined with vector FIR filters to produce the *vector median hybrid filter* [33]:

$$\mathbf{y}_i = \text{med}\left[\frac{1}{v}\sum_{j=1}^{v}\mathbf{x}_{i-j}, \ \mathbf{x}_i, \ \frac{1}{v}\sum_{j=1}^{v}\mathbf{x}_{i+j}\right] \tag{10.8.9}$$

where the median can be calculated as in (10.8.8).

10.9 NEURAL NETWORKS

Neural networks (connectionist models, parallel distributed processing models and neuromorphic systems) attempt to achieve a real-time response and a human-like performance using many simple processing elements operating in parallel as in biological nervous systems. Processing elements or nodes in neural networks are connected by branches with variable weights. Their elements are nodes that sum n weighted inputs and pass the result through a nonlinearity, as is seen in Figure 3.10.2:

$$y = f\left(\sum_{i=1}^{n} w_i x_i - \theta\right) \tag{10.9.1}$$

Three common types of nonlinearities $f(x)$ are shown in Figure 10.9.1.

There are numerous neural networks that have been investigated, such as the *Perceptron*, the *Hopfield network*, the *Multi-layer perceptron*, and *Kohonen's self-organizing feature maps* in [35,36]. One of the better known networks is the Perceptron. The single-layer perceptron is a neural net classifier that can be used with both continuous valued and binary inputs. A perceptron classifies an input into one of two classes as follows:

$$y = \begin{cases} 1 & \text{class A} \\ -1 & \text{class B} \end{cases} \tag{10.9.2}$$

(10.9.2) defines a decision boundary shown in Figure 10.9.2.

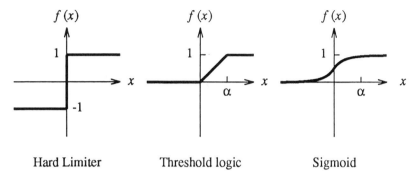

Figure 10.9.1: Nonlinearities used in neural network models.

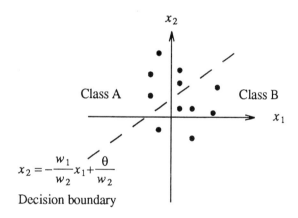

Figure 10.9.2: Pattern classification by using perceptron.

Multi-layer perceptrons are feed-forward networks with one or more layers of nodes between input and output nodes. These additional layers contain hidden units or nodes that are not directly connected to both input and output nodes. Multi-layers overcome many limitations of single-layer perceptrons, but were generally not used in the past because effective training algorithms were not available. This has been rectified with the development of the *backpropagation algorithm* [37,38]. Although it cannot be proved that these algorithms always converge, they have already found many successful applications. In fact, the back propagation algorithm is a generalization of Widrow's LMS algorithm [38]. It was a gradient search to minimize a cost function equal to the mean square difference between the desired output and actual nets output. The interesting point is that it has been shown that an arbitrary decision surface can be found in multi-layer neural nets [36].

Neural nets are attractive in digital signal and image processing for the following reasons: (1) an arbitrary decision surface can be formed in a multi-layered perceptron net so that any complex mapping from the set of noise signal to the noise-free signal can be realized; (2) neural nets are formed by a group of simple elements; (3) neural nets have attractive generalizing properties. However, the application of neural networks in this area is not mature yet and needs further study [40]. Neural nets have already been used for noise reduction in speech signals [41] and for speech recognition [42,43]. The relation of neural networks to adaptive filtering is described in [44]. Their applications in sonar signal classification, in brain research, and in image processing and analysis are presented in [45-48], respectively. Neural networks have found extensive applications in pattern recognition because of their learning capabilities. The work on neural networks originates from pattern recognition. Two of those applications are described in [49,50].

10.10 DISCUSSION

Some generalizations, extensions, and new trends have been presented in this chapter. A general structure for nonlinear filtering has been presented that is suitable for the implementation of a multitude of nonlinear filter classes. This structure is versatile and suitable for parallel computation. Its computational complexity is comparable to that of the median filter and the moving average filter. An overview of the performance of the various nonlinear filters described in this book is presented in section 10.5 and is summarized in Table 10.5.1. Several figures of merit are included in this table. The comparisons are qualitative, since no quantitative measure can represent fully the characteristics of the human eye, as has already been described in chapter 3. Finally, two new trends in nonlinear filtering have been described in this chapter. The first one is nonlinear color image processing. It is application-driven and arises from the need to use advanced image processing techniques for digital TV and video image processing applications. The second trend is the use of a new nonlinear tool, namely the neural networks, in digital signal and image processing. Although neural nets have not found yet extensive applications in this area, there is a great potential in their use, especially in certain areas, e.g., in adaptive filtering. Their great advantage comes from the fact that they have learning capabilities. Therefore, they can combine digital signal/image processing and analysis.

REFERENCES

[1]   J. Kittler, M. Duff (editors), *Image processing system architectures*, Research studies Press, 1985.

[2]   J. Alsford et al., "CRS image processing system with VLSI modules", in *Image processing system architectures*, Research Studies Press, 1985.

[3]   I. Pitas, A.N. Venetsanopoulos, "A new filter structure for the implementation of certain classes of image operations", *IEEE Transactions on Circuits and Systems*, vol. CAS-35, no. 6, pp. 636-647, June 1988.

[4]   K. Oflazer, "Design and implementation of a single chip 1-D median filter", *IEEE Transactions on Acoustics, Speech and Signal Processing*, vol. ASSP-31, pp. 1164-1168, Oct. 1983

[5]   I. Pitas, A.N. Venetsanopoulos, "Nonlinear order statistic filters for image filtering and edge detection", *Signal Processing*, vol. 10, pp.395-413, June 1986.

[6]   D.E. Knuth, *The art of computer programming*, vol. 3, Addison-Wesley, 1973.

[7]   Y.S. Fong, C.A. Pomalaza, X.H. Wang, "Comparison study of nonlinear filters in image processing applications", *Optical Engineering*, vol. 28, no. 7, pp. 749-760, July 1989.

[8]   R. Ding, A.N. Venetsanopoulos, "Generalized homomorphic and adaptive order statistic filters for the removal of impulsive noise and signal-dependent noise", *IEEE Transactions on Circuits and Systems*, vol. CAS-34, no. 8, pp. 948-955, Aug. 1987.

[9]   G. Sicuranza, A.N. Venetsanopoulos, "2-D quadratic filter implementation by a general-purpose nonlinear module", *IEEE Transactions on Circuits and Systems*, vol. CAS-36, no. 1, pp. 150-151, Jan. 1989.

[10]   A.K.Jain, *Fundamentals of digital image processing*, Prentice Hall, 1989.

[11]   R.C.Gonzalez, P.Wintz, *Digital image processing*, Addison- Wesley 1987.

[12]   D.L.MacAdam "Color essays", *Journal of the Optical Society of America*, vol.65, no.5, pp. 483-492, May 1975.

[13]   B.R.Hunt, "Karhunen-Loeve multispectral image restoration, part I: theory", *IEEE Transactions on Acoustics, Speech and Signal Processing*, vol. ASSP-32, no.3, pp.592-599, June 1984.

[14]   N.P. Galatsanos, R.T. Chin, "Digital restoration of multichannel images", *IEEE Transactions on Acoustics, Speech and Signal Processing*, vol. ASSP-37, no.3, pp.415-421, March 1989.

[15] G. Aggelopoulos, I. Pitas, "Least-squares multichannel filters in color image restoration", *Proc. European Conference on Circuit Theory and Design ECCTD89*, Brighton, England, September 1989.

[16] C.W.Therrien, "Multichannel filtering methods for color image segmentation", *Proceedings of the 1985 IEEE Conference on Computer Vision and Pattern Recognition*, pp. 637-639, 1985.

[17] N.Ohyama, M.Yachida, E.Badique, J.Tsujiuchi, T.Honda, "Least squares filter for color image segmentation", *Journal of the Optical Society of America A*, vol.5, no.1, pp.19-24, Jan. 1988.

[18] H.A.David, *Order statistics*, J.Wiley, 1980.

[19] G.A.F.Seber, *Multivariate observations*, J.Wiley, 1984.

[20] A.Papoulis, *Probability, random variables and stochastic processes*, McGraw Hill 1985.

[21] M.A.Chmielewski, "Elliptically symmetric distributions: a review and bibliography", *International Statistical Review*, vol.49, pp.67-74, 1981.

[22] R.J.Muirhead, *Aspects of multivariate statistical theory*, J.Wiley, 1982.

[23] V.Barnett, "The ordering of multivariate data", *Journal of the Royal Statistical Society A*,vol.139, pt.3, pp.318-354, 1976.

[24] F.P.Preparata, M.I.Shamos, *Computational Geometry*, Springer Verlag, 1986.

[25] G.A.Watterson, "Linear estimation in censored samples from multivariate normal populations", *Annals of Mathematical Statistics*, vol.30, pp.814-824, 1959.

[26] H.A.David, "Concomitants of order statistics", *Bulletin of the International Statistical Institute*, vol.46, pp.295-300, 1973.

[27] A.M.Mood, "On the joint distribution of the medians in samples from a multivariate population", *Annals of Mathematical Statistics*, vol.12, pp.268-279, 1941.

[28] C.M.Mustafi, "A recurrence relation for distribution functions of order statistics from bivariate distributions", *Journal of the American Statistical Association*, vol.64, pp.600-601, 1969.

[29] J.Galambos, "Order Statistics of samples from multivariate distributions", *Journal of the American Statistical Association*, vol.70, pp.674-680, 1975.

[30] P.S.Huber, *Robust statistics*, John Wiley, 1981.

[31] F.Hampel, E.Ronchetti, P.Rousseeuw, W.Stahel, *Robust statistics*, John Wiley, 1986.

[32] R. Gnanadesikan, J.R.Kettenring, "Robust estimates, residuals and outlier detection with multiresponse data", *Biometrics*, vol.28, pp. 81-124, March 1972.

[33] J. Astola, P. Haavisto, P. Heinonen, Y. Neuvo, "Median type filters for color signals", *Proc. IEEE Int. Symp on Circuits and Systems*, pp. 1753-1756, 1988.

[34] I. Pitas, "Marginal order statistics in multichannel and color image filtering", Technical report, University of Thessaloniki, Greece, 1989.

[35] R.P. Lippmann, "An introduction to computing with neural nets", *IEEE ASSP Magazine*, April 1987, pp. 4-22.

[36] R.P. Lippmann, "Neural nets for computing", *Proceedings of the Intl. Conference of Acoustics, Speech and Signal Processing*, New York City, April 1988, pp. 1-6.

[37] D.E. Rumelhart, G.E. Hinton, R.J. Williams, "Learning internal representations by error propagation", in *Parallel Distributed Processing: Explorations in the Microstructure of Cognition, vol. 1: Foundations*, D.E. Rumelhart and J.L. McClelland editors, MIT Press, 1986.

[38] B. Widrow, R. Winter, "Neural nets for adaptive filtering and adaptive pattern recognition", *Computer*, March 1988, pp. 25-39.

[39] N.B. Karayiannis, A.N. Venetsanopoulos, "Optimal least-squares training of associative memories: Learning algorithms and performance evaluation", *Neural Networks*, in press.

[40] "DARPA neural network study", ARCEA International Press, November 1988.

[41] S. Tamura, A. Waibel, "Noise reduction using connectionist models", *Proceedings of the Intl. Conference of Acoustics, Speech and Signal Processing*, New York City, April 1988.

[42] T. Kohonen, "The neural phonetic typewriter", *Computer*, vol. 21, no. 3, pp. 11-22, March 1988.

[43] D.J. Burr, "Experiments on neural net recognition of spoken and written text", *IEEE Transactions on Acoustics, Speech and Signal Processing*, vol. ASSP-36, no. 7, pp. 1162-1168, July 1988.

[44] B. Widrow, R. Winter, "Neural nets for adaptive filtering and adaptive pattern recognition", *Computer*, vol. 21, no. 3, pp. 25-39, March 1988.

[45] R.P. Gorman, T.J. Senjowski, "Learned classification of sonar targets using a massively parallel network", *IEEE Transactions on Acoustics, Speech and Signal Processing*, vol. ASSP-36, no. 7, pp. 1135-1140, July 1988.

[46] A.S. Gevins, N.H. Morgan, "Applications of neural network signal pro-
       cessing in brain research", *IEEE Transactions on Acoustics, Speech and
       Signal Processing*, vol. ASSP-36, no. 7, pp. 1152-1161, July 1988.

[47] Y.T. Zhou, R. Chellapa, A. Vaid, B.K. Jenkins, "Image restoration using a
       neural network", *IEEE Transactions on Acoustics, Speech and Signal
       Processing*, vol. ASSP-36, no. 7, pp. 1141-1151, July 1988.

[48] J.G. Daugman, "Complete discrete 2-D Gabor transforms by neural net-
       works for image analysis and compression", *IEEE Transactions on
       Acoustics, Speech and Signal Processing*, vol. ASSP-36, no. 7, pp. 1169-
       1179, July 1988.

[49] B. Widrow, R.G. Winter, R.A. Baxter, "Layered nets for pattern recogni-
       tion", *IEEE Transactions on Acoustics, Speech and Signal Processing*,
       vol. ASSP-36, no. 7, pp. 1109-1118, July 1988.

[50] K. Fukushima, "A neural network for visual pattern recognition", *Com-
       puter*, vol. 21, no. 3, pp. 65-75, March 1988.

CHAPTER 11

ALGORITHMS AND ARCHITECTURES

11.1 INTRODUCTION

One of the main problems in linear as well in the nonlinear image processing is the computational complexity and the execution speed of the image processing routines, especially for applications where real-time image processing is required. This problem mainly results from the amount of the data to be processed (e.g., 256K bytes for a single 512×512 8 bit image) as well as from the number of operations required per output pixel. Such operations are usually comparisons, additions, multiplications, and nonlinear function evaluations. Usually the number of operations required per output pixel is not very high. However, if this number is multiplied by the output image size (in pixels), the total amount of computations required is tremendous. Therefore, even simple nonlinear filters (e.g., median filters, erosion, dilation), which require only comparisons, are relatively slow for fast image processing applications. There are two solutions to the requirement for an increase of the speed of image processing routines. The first one is to construct fast algorithms for linear and nonlinear image processing. These algorithms can be relatively fast on general purpose computers for fast nonreal-time digital image processing. If real-time processing is a must, the only solution is to build image processors, whose architecture is optimized for image processing applications. If the performance of such image processors is not adequate, parallel image processing is the solution.

In this chapter, algorithms and structures for fast nonlinear image processing will be discussed. Special attention will be paid to the fast calculation of the order statistics of a set of numbers (called *selection*) (e.g., the maximum, minimum, and median), as well as to the *sorting* of a set of numbers. The reason for this is that many nonlinear filters, described in the previous chapters, are based on comparisons for the calculation of the order statistics of data sets.

11.2 SORTING AND SELECTION ALGORITHMS

Sorting and selection algorithms are a classical topic of computer science. Their performance, usually measured in number of comparisons, has been thoroughly analyzed. Several such algorithms (e.g., MERGESORT, QUICK-SORT) have become classical, and their description and analysis can be found in any computer science textbook on algorithms [1,2]. However, for convenience to the reader, a brief description of some sorting and selection algorithms will be included here. The algorithms that will be described in this chapter are those most commonly used in digital image processing applications.

```
#include <stdio.h>
#define n ...
int x[n];
/* x is an external array [1:n] of integers.  */

QUICKSORT(p,q)
/* sorts the elements x[i], i=p,..,q of the array  x[1:n]  */
int p, q;
{
 int j;
  if (p<q)
    {
      j=q+1;
      PARTITION(p,j);
      QUICKSORT(p,j-1);
      QUICKSORT(j+1,q);
    }
}

PARTITION(m,p)
int m,p;
{
  int i=m, v=x[m], temp;
/* x[m] is the partition element */
  for( ; ; ) {
        do i++; until (x[i] >= v);
        do p--; until (x[p] <= v);
        if (i<p) /* exchange x[i],x[p] */
            {temp=x[i]; x[i]=x[p]; x[p]=temp;}
        else break; }
  x[m]=x[p]; x[p]=v;
}
```

Figure 11.2.1: QUICKSORT and PARTITION algorithms.

Let us denote by $x[1:n]$ a vector $[x(1),..,x(n)]^T$ having n elements. The sorting of this vector is required. One of the most known and fast sorting algorithms is the QUICKSORT [1,2]. It is essentially a divide-and-conquer approach to sorting. The QUICKSORT algorithm is given in Figure 11.2.1. All algorithms in this section are in C-like language. The QUICKSORT algorithm operates as follows. An element of $x[1:n]$ is picked, e.g., $t=x(s)$, and all other elements of $x[1:n]$ are reordered so that all elements appearing before t in $x[1:n]$ are less than t and all elements appearing after t are greater or equal to t. This rearrangement is done by the PARTITION subroutine described in Figure 11.2.1 in C-like language. Thus the array $x[1:n]$ has been divided in two subarrays. The elements of the first subarray are less than the elements of second array. Now QUICKSORT can be applied to these two subarrays independently. Thus QUICKSORT is a recursive algorithm as it is shown in Figure 11.2.1. A thorough analysis of the performance of the QUICKSORT algorithm can be found in [2, pp. 123-127]. The time required by the QUICKSORT algorithm in the average is of the order $O\,(nlogn)$, whereas in the worst case it is $O\,(n^2)$.

```
int x[n], n;
SELECT(k)
/* finds the k-th smallest element in the array  x[1:n] */
int k;
{
 int m=1, r=n+1, j;   x[n+1] = ∞
 for ( ; ; ) {
        j=r;
        PARTITION(m,j);
        if (k<j)  r=j;
        else if (k>j)  m=j+1;
        else return;}
}
```

Figure 11.2.2: SELECTION algorithm.

The PARTITION subroutine can also be used to select the k-th smallest (k-th order statistic) of the elements of the array $x[1:n]$ in the following way. Let us suppose that the partitioning element t is positioned at $x(j)$ after the first partition. Then $j-1$ elements of the array $x[1:j-1]$ are smaller than or equal to $x(j)$ and $n-j$ elements of the array $x[j+1:n]$ are greater than or equal to $x(j)$. If $k=j$, the k-th order statistic is the element $x(j)$. If $k<j$, the k-th order statistic is the k-th smallest element of $x[1:j-1]$. If $k>j$, the k-th order statistic is the $(k-j)$-th smallest element in $\mathbf{x}[j+1:n]$. The resulting algorithm is SELECT, shown in Figure 11.2.2. SELECT places the k-th order statistic in the position $x(k)$ and partitions the remaining elements of $\mathbf{x}[1:n]$ accordingly: $x(i) \leq x(k)$, $1 \leq i < k$ $x(i) \geq x(k)$, $k < i \leq n$. If k is chosen to be $k = v+1$, $n = 2v+1$, SELECT computes the median. A thorough analysis of the performance of the SELECT algorithm as well as some modifications of the algorithm of Figure 11.2.2 can be found in [2, pp. 129-136]. It can be proven that the average time required by SELECT is of the order $O(n)$, which is not much less than the $O(n \log n)$ time required by the QUICKSORT algorithm, although SELECT provides much less information. Therefore, the QUICKSORT algorithm is often used for the calculation of the median in digital image processing applications.

The maximum and minimum selection is much easier than the calculation of the general k-th order statistic. It can be done in a straightforward way, as is described in Figure 11.2.3.

int x[n], n, max, min;

STRAIGHTMAXMIN()
/* finds the maximum and the minimum of the array x[1:n] */
{
 int i;
 max=x[1]; min=x[1];
 for(i=2;i=n;i++)
  {
  if(x[i]>max) max=x[i];
  if(x[i]<min) min=x[i];
  }
}

Figure 11.2.3: Straightforward algorithm for the calculation of maximum and minimum.

This algorithm requires $2(n-1)$ comparisons for the calculation of both maximum and minimum. The calculation of each of them separately requires only $n-1$ comparisons. Thus max/min filters are expected to be very fast. If both maximum and minimum are required simultaneously (e.g., for midpoint or for range calculation), the recursive algorithm of Figure 11.2.4 can be used. It requires only $3n/2-2$ comparisons. It can be proven that no algorithm based on comparisons can be built that uses smaller number of comparisons [2, p. 476].

int x[n], n;

```
MAXMIN(i,j,max,min)
/* finds the maximum and the minimum of the array x[i:j] */
int i, j, max, min;
{
 int fmax, fmin, gmax, gmin;
    if(i==j)      {max=x[i]; min=x[i];}
    else if(i==j-1) { if (x[i]<x[j]) {max=x[j]; min =x[i];}
               else      {max=x[i]; min =x[j];}}
    else        { mid=⌊(i+j)/2⌋;
                MAXMIN(i,mid,gmax,gmin);
                MAXMIN(mid+1,j,fmax,fmin);
                max=MAX(gmax,fmax);
                min=MIN(gmin,fmin); }
}
```

Figure 11.2.4: Recursive algorithm for the calculation of maximum and minimum.

The previously mentioned algorithms can be very easily adapted to image filtering applications. In this case, the array $x[1:n]$ contains the pixels that lie inside the filter window A centered at pixel (i,j). However, sorting or selection (e.g., median calculation) are *running* operations. Thus, the filter window centered at pixels (i,j) and $(i,j+1)$ overlap. This window overlapping cannot be taken into advantage if classical sorting and selection algorithms are used. The running sorting and selection algorithms described in the next section are based exactly on the window overlapping.

11.3 RUNNING MEDIAN ALGORITHMS

The first running algorithm has been proposed by Huang et al. [3] for median filtering. Let us suppose that the filter window has $m \times n$ pixels and that it is moving row-wise. When the window is moved from position (i,j) to the position $(i,j+1)$, $mn-2n$ pixels remain unchanged, n must be discarded and n new pixels must be inserted. Huang's algorithm is based on the local gray level histogram of the nm pixels and on its adaptation, as the window moves. The algorithm is shown in Figure 11.3.1.

```
/* MAIN PROGRAM */
define N,M;        /* image size N×M                    */
define n,m;        /* window size n×m                    */
int x[N][M];       /* image x(0:N-1,0:M-1)                    */
int hist [255];    /* histogram array                    */
int mdn;           /* median in a window                    */
int ltmdn;         /* number of pixels having value less than mdn */
int leftcol[n];    /* leftmost column of previous window        */
int rightcol[n];   /* rightmost column of current window        */
int th, m2, n2;
main ()
{
 th=(m*n)/2; n2=n/2; m2=m/2;
 for(i=n2;i<N-n2;i++)
   {
   find the histogram of the first window;
   find the median mdn of the first window and ltmdn;
   for(j=m2;j<M-m2;j++)
     {
     put leftmost column of the previous window in leftcol;
     put rightmost column of the current window in rightcol;
     runningmedian;
     }
   }
}

runningmedian
 {
 int g;
 /* update histogram and ltmdn */
 for(k=0; k<n; k++)
  {
  g=leftcol[k];
  hist[g]--;
  if (g<mdn) ltmdn--;
  g=rightcol[k];
  hist[g]++;
  if (g<mdn) ltmdn++;
  }
 /* find the median */
 if(ltmdn >th)
         do {
             mdn--;
```

$ltmdn=ltmdn-hist[mdn];$
    $}$
    until $(ltmdn <= th);$
else while$(ltmdn+hist[mdn] <=th)$
    $\{ ltmdn=ltmdn+hist[mdn]; mdn++; \}$
$}$

Figure 11.3.1: Running median algorithm.

It has the following steps:

Step-1:Calculate the local histogram of the nm pixels in the first window and find the median. Count the number $ltmdn$ of pixels with gray levels less than that of the median.

Step-2:Move to the next window by deleting n pixels and inserting the n new pixels. Update the histogram. Count the number $ltmdn$ of pixels whose gray level is less than the median of step 1.

Step-3:Starting from the median of the previous window, move up/down the histogram bins one at a time if the count $ltmdn$ is smaller/greater than $(mn+1)/2$ and update $ltmdn$ until the median is reached.

Step-4:If the line end has been reached, move to the next line and go to step 1. Otherwise go to step 2.

Huang's algorithm requires only $2n$ comparisons per output point, whereas the QUICKSORT algorithm requires $O(2n^2 logn)$ comparisons if $m \approx n$. Thus the running median algorithm is much faster than the QUICKSORT used in median filtering. This has been confirmed by simulation experiments [3,4].

Another approach for running median calculation is based on the partitioning of the pixels in the filter window in two subsets [5]. Let the window size be $m \times n$ and $W(i,j)$ be an $m \times n$ matrix containing a modified version of the data x_{kl} in filter window centered at the pixel (i,j), such that the elements of each column of $W(i,j)$ are in ascending order. The last $m-1$ columns of the window $W(i,j)$ are also included in the window $W(i,j-1)$. Only the rightmost column of $W(i,j)$ differs. This column can be obtained from the corresponding column of $W(i-1,j)$ by deleting the pixel $x(i-(m+1)/2,j+(n-1)/2)$ and by inserting the pixel $x(i+(m-1)/2,j+(n-1)/2)$ in such a way so that the rightmost column of $W(i,j)$ is ordered. The median at the pixel $(i,j-1)$ can be found in the following way. The pixels in the window $W(i,j-1)$ are partitioned into two subsets. Subset S_1 contains $(mn+1)/2$ elements and subset S_2 contains the rest $(mn-1)/2$ elements that are greater or equal to the largest element of S_1. The median at $(i,j-1)$ is the largest element in S_1. The median at position (i,j) can be obtained by updating $W(i,j-1)$ to the new window $W(i,j)$. This can be done in the following steps:

Step-1:Discard the leftmost column of $W(i,j-1)$.

Step-2:Obtain the rightmost column of $W(i,j)$ from the corresponding column of $W(i-1,j)$ as described previously.

Step-3:Include the elements of the new column that are less than or equal to the median at position $(i,j-1)$ into the subset S_1 and the rest of them to the subset S_2.

Step-4:(i) If the number of elements in S_1 is just one more than the number of elements in S_2, final partitioning has been obtained. (ii) If the number of elements in S_1 is less than $(mn+1)/2$, the smallest elements of S_2 are pushed to S_1 until the number of elements of S_1 becomes equal to $(mn+1)/2$. (iii) If the number of elements into subset S_1 is greater than $(mn+1)/2$, the largest elements of the subset S_1 are pushed into S_2 until the number of elements of S_1 becomes equal to $(mn+1)/2$. In all cases the new median is the largest element of S_1.

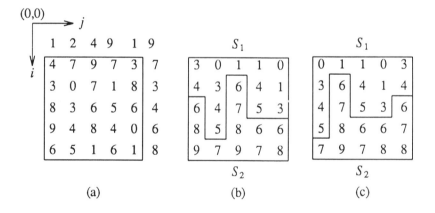

Figure 11.3.2: (a) Original two-dimensional sequence; (b) The partitioning of the window at the position (3,2); (c) The partitioning of the window at the position (3,3).

It should be noted that the subsets S_1, S_2 are separated by a border, as is seen in Figure 11.3.2. Finding the smallest number in S_2 or the largest number in S_1 involves searches along this border. Pushing elements from one subset to the other means changing the border. The above-mentioned algorithm can be illustrated be the example shown in Figure 11.3.2 for a 5×5 median filter. The original image pixels are shown in Figure 11.3.2a. The window $W(3,2)$ and its partition are shown in Figure 11.3.2b. Subset S_1 has 13 elements and subset S_2 has 12 elements. The median at the position $(3,2)$ is 5, i.e., the largest element of S_1 along its border. The rightmost column of the window $(2,3)$ is $[3\ 4\ 6\ 7\ 9]^T$. If the element 9 is discarded and the new element 8 is introduced the rightmost column of window $W(3,3)$ is obtained $[3\ 4\ 6\ 7\ 8]^T$. Thus the new window $W(3,3)$ is the one shown in Figure 11.3.2c. The elements 6,7,8 of the rightmost column are larger than the previous median 5 and it is introduced to subset S_2. The rest of the elements of the rightmost column of $W(i,j)$ are pushed to the subset S_1. This partition of the window $W(3,3)$ is indicated by the border in Figure 11.3.2c. The subset S_1 has now 13 elements and the subset S_2 12 elements. Thus the new median is 5 again, which is the largest element of S_1 along its border.

The algorithm involves basically three operations. The first one, which deletes an element from a list of m number and inserts a new number, requires m comparisons in the worst case. The second operation is to search for the position of the element in the rightmost column which is greater than or equal to the old median. This operations require at most m comparisons. The third operation is to update the two subsets. It involves the calculation of the smallest or the largest number along the border, i.e., from a set of n numbers. This process requires in the worst case $(2n-1)(m+1)/2$ comparisons. Thus in the worst case the number of operations required by this algorithm is of the order $O(n^2)$. However, it has been found that the actual number of comparisons in practical cases is much less and that the algorithm is 15%-25% faster than the Huang's algorithm [5] for window sizes up to 15×15. A list of the algorithm in C language can also be found in [5].

i	B_1	B_2	B_3	B_4	
1	0	0	1	1	$x_1=3$
2	1	1	0	0	$x_2=12$
3	1	0	0	0	$x_3=8$
4	1	0	1	1	$x_4=11$
5	0	1	1	1	$x_5=7$

Figure 11.3.3: Binary representation of pixels having $k=4$ bits each in a filter window of length 5.

Another approach to fast median filtering is to use the binary representations of the image pixels in the filter window, as it is seen, for example, in Figure 11.3.3. Rows represent pixels and columns represent bit planes. Let each pixel have k bits. The least significant bit plane is B_k and the most significant one is B_1. *Lexicographic sorting* uses the binary number representation [6]. In the first step, the sequences are sorted according to the most significant bit B_1. In the second step, they are ordered according to the next significant bit and the results of the first step. This process continues until the least significant bit is reached after k steps. The median is at row $v+1$ when the ordering is finished. Lexicographic ordering requires kn binary comparisons.

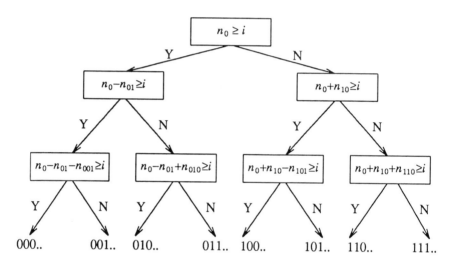

Figure 11.3.4: Binary tree for the selection of the i-th order statistic (adapted from [7]).

A modification of this algorithm has been proposed for the selection of the i-th order statistic [7]. Let n_0 be the number of pixels having the most significant bit B_1 equal to 0. If $n_0 \geq i$, then the i-th order statistic belongs to the partition of pixels that have $B_1=0$. The rest of the pixels can be discarded. If $n_0<i$, then the i-th order statistic belongs to the partition of pixels that have $B_1=1$. Let us consider the first case, i.e., $n_0 \geq i$. Let n_{01} be the number of pixels having 01 as the two most significant bits. If $n_0-n_{01}<i$, then the i-th order statistic lies in the partition of pixels having 01 as the two most significant bits. This procedure can be repeated until all bit planes are examined. Some pixels are discarded at each step, until only one pixel remains at the end. This pixel is the i-th order statistic. Thus, this algorithm determines the bit representation of the i-th order statistic

successively. The whole procedure can be described by a binary tree similar to that shown in Figure 11.3.4. Other algorithms using bit representation of the image pixels for the calculation of the median or other order statistics can be found in [8-12].

11.4 RUNNING ALGORITHMS FOR SORTING AND MAX/MIN SELECTION

Running algorithms for max/min selection are presented in [13]. The problem of running max/min selection can be formulated as follows. Let x_i, $i \in \mathbf{Z}$ be one-dimensional signal. The output of a max/min filter is given by:

$$y_i = T(x_i,..,x_{i-n+1}) \tag{11.4.1}$$

where n is the filter length and T is either the max or the min operator. Definition (11.4.1) is slightly different than the traditional one in that it introduces a constant time-shift at the output sequence, but it is used for simplicity reasons. If the filter length is a power of 2, i.e., $n=2^k$, a method similar to the FFT can be used for running max/min calculation:

$$y_i = T[T(x_i,..,x_{i-(n/2)+1}),T(x_{i-(n/2)},..,x_{i-n+1})] \tag{11.4.2}$$

The calculation of $T(x_i,..,x_{i-(n/2)+1})$ is common in the computation of y_i and $y_{i+(n/2)}$ and, therefore, it can be done only once. Similarly, the computation of $T(x_{i-(n/2)},..,x_{i-n+1})$ is common in the computation of y_i and $y_{i-(n/2)}$. Therefore, it can also be done only once. Once the terms $T(x_i,..,x_{i-(n/2)+1})$, $T(x_{i-(n/2)},..,x_{i-n+1})$ have been computed, only one extra comparison is needed for the calculation of y_i. In a similar way the max/min computation of $T(x_i,..,x_{i-(n/2)+1})$ can be further divided into the max/min computation of two subsequences having $n/4$ elements each:

$$y^{(1)}_i = T[T(x_i,..,x_{i-(n/4)+1}),T(x_{i-(n/4)},..,x_{i-(n/2)+1})] \tag{11.4.3}$$

where

$$y^{(l)}_i = T(x_i,..,x_{i-(n/2^l)+1}) \tag{11.4.4}$$

and $y^{(0)}_i=y_i$. The calculation of $T(x_i,..,x_{i-(n/4)+1})$ is common in the computation of $y^{(1)}_i$ and $y^{(1)}_{i+(n/4)}$, whereas the calculation of $T(x_{i-(n/4)},..,x_{i-(n/2)+1})$ is common in the computation of $y^{(1)}_i$ and $y^{(1)}_{i-(n/4)}$. This process can be repeated until subsequences of length 2 are reached. Therefore the computation of (11.4.1) is done in $\log_2 n$ steps. Only one extra comparison is needed at each step. Therefore the total number of comparisons required per output point is $\log_2 n$, whereas the conventional max or min calculation requires $n-1$ comparisons. The flow-diagram of this algorithm is shown in Figure 11.4.1. Similar algorithms can also be constructed for arbitrary filter lengths n. The algorithm can also be easily extended for the two-dimensional case. This algorithm leads also to a very efficient structure for the max/min filtering, shown in Figure 11.4.2. Its throughput delay is of the order $O(log_2 n)$.

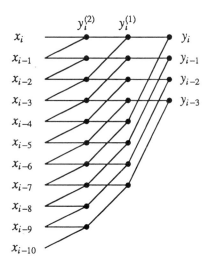

Figure 11.4.1: Flow diagram of the fast running max/min selection for $n=8$. Dots denote comparisons.

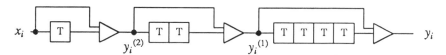

Figure 11.4.2: Structure of the running max/min selector. The triangle denotes a comparator.

The same approach can also be used for running sorting algorithms. The output of such an algorithm is a vector sequence $\mathbf{y}_i$ containing the sorted samples included in the filter window of size n at position i:

$$\mathbf{y}_i = S(x_i,..,x_{i-n+1}) \tag{11.4.5}$$

where x_i is the input sequence and S denotes a sorting operator for n numbers.

The computation of (11.4.5) can be decomposed to the merging of two sorted subsequences of length $n/2$:

$$\mathbf{y}_i = M \left[S \left(x_i, .., x_{i-(n/2)+1} \right), S \left(x_{i-(n/2)}, .., x_{i-n+1} \right) \right] \tag{11.4.6}$$

where $M[.,.]$ denotes the merging of two sorted subsequences. This procedure is repeated until subsequences of length 2 are reached. Thus the algorithm has again $\log_2 n$ steps. The only extra operations needed at each step l, $0 \leq l \leq \log_2 n - 1$ are the $n/2^l - 1$ comparisons needed for the calculation of the merging of two subsequences of length $n/2^{l+1}$ [2].

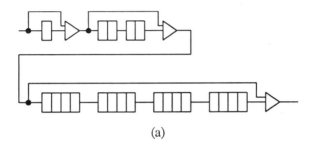

(a)

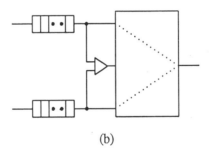

(b)

Figure 11.4.3: (a) Structure of a running sorter; (b) Structure of the merging element used in (a).

Thus the total number of comparisons required is given by:

$$C(n) = \sum_{l=0}^{\log_2 n - 1} \left(\frac{n}{2^l} - 1 \right) = 2(n-1) - \log_2 n \tag{11.4.7}$$

which is less than the $O(n\log n)$ required in the average case by the QUICK-SORT algorithm. This algorithm can also be easily extended to the two-dimensional case. It can also be used for the construction of a fast sorting structure, shown in Figure 11.4.3.

Another approach to running max/min calculation is the following [13]. If y_{i-1} is known and y_i must be computed, the window is moved only by one element (x_{i-n} is deleted and x_i is inserted in the window). If x_i is greater than y_{i-1} then the new element is the maximum, i.e., $y_i = x_i$. If x_i is less than or equal to y_{i-1}, the maximum is not affected by the introduction of the new element. In this case, if x_{i-n} is less than the maximum y_{i-1}, the new maximum is the same with the old one: $y_i = y_{i-1}$. If x_{i-n} is equal to the maximum y_{i-1}, a new max selection $y_i = T(x_i,..,x_{i-n+1})$ must be performed. This running max/min selection algorithm is described by the following formula:

$$y_i = \begin{cases} x_i & \text{if } x_i \geq y_{i-1} \\ y_{i-1} & \text{if } x_i < y_{i-1} \\ T(x_i,..,x_{i-n+1}) & \text{if } x_i < y_{i-1} \end{cases} \qquad (11.4.8)$$

A theoretical analysis of the average number of comparisons required by this algorithm is given in [13]. The average number of comparisons is approximately only 3 and it does not increase with the filter window size n. Therefore, it is much better than the straightforward max selection algorithm, which requires $n-1$ comparisons. The procedure MAXLINE, shown in Figure 11.4.4, is an implementation of the running max filter in PASCAL. Simulations have shown that this routine requires approximately 2 comparisons per output point [13].

```
procedure MAXLINE(finname:string; foutname:string);
 {* finname:  name of the input data file          *}
 {* foutname: name of the output data file          *}
 const order=5;
 var a:array[1..order] of real;
    old, max: real;
    fin, fout: file of real;
    n: integer;


procedure SHIFT;
 {* it shifts the contents of the array a          *}
 var i: integer;
 begin
  old=a[1]; for i:=2 to order do a[i-1]:=a[i];
 end;


procedure MAXIMUM(var max:real);
 { finds the maximum element of the array a         *}
```

```
var i: integer;
begin
 max=a[1]; for i:=2 to order do if a[i]>max then max:=a[i];
end;

procedure INITIALIZE;
 var i: integer;
 begin
 for i:=1 to order do read(fin,a[i]);
 maximum(max);
 write(fout, max);
end;

begin {MAXLINE}
 n:= order;
 assign(fin, finname); reset(fin);
 assign(fout, foutname); rewrite(fout);
 initialize;
 while not eof(fin) do
 begin
 shift; read(fin,a[n]);
 if a[n]>max then max:=a[n] else if old=max then maximum(max);
 write(fout, max);
 end;
end; {MAXLINE}
```

Figure 11.4.4: MAXLINE running one-dimensional max calculation routine.

This algorithm can be easily extended to image filtering if it is applied along rows and columns, respectively. In this case, the number of comparisons required per output pixel is approximately only 6, compared to $O(n^2)$ required by the classical algorithm.

A similar approach can be used for a running sorting algorithm. Its basic idea is the following. If y_{i-1} is known, y_i can be calculated as follows:

$$y_i = (y_{i-1} \ B\_del \ x_{i-n}) \ B\_ins \ x_i \qquad (11.4.9)$$

where B_del, B_ins denote deletion and insertion in the sorted array y_{i-1} [14,15]. Thus, the new output y_i of the running sorting algorithm can be done by using two binary searches [2, p. 100], one for deletion and one for insertion. Since the binary search requires $O(log_2 n)$ comparisons, the running sorting algorithm requires also the same order of the number of comparisons. An implementation of the running one-dimensional median filtering algorithm in PASCAL is the SORTLINE procedure, shown in Figure 11.4.5.

```
procedure SORTLINE(finname:string; foutname:string);
 {* finname:  name of the input data file          *}
 {* foutname: name of the output data file          *}
 var a, b:array[1..order] of real;
   old: real;
   fin, fout: file of real;
   in_pos, empty_pos: integer;

procedure SHIFT;
 {* it shifts the contents of the array b           *}
 var i: integer;
 begin
  old=b[1]; for i:=1 to n-1 do b[i]:=b[i+1];
 end;

procedure FIND_IN_POS(value: real; var in_pos: integer);
 var bottom, top, middle: integer;
   found: boolean;
 begin
  bottom:=1; top:=n; found:=false;
  if value>a[n] then
   begin
   in_pos:=n; found:=true;
   end
  else if value<a[1] then
   begin
   in_pos:=0; found:=true;
   end
  else
   begin
   while (not found) and (bottom <=top) do
    begin
    middle:=(bottom+top) div 2;
    if(a[middle]<=value) and (a[middle+1]>=value)
     then found:=true
    else if a[middle]>value then top:=middle-1
    else bottom:=middle+1;
    end;
```

```
   if found=true then in_pos:=middle
   else writeln('Error in FIND_IN_POS');
  end;
end;

procedure B_DELETE(value: real; var empty_pos: integer);
 var bottom, top, middle: integer;
   found: boolean;
 begin
  bottom:=1; top:=n; found:=false;
  while (found=false) and (bottom <=top) do
  begin
   middle:=(bottom+top) div 2;
   if(a[middle]=value) then found:=true
   else if a[middle]>value then top:=middle-1
   else bottom:=middle+1;
  end;
  if found=true then empty_pos:=middle
   else writeln('Error in B_DELETE');
 end;

procedure B_INSERT(value: real ; empty_pos: integer);
 var in_pos, i:integer;
 begin
  find_in_pos(value, in_pos);
  if empty_pos <= in_pos then
   begin
   for i:=empty_pos to in_pos-1 do a[i]:=a[i+1];
   a[in_pos]:=value;
   end
  else if empty_pos > in_pos then
   begin
   for i:=empty_pos downto in_pos+2 do a[i]:=a[i-1];
   a[in_pos+1]:=value;
   end;
 end;

procedure SORT;
 var i: integer;
   sorted: boolean;
   r: real;
 begin
  for i:=1 to n do a[i]:=b[i]; sorted:=false;
```

```
while sorted=false do
begin
 sorted:=true;
 for i:=1 to n-1 do
 if a[i]>a[i+1] then
  begin
   sorted:=false; r:=a[i+1]; a[i+1]:=a[i]; a[i]:=r;
  end;
 end;
end;

procedure SEND_OUT;
 var i: integer;
 begin for i:=1 to n do write(fout,a[i]); end;

procedure INITIALIZE;
 var i: integer;
 begin
 for i:=1 to n do read(fin, b[i]);
 sort;
 send_out;
 end;

begin {SORTLINE}
n:=order;
assign(fin, finname); reset(fin);
assign(fout, foutname); rewrite(fout);
initialize;
while not eof(fin) do
begin
 shift
 read(fin, b[n]);
 b_delete(old, empty_pos);
 b_insert(b[n], empty_pos);
 send_out;
 end;
end {SORTLINE}
```

Figure 11.4.5: SORTLINE running one-dimensional running sorting routine.

These algorithms can also be easily extended to the two-dimensional case. In this case, the binary search routine will be used to delete n elements from the sorted output y_{ij} and to insert the n new elements. Thus, the two-dimensional sorting algorithm requires $O(2n\log_2(mn))$ comparisons, which is much lower than the $O(2n^2\log n)$ required by the QUICKSORT algorithm if $m \simeq n$. It is also close to the $O(n)$ number required by the Huang's algorithm, although it provides much more information. It has been proven by simulation that an algorithm similar to SORTLINE, called NOSORT (NS) [15], is much faster than the QUICKSORT algorithm both for one-dimensional and for two-dimensional filtering [4].

11.5 FAST STRUCTURES FOR MEDIAN AND ORDER STATISTICS FILTERING

In the previous sections we were mostly interested in the construction of fast algorithms for median and order statistics filtering mainly by minimizing the number of comparisons required. No special attention has been paid for data movements or for the complexity of the final decision structure. However, sorting algorithms can be constructed that have a very regular structure and which can be used for the development of very regular and parallel architectures for sorting. Such algorithms are the so-called *sorting networks* [1, pp. 200-246]. Such a sorting network is shown in Figure 11.5.1. The data enter at the left side of the network. Comparators are represented by vertical connections (bars) between two network lines. Each comparator can shuffle its inputs, if necessary, so that the large number appears on the lower output of the comparator and the small number appears to its upper output. Sorting networks have many advantages apart from their regularity and parallelism. They always do *in-place calculations*, i.e., the data can be maintained in the same n locations. Also, no storage of the results of previous comparisons is required. One simple way to construct a sorting network is by *insertion*, as is shown in Figure 11.5.1a. The $(n+1)$-st element can be inserted to its proper place, after n elements have been sorted. The sorting network for $n=5$ number is shown in Figure 11.5.1b. Another way to construct a sorting network is by *selection*, as shown in Figure 11.5.2. In this case first the maximum element $x_{(n+1)}$ is calculated. Repeated application of this maximum selection leads to the *bubble sorting network* [1]. If parallelism is allowed, straight insertion also leads to bubble sorting network, as is seen in Figure 11.5.1b [1, p.224]. Networks can be constructed that require the minimum number of comparators, if hardware restrictions exist. Such a network for sorting $n=9$ numbers requires only 25 comparators and its time delay is 9 [1, p.228]. If parallel computation is possible, sorting networks can be constructed that require the minimum delay. Such a network for $n=9$ numbers requires 25 comparators and its time delay is only 8 [1, p. 231].

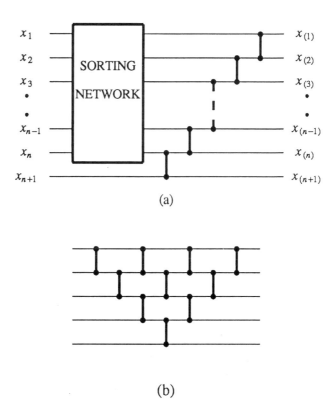

(a)

(b)

Figure 11.5.1: (a) Sorting by insertion; (b) Sorting network for $n=5$ numbers.

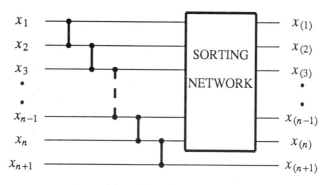

Figure 11.5.2: Sorting by selection.

Finally, a network that can be easily be implemented in hardware is the *odd-*

even transportation network shown in Figure 10.4.1. It requires $n(n-1)/2$ comparators in a brick-like pattern. Its simplicity lies in the fact that only two operations alternate. This network has been proposed for a VLSI implementation of an one-dimensional median filter of length $n=5$, which is shown in Figure 11.5.3 [16]. The two blocks used in this implementation are delays (T) and compare-and-swap (CS) circuits. A smaller odd-even transportation network for $n=3$, which has been used as a building block in a sorting network of $n=9$ numbers, is shown in Figure 11.5.4 [17]. The odd-even transportation network has been proposed for the implementation of a 3×3 nonlinear filter module, as is shown in Figure 10.4.3 [14]. This filter module can be used for the implementation of a variety of nonlinear digital filters (e.g., median filters, L-filters, morphological filters, homomorphic filters), as has already been described in chapter 10.

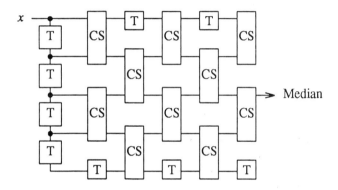

Figure 11.5.3: A VLSI structure for median filtering for n=5 filter size.

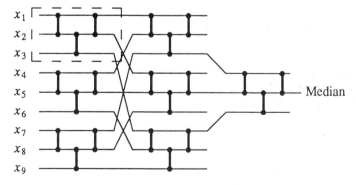

Figure 11.5.4: Median filtering structure for $n=9$ numbers, using the odd-even transportation network for $n=3$.

Simpler networks can be constructed for the implementation of the max/min (erosion/dilation) filters [14]. Such a network for the calculation of the minimum and the maximum of $n=8$ numbers is shown in Figure 10.4.4. The erosion/dilation filter has delay of the order $O(log_2 n)$ and it requires $3n/2-2$ comparators.

With the development of the technology for Switched Capacitor (SC) filters [18] and Charge Coupled Device (CCD) [19] filters and delay lines, analog techniques have been proposed for the implementation of some nonlinear filters. The analog implementation of a median filter is described in [20]. Also analog/digital and Switched Capacitor implementations of the median filters and of the median hybrid filters have been proposed recently [21-23].

11.6 MORPHOLOGICAL IMAGE PROCESSORS

The main characteristic of the morphological operations is their simplicity. The basic operations involved are the erosion and the dilation. If binary images are involved, the following definition of dilation and erosion can be used for their implementation on a computer:

$$X \oplus B^s = \{z \in E : B_z \cap X \neq \varnothing\} \tag{11.6.1}$$

$$X \ominus B^s = \{z \in E : B_z \subset X\} \tag{11.6.2}$$

where z is and arbitrary position on the image plane and B_z is the translate of the structuring element B in that position. Both definitions of erosion and dilation suggest that local operations around the position $z=(i,j)$ of the image plane are required for the calculation for both erosion and dilation. Similarly the definitions of dilation and erosion of a function by function:

$$[f \oplus g^s](x) = \max_{\substack{z \in D \\ z-x \in G}} \{f(z) + g(z-x)\} \tag{11.6.3}$$

$$[f \ominus g^s](x) = \min_{\substack{z \in D \\ z-x \in G}} \{f(z) - g(z-x)\} \tag{11.6.4}$$

suggest that the computation of the local maximum and minimum at each point z in the domain D of function f is required for the calculation of dilation and erosion, respectively. Thus *cellular* computer architectures [24-31] can be used for the implementation of the morphological operations. Such a cellular array architecture operating on a 3×3 image neighborhood is shown in Figure 11.6.1. Each of its cell processors has the structure of Figure 11.6.2. Each cell has a register for storing the state of the cell and a logic module which calculates the new state of the module as a function of its current state and the states of the neighboring cells.

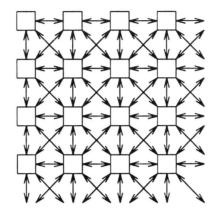

Figure 11.6.1: Cellular array architecture (adapted from [30]).

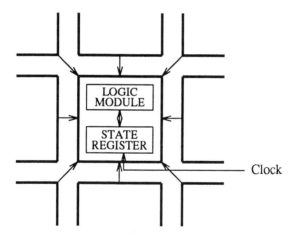

Figure 11.6.2: Structure of a cell (adapted from [30]).

An example of cellular image processor is CLIP (Cellular Logic Image Processor) [25]. CLIP4 has an array of 96×96 processors. Each cell has the possibility of storing a large stack of binary images. The processing element can compute two independent binary outputs. One is transmitted to all neighboring processors, and the other corresponds to the new local image element. The computation time per image element is one nanosecond. CLIP can implement different programmable neighborhoods that have square or hexagonal shape. Several other cellular array image processors exist, e.g., the Binary Image Processor

(BIP) [26], Preston- Herron Processor [27], PICAP system [28], Massively Parallel Processor [29] and Cytocomputer [30]. One of the most important problems with the cellular arrays is that their current size is of the order 100×100, whereas the image sizes are 512×512 or 1024×1024 pixels. Therefore, the image must be partitioned into segments before it is processed, and each segment must be processed independently. This approach causes border effects to propagate in the segments when multiple local operations are applied to the image. It also introduces a heavy I/O load to the system. Another approach that does not have such problems is pipelining, when the nature of image processing operations permit such an approach.

One of the advantages of the morphological operations is not only that they permit pipelining but the use of pipelining reduces the computational complexity of the operations. Let us use the following definitions of the dilation and erosion:

$$X \oplus B^s = \bigcup_{b \in B} X_{-b} \qquad\qquad (11.6.5)$$

$$X \ominus B^s = \bigcap_{b \in B} X_{-b} \qquad\qquad (11.6.6)$$

where X_{-b} denotes translation of the image X. Therefore, for binary images, erosions and dilations can be computed by calculating the unions or the intersection of $N(B)$ translates of the image X, where $N(B)$ is the number of elements in the structuring element B. Specialized pipeline computer architectures can compute erosion and dilation in much less than $N(B)$ operations. Pipelining is based on the decomposition property of dilation and erosion. Let the structuring element B^s be decomposed as follows:

$$B^s = B_1{}^s \oplus B_2{}^s \ldots \oplus B_K{}^s \qquad\qquad (11.6.7)$$

The dilation and the erosion can be implemented in K steps, as follows:

$$X \oplus B^s = (..((X \oplus B_1{}^s) \oplus B_2{}^s).. \oplus B_K{}^s) \qquad\qquad (11.6.8)$$

$$X \ominus B^s = (..((X \ominus B_1{}^s) \ominus B_2{}^s).. \ominus B_K{}^s) \qquad\qquad (11.6.9)$$

An example of the decomposition of a structuring element having 21 points is shown in Figure 6.2.7. The implementation of dilation by using definition (11.6.5) requires 20 translations of the object X and unions. The structuring element B can be decomposed in three smaller structuring elements. Therefore the dilation can be implemented in 3 steps. These dilations can be done in pipeline. Thus the decomposed implementation of dilation requires only 8 translations and unions instead of 20. The saving is much more when large structuring elements are used. The decomposed implementation of erosion and dilation have been used in the first morphological image processor called *Texture Analyzer* [31]. A similar approach is used in the *Cytocomputer* [30,32,33]. It is a pipeline system, whose structure is shown in Figure 11.6.3.

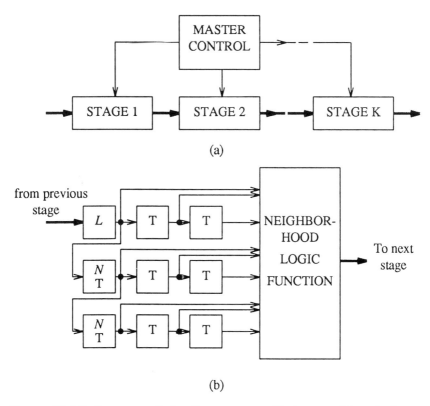

Figure 11.6.3: Structure of the Cytocomputer. T denotes delay. *L* denotes point-to-point logical operation (adapted from [30]).

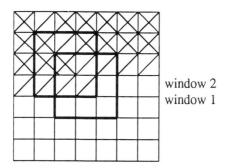

window 2
window 1

Figure 11.6.4: Sequential scanning through a digital image in two stages (adapted from [30]).

Each stage of the pipeline performs one type of point-by-point or neighborhood logic operation. Operations are performed either on a 3×3 square neighborhood or in a 7-element hexagonal neighborhood. Each stage can store two image lines having N pixels each. It has also nine registers which contain the pixels in an image neighborhood. These pixels are the input to the neighborhood logic function. This function performs morphological operations both for binary and for gray-scale images. All data transformations and data transfers are computed in a single step. The output of a stage is input to the subsequent stage. The operation of two stages working on 3×3 neighborhoods is shown in Figure 11.6.4. The operation of a morphological processor like the Cytocomputer can be complemented by a *point processor* performing arithmetic and logical operations between two images, by a *connectivity processor* performing analysis of the connected components of an image and a *pixel location and histogram processor* which evaluates histograms and counts pixels in an image window [33]. The pipelining of these four processors leads to an *image flow architecture*, which can perform a variety of image processing operations. A special language, called *Basic Language of Image X-formation (BLIX)*, has been implemented for the programming of this image flow processor for machine vision applications [33].

11.7 IMPLEMENTATION TECHNIQUES FOR QUADRATIC DIGITAL FILTERS

As we explained in chapter 8, a nonlinear polynomial filter is reduced to a quadratic one, if the input-output relation is restricted to the second order (third) term of the discrete Volterra series. Quadratic digital filters are the simplest non-linear time-invariant systems and constitute an efficient class of nonlinear filters in a variety of applications of digital signal processing, such as in the optimal detection of a Gaussian signal in Gaussian noise [34].

For the realization of finite extent quadratic digital filters the direct approach [35], which requires a large number of operations, and the 2-d convolution approach, which is not in general very efficient, have been suggested [34]. Also another more efficient approach, which is based upon the distributed arithmetic technique, for filters having fixed coefficients has been proposed [36]. This latter approach has been extended in order to exploit memory-oriented structures that properly match the characteristics of distributed arithmetic [37]; the penalty paid, however, is the requirement of very large memory and the low throughput rate for higher order filters.

Moreover, a realization structure for quadratic digital filters has been proposed, based on matrix decomposition [38]; this structure resulted in a parallel configuration of linear finite impulse response (FIR) filters in cascade with a set of square-in add-out type of operations. Specifically, the lower upper (LU) triangular decomposition, in conjunction with the singular value (SV) decomposition, was considered. Also the 2-d finite extent quadratic digital filters were implemented with the matrix decomposition approach [39]. The above matrix decomposition approaches are based on the idea of the exact expansion of a general m-d real rational function (i.e., a real rational function of M independent variables) in terms of polynomials of order one, each one of which is a function of one of the M variables [40], [41].

The increasing interest in real-time signal processing is mainly due to the enormous growth in both signals to be processed and need for fast processing. Therefore, the main objective in the implementation of digital filters is the increase of the throughput rate. Presently a number of hardware design modifications that serve to increase computation speed and control system throughput are available. Among the most efficient of those techniques is one that utilizes array processors (APs) to reduce computation time. In particular, the VLSI APs are special-purpose, locally interconnected computing networks that maximize concurrency by exploiting parallelism and pipelining. The two most dominant VLSI APs are the systolic and wavefront arrays, which take advantage of both parallelism and pipelining by using the concept of computational wavefront [42-46]. A systolic array is a network of elementary processing elements (PEs) that rythmically compute and pass data through the system. A wavefront array may be seen as a systolic array, which is characterized by data flow computing. The existence of data flow implies that the requirement for correct timing in the systolic arrays is replaced by the requirement for correct sequencing in the wavefront arrays.

Sections 11.8 - 11.10 address the implementation of quadratic digital filters via systolic and wavefront VLSI APs [47]. The structures obtained permit very high sampling rates and are characterized by parallelism pipelining, modularity and regularity, local interconnections, absence of time-shared buses, and the highest possible speed-up factor [43].

In section 11.8, the quadratic digital filter is described via a quadratic form, where the coefficient matrix is triangular; this is achieved by exploiting the existing symmetry and simplifies substantially the associated implementation. In sections 11.9 and 11.10 the systolic and wavefront array implementations are presented, respectively. The combined parallelism and pipelining, which results from the proper organization of computations, offers high concurrency and full processor efficiency that corresponds to the highest possible speed-up factor. In section 11.11 the matrix description is introduced for 2-d finite support quadratic filters. Section 11.12 presents realizations of 2-d quadratic filters based on the LU and the LU-SV decompositions.

11.8 MATRIX DESCRIPTION OF QUADRATIC DIGITAL FILTERS

A class of nonlinear, causal, shift-invariant digital filters is defined by using kernels with finite dimensions. Specifically the input-output relationship of these nonlinear digital filters is given by:

$$y(n) = h_0 + \sum_{k=1}^{K} \bar{h}_k[x(n)] \qquad (11.8.1)$$

where h_0 is the constant term, K is the order of nonlinearity and $\bar{h}_k[x(n)]$ is defined as:

$$\bar{h}_k[x(n)] = \sum_{i_1=0}^{N-1} \sum_{i_2=0}^{N-1} \cdots \sum_{i_k=0}^{N-1} h_k(i_1,i_2,...,i_k)x(n-i_1)x(n-i_2)...x(n-i_k) \qquad (11.8.2)$$

Clearly the term $\bar{h}_k[x(n)]$ includes N^k terms. The linear term of the Volterra expansion (11.8.2) is written as:

$$y_1(n) = \bar{h}_1[x(n)] = \mathbf{X}_1^T(n)\mathbf{H}_1 \qquad (11.8.3)$$

where

$$\mathbf{X}_1^T(n) = [x(n)\ x(n-1)\ \cdots\ x(n-N+1)] \qquad (11.8.4)$$

and

$$\mathbf{H}_1^T = [h_1(0)\ h_1(1)\ \cdots\ h_1(N-1)] \qquad (11.8.5)$$

The general nonlinear term of order $k>1$ may be written in the form [37]

$$y_k(n) = h_k[x(n)] = \mathbf{X}_k^T(n)\mathbf{H}_k \qquad (11.8.6)$$

where

$$\mathbf{X}_k(n) = [\mathbf{X}_1(n)\otimes\mathbf{X}_{k-1}(n)] \qquad \in \mathbf{R}^{N^k} \qquad (11.8.7)$$

and $\otimes$ denotes the Kronecker product of matrices [48].

The vector $\mathbf{H}_k \in \mathbf{R}^{N^k}$ is formed as follows:

$$\mathbf{H}_k = [(\mathbf{H}_k)_\tau] = [h_k(i_1,i_2,\ldots,i_k)] \qquad (11.8.8)$$

where $(\mathbf{H}_k)_\tau$ denotes the τ-th, $\tau=1,2,...,N^k$ element of $\mathbf{H}_k$ and:

$$i_m = \left[\frac{\tau-1}{N^{k-m}}\right] \bmod N^{k-m+1}, \qquad m=1,2,...,k \qquad (11.8.9)$$

Note that $[q]$ denotes the integer part of the nonnegative number q.

Each coefficient $h_k(i_1,i_2,...,i_k)$ corresponds to the product $x(n-i_1)x(n-i_2)...x(n-i_k)$ independently of the ordering of the indices $i_1,i_2,...,i_k$. Therefore, we can retain only those terms $h_k(i_1,i_2,...,i_k)$, which have different combinations of the indices $i_1, i_2, ..., i_k$. Thus the dimension of $\mathbf{H}_k$ may be substantially reduced and the k-th nonlinear term may be written in the form:

$$y_k(n) = \overline{h}_k[x(n)] = \hat{\mathbf{X}}_k^T(n)\hat{\mathbf{H}}_k \tag{11.8.10}$$

where $\hat{\mathbf{X}}_k$ includes the products $x(n-i_1)x(n-i_2)...x(n-i_k)$ only once. It may be shown by induction that the dimension of $\mathbf{H}_k$ is:

$$d_k = \frac{1}{(k-1)!} \sum_{i=1}^{N} \prod_{j=0}^{k-2} (i+j) \tag{11.8.11}$$

For example, let $N=3$ and $k=4$. Then, while the dimension of $\mathbf{H}_4$ is $N^k = 3^4 = 81$, the dimension of $\hat{\mathbf{H}}_4$ is given by:

$$d_4 = \frac{1}{3!}(1 \cdot 2 \cdot 3 + 2 \cdot 3 \cdot 4 + 3 \cdot 4 \cdot 5) = 15 \tag{11.8.12}$$

The nonlinear quadratic term is given by:

$$y_2(n) = \sum_{i_1=0}^{N-1} \sum_{i_2=0}^{N-1} h_2(i_1, i_2)x(n-i_1)x(n-i_2) = \mathbf{X}_2^T \mathbf{H}_2 \tag{11.8.13}$$

where $\mathbf{H}_2 \in \mathbf{R}^{N^2}$ is given by:

$$\mathbf{H}_2 = [(\mathbf{H}_2)_\tau] = [h_2(i_1, i_2)], \quad \tau = 1, 2, ..., N^2 \tag{11.8.14}$$

Applying (11.8.9) for $k=2$ and $m=1,2$, we obtain:

$$i_1 = \left\lceil \frac{\tau-1}{N} \right\rceil, \qquad i_1 = 0, 1, ..., N-1 \tag{11.8.15}$$

$$i_2 = (\tau-1) \mod N, \quad i_2 = 0, 1, ..., N-1 \tag{11.8.16}$$

Moreover, according to (11.8.10), $y_2(n)$ may be written in the form:

$$y_2(n) = \overline{h}_2[x(n)] = \hat{\mathbf{X}}_x^T(n)\hat{\mathbf{H}}_2 \tag{11.8.17}$$

where $\hat{\mathbf{H}}_2 \in \mathbf{R}^{d_2}$ and

$$d_2 = 1+2+...+N = N(N+1)/2 \tag{11.8.18}$$

It is easily seen that the quadratic term $y_2(n)$ may be written in the quadratic form:

$$y_2(n) = \mathbf{X}_1^T(n)\mathbf{R}\mathbf{X}_1(n) \tag{11.8.19}$$

where $\mathbf{R}$ is a lower (it may be also upper) triangular matrix of the form:

$$\mathbf{R} = [r_{ij}] = \begin{bmatrix} h_2(0,0) & & & 0 \\ h_2(0,1) & h_2(1,1) & & \\ \vdots & \vdots & & \\ h_2(0,N-1) & h_2(1,N-1) & \cdots & h_2(N-1,N-1) \end{bmatrix} \tag{11.8.20}$$

Note that $\mathbf{R}$ includes only $d_2=N(N+1)/2$ nonzero elements which are the elements of $\hat{\mathbf{H}}_2$. It is seen from (11.8.20) that the (i,j)th element of the coefficient matrix $\mathbf{R}$ is given by:

$$r_{ij} = \begin{cases} h_2(j-1,i-1) & \text{if } i \geq j \\ 0 & \text{if } i < j \end{cases} \qquad (11.8.21)$$

11.9 SYSTOLIC ARRAY IMPLEMENTATION

In this section the systolic implementation of the quadratic digital filter will be described. The quadratic digital filter is described by (11.8.19). At first we implement the row-matrix multiplication (RMM) $\mathbf{X}_1^T(n)\mathbf{R}$, which is actually a matrix vector multiplication [47,49]. In the sequel we apply n more different PEs in the n outputs of the RMM, in order to implement the row-vector multiplication (RVM) $[\mathbf{X}_1^T(n)\mathbf{R}]\mathbf{X}_1(n)$. The block and the layout diagrams of the proposed systolic implementation are given in Figures 11.9.1 and 11.9.2, respectively. The systolic arrays implementation is characterized by a time-scaling factor $\alpha=1$, which is the more advantageous for the pipelining rate [43,44], since no blanks are required to be placed among the input samples. This is due to the fact that both the RMM and the RVM have time-scaling factors $\alpha=1$, and they are properly synchronized to each other.

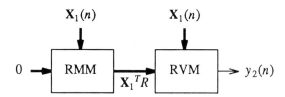

Figure 11.9.1: Block diagram of the systolic array implementation (adapted from [43]).

The execution time of the operations in each type of the used PEs equals to:

$$\tau = Mu + Ad \qquad (11.9.1)$$

where Mu and Ad denote the time needed to perform one multiplication and one addition, respectively.

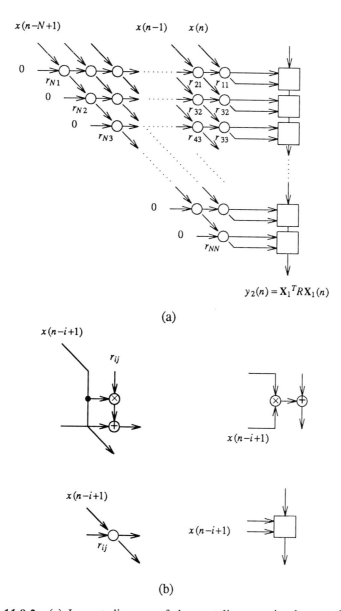

$$y_2(n) = \mathbf{X}_1{}^T R \mathbf{X}_1(n)$$

(a)

(b)

Figure 11.9.2: (a) Layout diagram of the systolic array implementation; (b) The structures and the symbols of the used PEs.

The throughput rate of the whole implementation structure is therefore given by:

$$R = \frac{1}{\alpha\tau} = \frac{1}{\tau} = \frac{1}{Mu+Ad} \tag{11.9.2}$$

which is very high and meets the usual requirements for real-time processing applications. For example, using 16 bit multipliers and adders, we have Mu=115 ns and Ad=19 ns, i.e., τ=134 ns and:

$$R = \frac{1}{134\times10^{-9}} = 7.46\times10^6 \text{OP/s} = \text{MOP/s}$$

Note that the structure considered is nonrecursive, and therefore there are no feedback loops that impose iteration bounds [50]. The latency is also called by some authors *throughput delay* and is defined as the time interval separating the appearance of an input sample at the input port, from the appearance of the corresponding output sample at the output port. In a systolic array implementation the latency is determined by the number of cutsets, i.e., the number of latches separating the first input sample from the last output sample in a snapshot. It is seen from the layout diagram (Figure 11.9.2a) that the latency is given by:

$$L_s = 2N\tau \tag{11.9.3}$$

since $2N$ cutsets separate the first input sample from the output sample.

11.10 WAVEFRONT ARRAY IMPLEMENTATION

The signal-flow graph (SFG), which is associated with the systolic array implementation, may be transformed to the data-flow graph (DFG) by substituting every time-delay operator D (controlled by a global clock and denoted by the intersection of the graph with the cutset) in the SFG, with a separator (*diamond*) that is locally controlled by handshaking [45]. Thus the requirement of global synchronization is not needed any more. The wavefront array implementation has the same throughput rate with the systolic one, which is given by (11.9.1), since the data transfer time Δ is negligible [43]. The DFG of the wavefront array implementation is shown in Figure 11.10.1. However, due to the data flow timing synchronization, the latency is highly reduced and is now given by:

$$L_w = (N+1)\tau \tag{11.10.1}$$

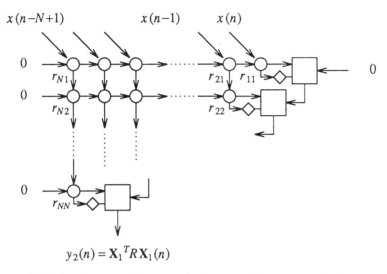

$$y_2(n) = \mathbf{X}_1^T R \mathbf{X}_1(n)$$

Figure 11.10.1: Layout diagram of the wavefront array implementation (adapted from [43]).

11.11 MATRIX DESCRIPTION OF 2-D FINITE-SUPPORT QUADRATIC FILTERS

A 2-d finite-support quadratic digital filter is described by the discrete 2-d Volterra series truncated to the second-order term:

$$y(n_1,n_2) = h_0 + \sum_{i_{11}=0}^{N_1-1} \sum_{i_{12}=0}^{N_2-1} h_1(i_{11},i_{12})x(n_1-i_{11},n_2-i_{12}) \qquad (11.11.1)$$

$$+ \sum_{i_{11}=0}^{N_1-1} \sum_{i_{12}=0}^{N_2-1} \sum_{i_{21}=0}^{N_1-1} \sum_{i_{22}=0}^{N_2-1} h_2(i_{11},i_{12},i_{21},i_{22})x(n_1-i_{11},n_2-i_{12})x(n_1-i_{21},n_2-i_{22})$$

where h_0 is a constant term, and $h_1(i_{11},i_{12})$ and $h_2(i_{11},i_{12},i_{21},i_{22})$ are the filter coefficients of the linear and the quadratic terms, respectively. We make here the usual assumption, that the quadratic kernel h_2 is a symmetric function of the pairs of indices (i_{11},i_{12}) and (i_{21},i_{22}), so that, as directly seen from (11.11.1), $h_2(i_{11},i_{12},i_{21},i_{22}) = h_2(i_{21},i_{22},i_{11},i_{12})$. The quadratic part of the output may be written in the form [39,51]:

$$y_q(n_1,n_2) = \mathbf{X}^T(n_1,n_2)\mathbf{H}\mathbf{X}(n_1,n_2) \qquad (11.11.2)$$

where $\mathbf{X}(n_1,n_2)$ is the $N \times 1$ $(N=N_1N_2)$ vector:

$$\mathbf{X}(n_1,n_2) = [x(n_1,n_2)x(n_1-1,n_2)$$
$$\cdots x(n_1-N_1+1,n_2)| \cdots |x(n_1,n_2-N_2+1)$$
$$x(n_1-1,n_2-N_2+1)$$
$$\cdots x(n_1-N_1+1,n_2-N_2+1)]^T \qquad (11.11.3)$$

The vector $\mathbf{X}(n_1,n_2)$ may be written in a compact form as follows:

$$\mathbf{X}(n_1,n_2) = [\mathbf{X}_\tau(n_1,n_2)] = x(n_1-e,n_2-f) \qquad (11.11.4)$$

where $\mathbf{X}_\tau(n_1,n_2)$ denotes the τ-th, element of $\mathbf{X}(n_1,n_2)$ $(\tau=1,2,...,N)$ and:

$$e = (\tau-1) \ mod \ N_1, \qquad e = 0,1,...,N_1-1 \qquad (11.11.5a)$$

$$f = [(\tau-1)/N_1], \qquad f = 0,1,...,N_2-1 \qquad (11.11.5b)$$

The symbol $[x]$ denotes the integer part of $x \geq 0$, and the index τ is determined in terms of e, f by the relation:

$$\tau = fN_1 + e + 1 \qquad (11.1.6)$$

The $N \times N$ matrix $\mathbf{H}$ is partitioned in submatrices $\mathbf{H}(i,j)$ of dimensions $N_1 \times N_1$ as follows:

$$\mathbf{H} = \begin{bmatrix} \mathbf{H}(0,0) & \mathbf{H}(0,1) & \cdots & \mathbf{H}(0,N_2-1) \\ \mathbf{H}(1,0) & \mathbf{H}(1,1) & \cdots & \mathbf{H}(1,N_2-1) \\ \vdots & \vdots & & \vdots \\ \mathbf{H}(N_2-1,0) & \mathbf{H}(N_2-1,1) & \cdots & \mathbf{H}(N_2-1,N_2-1) \end{bmatrix} \qquad (11.11.7)$$

Each submatrix $\mathbf{H}(i,j)$, $i,j=0,1,...,N_2-1$ has the form:

$$\mathbf{H}(i,j) = \begin{bmatrix} h(0,i,0,j) & h(0,i,1,j) & \cdots & h(0,i,N_1-1,j) \\ h(1,i,0,j) & h(1,i,1,j) & \cdots & h(1,i,N_1-1,j) \\ \vdots & \vdots & & \vdots \\ h(N_1-1,i,0,j) & h(N_1-1,i,1,j) & \cdots & h(N_1-1,i,N_1-i,j) \end{bmatrix} \qquad (11.11.8)$$

Note that according to (11.11.7) and (11.11.8) and the symmetry of h_2, the square matrix $\mathbf{H}$ is symmetric and can be written in a compact form as follows:

$$\mathbf{H} = [h_{\alpha\beta}] = h_2(k,i,l,j) \qquad (11.11.9)$$

where $h_{\alpha\beta}$ denotes the $\alpha\beta$-th element of $\mathbf{H}$ $(\alpha,\beta=1,2,...,N)$ and:

$$k = (\alpha-1) \ mod \ N_1, \qquad k = 0,1,...,N_1-1$$
$$l = (\beta-1) \ mod \ N_1, \qquad l = 0,1,...,N_1-1$$
$$i = [(\alpha-1)/N_1], \qquad i = 0,1,...,N_2-1$$
$$j = [(\beta-1)/N_1], \qquad j = 0,1,...,N_2-1$$

The quadratic part of the filter (11.11.1) may be realized in a decomposed form [52] by applying a decomposition on the coefficient matrix $\mathbf{H}$. This decomposition may be accomplished through a number of various approaches, i.e., general decomposition, lower-upper triangular (LU) decomposition, singular value decomposition (SVD), Jordan (J) decomposition, Walsh-Hadamard transform (WHT) decomposition, etc. [52-54].

Since the matrix $\mathbf{H}$ is square and symmetric, it may be decomposed exactly into a finite sum of rank one symmetric matrices as follows [54]:

$$\mathbf{H} = \sum_{i=1}^{r} q_i \mathbf{R}_i \mathbf{R}_i^T \tag{11.11.10}$$

where $r = \text{rank } \mathbf{H} \leq N$. The q_i's are real scalars and the $\mathbf{R}_i$'s are $N \times 1$ vectors of the form:

$$\mathbf{R}_i = [(\mathbf{R}_i)_\tau] = r_i(e,f), \qquad \tau = 1,2,...,N \tag{11.11.11}$$

where $(\mathbf{R}_i)_\tau$ denotes the τ-th element of $\mathbf{R}_i$ and e, f are given by (11.11.5a) and (11.11.5b), respectively. The substitution of (11.11.10) in (11.11.2) leads to the equation:

$$y_q(n_1,n_2) = \sum_{i=1}^{r} q_i [\mathbf{X}^T(n_1,n_2)\mathbf{R}_i][\mathbf{R}_i^T \mathbf{X}(n_1,n_2)]$$

$$= \sum_{i=1}^{r} q_i y_i^2(n_1,n_2) \tag{11.11.12}$$

where

$$y_i(n_1,n_2) = \mathbf{R}_i^T \mathbf{X}(n_1,n_2) = \sum_{k=0}^{N_1-1}\sum_{i=0}^{N_2-1} r_i(k,l)x(n_1-k,n_2-l) \tag{11.11.13}$$

Clearly, (11.11.13) represents a 2-d FIR digital filter of size $N_1 \times N_2$. Hence, the overall output $y_q(n_1,n_2)$, which is given by (11.11.12), may be computed by the sum of the r parallel square-in add-out operations. In other words, the quadratic 2-d finite-support filter (11.11.2) is realized by Figure 11.11.1,

i)   combining the r 2-d linear FIR filters of the form (11.11.13), where in each branch a square function is also placed, and

ii)   summing all the weighted partial results.

From Figure 11.11.1, it is seen that the general structure derived is characterized by a high degree of inherent parallelism; each parallel branch operates simultaneously and independently on a common input array. Moreover, this structure requires only a single sharing bus, where r communication paths are connected.

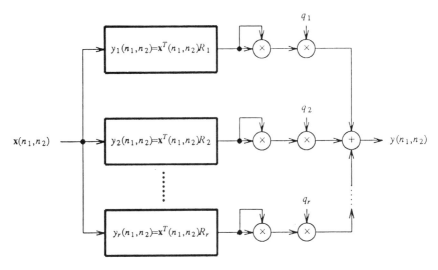

Figure 11.11.1: The general matrix decomposition structure of a 2-d finite-support quadratic digital filter (adapted from [39, 51]).

11.12 REALIZATION OF 2-D QUADRATIC FILTERS BASED ON THE LU AND THE LU-SV DECOMPOSITIONS

According to the LU matrix decomposition [54], a square $N \times N$ constant symmetric matrix $\mathbf{H}$ may be decomposed into a finite sum of rank one symmetric matrices, in the form:

$$\mathbf{H} = \sum_{i=1}^{r} d_i \mathbf{L}_i \mathbf{L}_i^T \qquad (11.12.1)$$

where $r = \text{rank } \mathbf{H}$ and $\mathbf{L}_i$ are $N \times 1$ constant vectors with $i-1$ leading zeros [53, 54]. The substitution of $\mathbf{H}$, in the form of (11.12.1), in (11.11.2) results in:

$$y_q(n_1, n_2) = \sum_{i=1}^{r} d_i [\mathbf{X}^T(n_1, n_2) \mathbf{L}_i][\mathbf{L}_i^T \mathbf{X}(n_1, n_2)]$$

$$= \sum_{i=1}^{r} d_i y_i^2(n_1, n_2) \qquad (11.12.2)$$

$$y_i(n_1, n_2) = \mathbf{X}^T(n_1, n_2) \mathbf{L}_i = \sum_{j=1}^{N} l(j, i) \mathbf{X}_j(n_1, n_2) \qquad (11.12.3)$$

$\mathbf{X}_\tau(n_1, n_2) = x(n_1 - e, n_2 - f)$ denotes the τ-th element of $\mathbf{X}(n_1, n_2)$ and $l(j, i)$ are the nonzero elements of the vectors $\mathbf{L}_i$.

(11.12.2) and (11.12.3) lead to a more advantageous realization which is shown in Figure 11.12.1. Specifically, the set of the parallel square-in add-out type operations (Figure 11.11.1) is reduced now to a set of sequential square-in add-out operations. This is due to the structure of the vectors $\mathbf{L}_i$, which result in the reduction of the throughput delay of each 2-d FIR filter, as i increases. Specifically, the number of nonzero elements of the FIR filter at the ith stage is one less than that at the $(i-1)$th stage, i.e., $y_i(n_1, n_2)$ is produced earlier than $y_{i-1}(n_1, n_2)$. Each parallel branch consists of one 2-d FIR filter cascaded with a square function and a multiplier. Clearly, different realizations of these linear 2-d FIR filters result in different realizations of the whole 2-d quadratic filter.

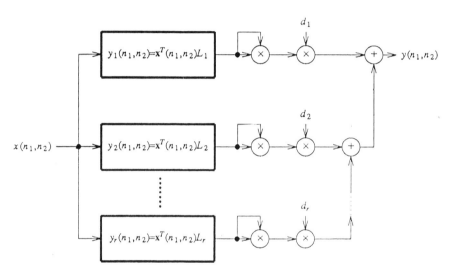

Figure 11.12.1: The LU decomposition structure of a 2-d finite support quadratic digital filter (adapted from [39, 51]).

The LU decomposition does not imply any sharing buses and thus is well suited for VLSI implementation [53, 55]. It appears also to be the most advantageous decomposition structure with respect to the hardware cost expressed as numbers of delays, multipliers, and adders.

In order to further reduce the structural complexity of the realization, it is often convenient to apply first the SV decomposition to the $\mathbf{H}$ matrix so that an approximate expression:

$$\mathbf{H}_p = \sum_{i=1}^{p} \delta_i \mathbf{W}_i \mathbf{W}_i^T \tag{11.12.4}$$

can be found, where W_i are the normalized eigencolums associated with the corresponding eigenvalues δ_i ($i=1,...,p\leq r$) of H, arranged in a decreasing order of magnitude. Then the application of the LU decomposition to the approximation (11.12.4) leads to a realization of the type of Figure 11.12.1, possibly with $p<r$. Thus, the SV-LU decomposition combines the inherent computational advantages of the LU decomposition with the savings that result from the MSE approximation of the SV decomposition.

11.13 DISCUSSION

Several algorithms and architectures for fast nonlinear filtering have been described in this chapter. Classical sorting and selection algorithms, e.g., QUICKSORT, have been analyzed and compared to fast running median and running ordering algorithms. Running algorithms take into account the fact that the filter windows in consecutive positions overlap. Thus they are expected to have better preformance than the classical algorithms. This fact has been proven by simulations on general purpose computers. Special algorithms for median filtering and ranked order filtering have been described that operate on the binary level of the image pixels. Their complexity is usually linearly related with the number of bits per pixel. They are are better suited for filter realizations in hardware or in assembly language. Finally hardware structures have been presented for nonlinear filter realization. Most of these structures are based on the sorting networks.

Some parallel and pipelined structures have been presented for the implementation of morphological operations. Morphological operations are performed locally and they can be decomposed in series of consecutive operations. Therefore, they are very well suited for such structures. Almost all nonlinear filters described in this book perform locally. Their operation is uniform throughout the image. Thus most of them can take advantage of the recent advances in parallel computation.

Systolic and wavefront array implementations of finite extent quadratic digital filters were also presented. The underlying realization structure is based on a matrix description of the quadratic filter, where the associated coefficient matrix has been reduced to triangular form by exploiting the symmetry in the Volterra kernel. The resulting implementations permit very high sampling rates, which are suitable for real-time digital signal processing applications and are characterized by modularity, regularity, and high concurrency, due to the combined parallelism and pipelining. Thus full processor efficiency is attained that corresponds to the highest possible speed-up factor. Realization structures of two-dimensional finite-support quadratic digital filters, which are based on matrix decomposition, were also introduced. The resulting structures consist generally of a number of parallel branches that operate simultaneously and independently on a common input array. By proper decomposition of the coefficient matrix, each parallel branch may be realized by a two-dimensional finite impulse response digital filter in cascade with a squaring operator. The lower-upper (LU) triangular decomposition and the singular value lower-upper (SV-LU) decomposition is used in order to arrive at efficient structures with computational and hardware advantages. Work on parallel implementations of quadratic digital filters shift invariant or adaptive continues [56,57]. Higher order or dimensionality polynomial filters can be implemented through extended implementations based on the approaches described in sections 11.8 to 11.11.

REFERENCES

[1]   D.E. Knuth, *The art of computer programming*, vol. 3, Addison-Wesley, 1973.

[2]   E.Horowitz, S. Sahni, *Fundamentals of Computer Algorithms*, Computer Science Press, 1984.

[3]   T.S. Huang, G.J. Yang, G.Y. Tang, "A fast two-dimensional median filtering algorithm", *IEEE Transactions on Acoustics, Speech and Signal Processing*, vol. ASSP-27, no. 1, pp. 13-18, 1979.

[4]   F. Pasian, "Sorting algorithms for filters based on ordered statistics: performance considerations", *Signal Processing*, vol. 14, pp. 287-293, 1988.

[5]   M.O. Ahmad, D. Sundararajan, "A fast algorithm for two-dimensional median filtering", *IEEE Transactions on Circuits and Systems*, vol. CAS-34, no. 11, pp. 1364-1374, Nov. 1987.

[6]   A.V. Aho, J.E. Hopcroft, J.D. Ullman, *The design and analysis of computer algorithms*, Addison-Wesley, 1974.

[7]   P.E. Danielsson, "Getting the median faster", *Computer Graphics and Image Processing*, vol. 17, pp. 71-78, 1981.

[8]   E. Ataman, V.K. Aatre, K.M. Wong, "A fast method for real time median filtering", *IEEE Transactions on Acoustics, Speech and Signal Processing*, vol. ASSP-28 , no. 4, pp. 415-421, Aug. 1980.

[9]   R.T. Hoctor, S.A. Kassam, "An algorithm for order statistic determination and L-filtering", *Proc. IEEE Int. Conf. on Acoustics, Speech and Signal Processing*, pp. 1686-1689, 1988.

[10] O. Yli-Harja, J. Astola, Y. Neuvo, "Generalization of the radix-method of finding the median to weighted median, order statistic and weighted order statistic filters", *SPIE Visual Communication and Image Processing*, vol.

1001, pp. 69-75, 1988.

[11]   V.V.B. Rao, K.S. Rao, "A new algorithm for real-time median filtering", *IEEE Transactions on Acoustics, Speech and Signal Processing*, vol. ASSP-34 , no. 6, pp. 1674-1675, Dec. 1986.

[12]   R.T. Hoctor, S.A. Kassam, "An algorithm and a pipelined architecture for order statistic determination and *L*-filtering", *IEEE Transactions on Circuits and Systems*, vol. CAS-36, no. 3, pp. 344-352, March 1989.

[13]   I. Pitas, "Fast algorithms for running ordering and max/min calculation", *IEEE Transactions on Circuits and Systems*, vol CAS-36, no. 6, pp. 795-804, June 1989.

[14]   I.Pitas, A.N. Venetsanopoulos, "A new filter structure for the implementation of certain classes of image processing operations", *IEEE Transactions on Circuits and Systems*, vol. CAS-35, no. 6, pp. 636-647, June 1988.

[15]   J. Bee Bednar, T.L. Watt, "Alpha trimmed means and their relationship to median filters", *IEEE Transactions on Acoustics, Speech and Signal Processing*, vol. ASSP-32 , no. 1, pp. 145-153, Feb. 1984.

[16]   K. Oflazer, "Design and implementation of a single-chip 1-D median filter", *IEEE Transactions on Acoustics, Speech and Signal Processing*, vol. ASSP-31, no. 5, pp. 1164-1168, Oct 1983.

[17]   J. Alsford, P. Dewer, J. Illingworth, J. Kittler, J. Lewis, K. Paler, P. Wilde, W. Thomas, "CRS image processing system with VLSI modules", in *Image processing system architectures*, J. Kittler, M. Duff editors, Research Studies Press, 1985.

[18]   P.E. Allen, *Switched capacitor circuits*, Van Nostrand Reinhold, 1984.

[19]   D.F. Franke editor, *Charge coupled devices*, Springer Verlag, 1980.

[20]   D.R. Morgan, "Analog sorting networks ranks inputs by amplitudes and allows selection", *Electronic Design* 2, Jan. 18, pp. 72-74, Jan. 1973.

[21]   T. Jarske, P. Heinonen, Y. Neuvo, "Switched-capacitor linear median hybrid filters", *Proc. IEEE Int. Symp. on Circuits and Systems*, pp. 260-263, 1987.

[22]   J.S. Jimmy Ly, W. Harney Holmes, "Analog implementation of median filters", *IEEE Transactions on Circuits and Systems*, vol. CAS-35, no. 8, pp. 1032-1033, Aug. 1988.

[23]   K. Chen, P. Heinonen, Q. Ye, Y. Neuvo, "Analog/Digital hybrid median filter realizations", *Proc. 1987 Int. Conf. on Digital Signal Processing*, pp. 349-352, Florence, Italy, 1987.

[24]   K. Preston, "Cellular logic computers for pattern recognition", *Computer*, vol. 16, no. 1, pp. 36-47, Jan. 1983.

[25] M.J.B. Duff, "A cellular logic array for image processing", *Pattern Recognition*, vol. 5, pp. 229-247, 1973.

[26] S.B. Gray, "Local properties of binary images in two dimensions", *IEEE Transactions on Computers*, vol. C-20, no. 5, p. 551, May 1971.

[27] J.M. Herron et al., "A general purpose high speed logical transform processor", *IEEE Transactions on Computers*, vol. C-31, no. 8, pp. 795-800, Aug. 1982.

[28] B. Kruse, "A parallel picture processing machine", *IEEE Transactions on Computers*, vol. C-22, no. 12, pp. 1075-1087, Dec. 1973.

[29] K.E. Batcher, "Design of a massively parallel processor", *IEEE Transactions on Computers*, vol. C-29, no. 9, pp. 836-840, Sept. 1980.

[30] S.R. Sternberg, "Biological image processing", *Computer*, vol. 16, no. 1, pp. 22-34, Jan. 1983.

[31] J.C. Klein, J. Serra, "The texture analyzer", *Journal of Microscopy*, vol. 95, pt. 2, pp. 349-356, Apr. 1972.

[32] S.R. Sternberg, "Pipeline architectures for image processing" in *Multi-computers and image processing algorithms and programs*, Academic Press, 1982.

[33] S.R. Sternberg "Machine vision non-contact gaging", *Proc. Applied Machine Vision Conference*, Chicago, 1984.

[34] B. Picinbono, "Quadratic filters", *Proc. IEEE Int. Conf. ASSP*, Paris, France, pp. 298-301, 1982.

[35] E. Biglieri, "Theory of Volterra processors and some applications", *Proc. IEEE Int. Conf. ASSP*, Paris, France, pp. 294-297, 1982.

[36] G.L. Sicuranza, "Nonlinear digital filter realization by distributed arithmetic", *IEEE Transactions on Acoustics, Speech and Signal Processing*, vol. ASSP-33, no.4, pp. 939-945, Aug. 1985.

[37] G.L. Sicuranza, G. Ramponi, "Adaptive nonlinear digital filters using distributed arithmetic", *IEEE Transactions on Acoustics, Speech and Signal Processing*, vol. ASSP-34, no.3, pp. 518-526, June 1986.

[38] H.H. Chiang, C.L. Nikias, A.N. Venetsanopoulos, "Efficient implementations of quadratic digital filters", *IEEE Transactions on Acoustics, Speech and Signal Processing*, vol. ASSP-34, no.6, pp. 1511-1528, Dec. 1986.

[39] B.G. Mertzios, G.L. Sicuranza, A.N. Venetsanopoulos, "Efficient structures of two-dimensional quadratic digital filters", *IEEE Transactions on Acoustics, Speech and Signal Processing*, vol. ASSP-37, no. 5, pp. 765-768, May 1989.

[40] A.N. Venetsanopoulos, B.G. Mertzios, "A decomposition theorem and its implications to the design and realization of two-dimensional filters", *IEEE Transactions on Acoustics, Speech and Signal Processing*, vol. ASSP-33, no. 6, pp. 1562-1574, Dec. 1985.

[41] B.G. Mertzios, A.N. Venetsanopoulos, "Modular realization of m-dimensional filters", *Signal Processing*, 7, pp. 351-369, 1984.

[42] H.T. Kung, "Why systolic architectures", *Computers*, 15, pp. 37-46, 1982.

[43] S.Y. Kung, "On supercomputing with systolic/wavefront array processors", *Proc. IEEE*, 72, pp. 867-884, 1984.

[44] S.Y. Kung, *VLSI array processors*, Prentice Hall, 1987.

[45] K. Hwang, F.A. Briggs, *Computer architectures and parallel processing*, McGraw-Hill, New York, 1984.

[46] H.H. Lu, E.A. Lee, D.G. Messerschmitt, "Fast recursive filtering with multiple slow processing elements", *IEEE Transactions on Circuits and Systems*, vol. CAS-32, no. 11, pp. 1119-1129, Nov. 1985.

[47] B.G. Mertzios, A.N. Venetsanopoulos, "Implementation of quadratic digital filters via VLSI array processors", *Archiv fur Elektronik und Ubertrangungstechnik* (Electronics and Communication), vol. 43, no. 3, pp. 153-157, 1989.

[48] D.F. Elliot, R. Rao, *Transforms, algorithms, analyses and applications*, Academic Press, 1982.

[49] B.G. Mertzios, V.L. Syrmos, "Implementation of digital filters via VLSI array processors", *IEE Proc.*, Part G, 135, pp. 78-82, 1988.

[50] K.K. Parhi, D.G. Messerschmitt, "Concurrent cellular VLSI adaptive filter architectures", *IEEE Transactions on Circuits and Systems*, vol. CAS-34, no.10, pp. 1141-1151, Oct. 1987.

[51] V.G. Mertzios, G.L. Sicuranza, A.N. Venetsanopoulos, "Efficient structures for two-dimensional quadratic filters", *Photogrammetria*, vol. 43, no. 3/4, pp. 157-166, 1989.

[52] A.N. Venetsanopoulos, B.G. Mertzios, "A decomposition theorem and its implications to the design and realization of two-dimensional filters", *IEEE Trans. Acoust., Speech, Signal Processing*, vol. ASSP-33, no.6, pp. 1562-1574, Dec. 1985.

[53] C.L. Nikias, A.P. Chrysafis, A.N. Venetsanopoulos, "The LU decomposition and its implications to the realization of two dimensional digital filters, *IEEE Trans. Acoust., Speech, Signal Processing*, vol. ASSP-33, no.3, pp. 694-711, June 1985.

[54] F.R. Gantmacher, *The theory of matrices*, vol. 1, Chelsea, New York, 1974.

[55] H.H. Chiang, C.L. Nikias, A.N. Venetsanopoulos, "Efficient implementation of quadratic digital filters", *IEEE Trans. Acoust., Speech, Signal Processing*, vol. ASSP-34, no. 6, pp. 1511-1528, Dec. 1986.

[56] Y. Lou, C.L. Nikias, A.N. Venetsanopoulos, "Efficient VLSI array processing structures for adaptive quadratic digital filters", *Circuits, Systems and Signal Processing*, vol. 7, no. 2, pp. 253-273, Feb. 1988.

[57] D. Hatzinakos, C.L. Nikias, A.N. Venetsanopoulos, "Massively parallel architecture for quadratic digital filters", *IEEE Trans. on Circuits and Systems*, in press.

INDEX